INDUSTRIAL ODOR TECHNOLOGY ASSESSMENT

INDUSTRIAL ODOR TECHNOLOGY ASSESSMENT

Paul N. Cheremisinoff, P.E.
Associate Professor of Environmental Engineering
New Jersey Institute of Technology
Newark, New Jersey

Richard A. Young
Editor, Pollution Engineering
Technical Publishing Company
Barrington, Illinois

Editors

ann arbor science PUBLISHERS INC.
POST OFFICE BOX 1425 • ANN ARBOR, MICH. 48106

P. O. Box 1425, Ann Arbor, Michigan 48106

Library of Congress Catalog Card No. 74-28656
ISBN 0-250-40093-6

Manufactured in the United States of America

FOREWORD

Detection and discrimination of odorants provide man with a wealth of sensory information about the ambient chemical environment. In many cases an air contaminant may be immediately detected and identified by odor. Odors have many different qualities, described by such terms as "putrid," "pungent," or "pleasant." The word "odor" itself can evoke the memory of a sensation. A major factor in the total air pollution picture, odor is easily noticed and frequently the source of complaints to air pollution authorities.

Odor emissions that may be objectionable are not limited to offensive or unpleasant smells; any odor can be deemed a nuisance and subject to complaints. Chemical compounds are often the cause of odor problems, varying widely from source to source comprising a wide range of activities—industries such as plastics, pulp and paper, glass, rendering, foods and chemicals. Manufacturing is usually thought of as a principal source of odor—the nature of the enterprise, its raw materials and finished product all determine the extent of the problem.

The preparation of this volume is devoted entirely to practical understanding and control of odors. As wide as possible a range of industries and sources, as well as state-of-the-art working techniques, are presented by experts in their respective fields. In addition, attention has been given to areas touching economics, planning, social and legal aspects essential in such a technology assessment.

The book attempts to stress practical engineering, written in general technical language for expert, student or novice. It should be useful to the pollution engineer; control official; plant, chemical, and process engineer; sewage treatment operator; and to consultants in their respective fields.

Paul N. Cheremisinoff
Richard A. Young

CONTENTS

LIST OF CONTRIBUTORS

Robert M. Bethea, Institute for Environmental Technology, Texas Tech University, Lubbock, Texas 79409
Mahesh V. Bhatia, Engelhard Minerals & Chemicals Corp., East Newark, N.J. 07029
Carleton H. Boll, RaySolv., Inc., Highland Park, N.J. 08904
William A. Cheney, United Air Specialists Inc., Cincinnati, Ohio 45200
Nicholas P. Cheremisinoff, Clarkson College of Technology, Potsdam, N.Y. 13676
Paul N. Cheremisinoff, New Jersey Institute of Technology, Newark, N.J. 07102
Clifton W. Cooke, E. I. duPont de Nemours & Co., Newport, Del. 19804
John R. Dervay II, Air Preheater Co., Wellsville, N.Y. 14895
W. P. Dolan, Rossnagel & Associates, Cherry Hill, N.J. 08003
Andrew Dravnieks, Odor Sciences Center, IIT Research Institute, Chicago, Ill. 60616
R. David Flesh, Copley International Corp., La Jolla, Calif. 92037
Kenneth J. Ficek, Carus Chemical Co., La Salle, Ill. 61301
Stanley E. Gifford, Regenerative Heat Corp., Middletown, Conn. 06457
Ronald D. Gottschling, Welsbach Ozone Systems Corp., Philadelphia, Pa. 19100
Yousuf H. Habib, Foster Wheeler Energy Corp., Livingston, N.J. 07039
Jerry Harris, Harris County Stream & APCD, Pasadena, Texas 77502
Thomas M. Hellman, Allied Chemical Corp., Morristown, N.J. 07960
George A. Herr, Eli Lilly & Co., Lafayette, Ind. 47900
A. T. Hugh, Harris County Stream & APCD, Pasadena, Texas 77502
Oscar Bud Lauren, Environmental Odor Consultant, Countryside, Ill. 60525
W. D. Lovett, Calgon Corp., Pittsburgh, Pa. 15230
Angelo C. Morresi, New Jersey Institute of Technology, Newark, N.J. 07102
Carl Nebel, Welsbach Ozone Systems Corp., Philadelphia, Pa. 19100
Thomas J. Olenik, New Jersey Institute of Technology, Newark, N.J. 07102
John Polhemus, Ball Corp., Muncie, Ind. 47302
R. L. Poltorak, Calgon Corp., Pittsburgh, Pa. 15230
W. A. Quebedeaux, Harris County Stream & APCD, Pasadena, Texas 77502
Charles W. Raleigh, FMC Corp., Princeton, N.J. 08540
W. B. Rossnagel, Rossnagel & Associates, Cherry Hill, N.J. 08003
Stanley W. Schroeder, Lamb, Eastman & Keats, Washington, D.C.
Richard H. Snow, IIT Research Institute, Chicago, Ill. 60616

M. F. Terraso, Harris County Stream & APCD, Pasadena, Texas 77502
Amos Turk, The City College of the City University of New York, N.Y. 10031
N. W. Tyler, Harris County Stream & APCD, Pasadena, Texas 77502
Michael Y. Yang, W. R. Grace & Co., Fords, N.J. 08863
Harry A. Young, Jr., Neistein, Richman, Hauslinger and Young, **Ltd.**, Chicago Ill.
Richard A. Young, *Pollution Engineering*, Barrington, Ill. 60010

CHAPTER 1

HUMAN RESPONSE AND EFFECTS OF ODORS

Angelo C. Morresi, Paul N. Cheremisinoff

New Jersey Institute of Technology
Newark, N.J. 07102

Richard A. Young

Pollution Engineering
Barrington, Ill. 60010

> O, how much more doth beauty beauteous seem
> By that sweet ornament which truth doth give.
> The rose looks fair, but fairer we it deem
> For that sweet odour which doth in it live.
>
> Shakespeare

The sense of smell seems to add a fourth dimension to our environment. We are able to grasp an object, inspect it thoroughly, and speak objectively of it. However, the object's odor forces us to make a decision on whether we want it near or away from us. Its odor contributes an indescribable feeling that adds a finer quality to its beauty or ugliness. Where sight can tell us if a person is beautiful and neat or ugly and sloppy, smell can give us an insight to a person's personal hygienic habits by, say, his rancid body odor or halitosis.

The rose Shakespeare writes about can be duplicated to the extent that the eye cannot distinguish between real and artificial. The texture of the impostor has been developed to the "feel" of the rose. Yet the rose's fragrance goes unreproduced. These natural scents are not duplicated because of their complexity, sometimes containing as many as 40 chemical entities. We often know only a few trace constituents of the odor in question and thereby produce a mixture of lesser quality.[1] It is almost as if God let man play around and manufacture a rose to a certain extent, but only He could add the finishing touches or secret ingredient to cross the line from make-believe to reality.

Plato noted the complexities of odors and formulated his own theory on olfaction. The four primary kinds of matter—earth, fire, air, and water—have no odor and only smell when changing phases, that is, air changing into water (mist) or water changing into air (smoke).[2]

Aristotle, Plato's formidable disciple, followed the above theory, but went on to classify smells—rich and astringent, pungent, sweet, harsh and fetid (bitter). People reacted to an odor in one of two manners, and he referred to two species of smellable objects. Pleasantness and unpleasantness of a nutritive substance are relative to one's degree of hunger; therefore they are of incidental concern. Then there are those which are pleasant or unpleasant in themselves, such as flowers or feces. Thus we might conclude some odors are compatible with particular functions[3] and that our reaction to them is based on our compatibility with that function.

Concerning Odours was written in the third century B.C. by Theophrastus. He maintained that all entities have their own peculiar odor: plant, animal, object. He referred to odors as "due to [a] mixture, for anything which is uncompounded has no smell."[3,4] His detailed classification of odors was the first of its kind and is considered the most important contribution to olfaction from the classical period to the Middle Ages.[3]

Research over the years into the science of odors has been limited, first by the lack of interest of researchers in examining the role smells exhibit in our lives, both consciously and subconsciously, and second, by the lack of a useful nomenclature to describe the systems being examined. It was not until 1895 that Zwaadermaker made his system of odor classification, incorporating in it the works of Linnaeus (1752), Lorry (1784), and Haller (1763) to produce nine categories.[1,3] Table 1-1 shows Zwaadermaker's nine categories.[1] The first four are classed as pleasant and the last four as unpleasant. The middle class, alliaceous odors, is unpleasant when associated with food smells.

As proof of the lack of a useful nomenclature, in describing Zwaadermaker's system Harper, Bate Smith, and Land[3] used the word "foul" when considering category number 8 and "fragrant" for number 4, whereas Moncrieff[1] notes "balsamic" could be used for number 4, as well as "caprilic" and "fetid" for 7 and 9 respectively. He uses "repulsive" for category 8.

These variations exemplify the fact that there is no science of odor. Bienfang[5] asks "How does one measure smell?" "What is the difference between one kind of smell and another?" Much research is now being

Table 1-1. Nine Categories of Odor

Odor Classification	Description
1. Etheral	Fruity ethers or perfumes, beeswax, ether, etc.
2. Aromatic	Camphor, spices, anise, citron, almond, etc.
3. Fragrant	Flowers, vanilla, balsam, etc.
4. Ambrosiac	Amber, musk, etc.
5. Aliaceous	Onion, acetylene, iodine, etc.
6. Enpyreunatic	Roasted coffee, tobacco smoke, xylol, naphthalene, etc.
7. Hircine	Goaty odors, cheese, sweat, chestnuts, etc.
8. Foul	Narcotics, some bugs, coriander, flowers, etc.
9. Nauseous	Carrion, carrion flowers, feces, etc.

done because these questions are beginning to be asked more and more, as our society becomes overly odor conscious what with the deodorant and mouthwash business grossing billions of dollars each year.

RESPONSE TO ODORS

Alland[6] discusses human response to stimuli and notes that the ability of an organism, such as the olfactory system, to respond to a stimuli (odor) depends first upon its ability to perceive it. This perception varies with the prior experience of the individual, the degree of the odor threshold, and frequency of encounters.

Cassel,[7] in defining perception, outlines an individual's reaction to an odor step by step. "Perception is defined as the process of translating into mental imagery or concepts a vast array of impinging dynamics (odors) from both the psychological and physical domains of an individual." As an odor is perceived, the following structural functions occur:

1. Stimulation–the stimuli or odor is present
2. Discrimination–determination of what an odor is
3. Association–cue reduction
4. Mediation–autonomic involvement with affect
5. Inference–short memory involvement with odor
6. Subception–unconscious involvement with odor
7. Adaptation–determination of odor as a relevant stimuli.

We can conclude that smelling is a complex system of nose and nostril. The sense of smell, although it has declined from its once most important status in primitive man[4] to the least appreciated sense, is a highly complex microscopic system of interactions between primary receptor neurons reduced a thousandfold at the glomeruli layer, branching into mitral cells and the brain. The brain is cognizant of a particular odor through experience, learning, and inborn defenses (as with the foul odors of some poisons).

However, we cannot determine an odor's origin or source without experience. Bartley[8] notes that this sensory experience is not a matter of imputation, inference, or conceptualization. We can only experience, and therefore recognize an odor through something that has happened.

Through our experiences we incorporate into our lives a "library" of smells and smell encounters. Hartshorne[9] explains that smells are either the inevitable effects of the necessary body chemistry–sex lures, warning signals against attack–or deterrents of certain entities for the consumption of food, as with the skunk or stinkbug. The primary function of smell must be to detect the presence of the chemically inevitable body odors, and therefore to identify friend, foe, or quarry, and further to differentiate between edible and nonedible foods.

This primary or practical role of the sense of smell is probably the main reason that it is "often neglected" and considered just a "useful auxiliary" sense.[10] Ellis[10] writes of the highly emotional nature of smells through previous associations. Specific odors stir the imagination and motivate the recalling of distant memories.[5,10] These memories are often tied in with deep emotional experiences such as mom's baking lasagne and its characteristic odor. Smelling this odor, or any other emotional odor, forces us to remember "the good old days" (there is nothing to stir nostalgia like odors)[5] and furnishes impressions that can easily change our emotional state. If the odors bring back fond memories, we are likely to become carefree and bubbly; however, the odors might bring back adverse thoughts, leaving us in a deep state of depression. Thus the sense of smell is a highly personal possession, which in our society has greater significance in the imaginative or power-of-suggestion arena than the so-called primary function of a warning device.

The entire nervous system is affected by odors. Wouters and Koster[11] discuss the influence of odors on heartbeat, respiration, and other reflexes. Ellis[10] further describes odors as "powerful stimulants" causing increased body energy which can result in nervous exhaustion if allowed to be excessive and prolonged.

McCartney[4] cites the works of Alexander (1951) and Kistiakowsky (1950) in formulating a theory as to part of the cause of an odor reaction. The enzyme balance of the olfactory cells might be affected by odorants (1) by modifying existing enzymes, (2) by forming new enzymes, (3) by inhibiting all or part of the activity of normal cellular enzymes, or (4) by all the above. Furthermore, it may be assumed that the odorants inhibit enzymic reactions, thus proving the stimulatory effect of odors to be physical as well as imaginary. Enzymes control the rate, quantity, and selectivity of the body reactions. Therefore when Ellis[10] concurs with Wouters and Koster[11] by describing the effects of certain volatile oils such as antipasnodics and anesthetics, and notes that "they stimulate digestion, circulation, and the nervous system," and in large doses produce depression, we realize that the sense of smell is far from an "auxiliary sense."

Smith, Allison, Goff, and Principato[12] experimented with human odorant-evoked responses (OER). They investigated whether the OER is due to the stimulation of the olfactory receptors or stimulation of the nasal trigeminal afferents in the nose. Through the use of anesthesia the subjects were tested and it was found that the absence of trigeminal innervation produced an absence of OER, while the loss of olfactory function produced no OER change. Thus the question of how human response occurs is still with us, as it was when Plato[2] described it as small particles swirling around a nasal vein to the brain.

Our response to odors is now conclusive; the anatomy of the human olfactory system is not our concern.

With regard to human responses to odors, odors play an important role as sexual stimuli. Men and women exhibit a difference in body odors,[9] and Morris[13] suggests this is part of the pair-formation process: each individual imprints on his or her partner a specific body odor and, as a result, we become accustomed to and compatible with the odor as well as the person. As we grow older, we transmit different odors. Through puberty, adolescence, and early manhood and womanhood there is the gradual development of the adult odor of skin and excreta.[10] We possess a large number of scent glands–apocrine glands, as well as a large number of sweat-scent-producing glands, in high concentrations in the armpits and sexual regions of the body.

Significant in this aspect are the changes in odor preferences as puberty is attained.[1,4,10,13] Morris[13] notes strong preferences for sweet and fruity odors before puberty. However, this response declines with an increased interest in flowery, oily, and musky odors as sexual maturity is reached.

Figure 1-1 shows the variation with age of the liking for musk lactone for both sexes.[1] Males and females were tested. There is a marked increase in the curve as adolescence is approached, a sharp fall at the ages 18 to 20, and then a steady increase until the age of 45. The graph's significance is the maximum liking of musk lactone at the advent of our sexual lives and also at their decline. This curve applies to both sexes, but for males the increase in responsiveness to musk lactone is greater than for females.[1,13]

McCartney[4] cites work done with a large number of subjects using benzene, coffee, peppermint, and other odorants. It shows that with increased age, olfactory acuity decreased, and that nonsmokers over the age of 15

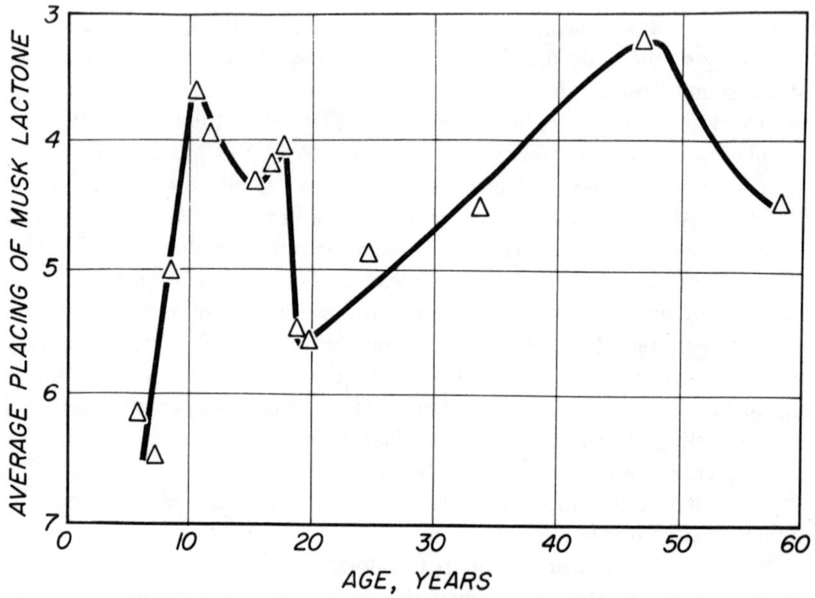

Figure 1-1. Preference for musk lactone with respect to age (both sexes together).

had greater acuity than smokers of all ages. Although no follow-up work has been done, this might be the reason why the greater changes in olfactory preferences take place in the first 20 years of life, and therefore the changes are less conspicuous.

Table 1-2 shows the placing of 10 odorants by subjects according to the subjects' sex and age.[1] Males and females agree on what they like best or dislike the most; however, the intermediate odors show many differences. There is a sex difference at a very early age (under eight) of olfactory preferences, and this continues up to about 40 years. After 40, there does not seem to be a great difference, possibly because of the decline in acuity.

From the data thus far published it would seem that neither sex has a definite edge in olfactory acuity. Thus when it is noted that the olfactometric sensitivity varies from man to woman appreciably, it is probably due more to the differences in preferences for a variety of odors than to the actual threshold limits. The notion that women are able to smell certain substances such as foods and flowers to a greater degree than men, and that men are more adept at smelling others, such as chemicals and pharmaceuticals, is more likely attributable to the fact that women and men come in contact to a greater extent with the items they have a greater sensitivity toward and thus recognize them more easily—rather than to a scientific explanation.

Table 1-2. Odorant Classification by Sex and Age

Odorant	Average Placing by Males (in Years)						Average Placing by Females (in Years)						Difference in Average Placings of Males and Females
	0-7	8-14	15-19	20-40	Over 40	All Groups	0-7	8-14	15-19	20-40	Over 40	All Groups	
1 (Strawberry essence)	1.53	1.86	1.89	2.86	4.14	2.46	1.47	1.30	2.37	3.58	3.96	2.54	–0.08
2 (Spearmint oil)	5.25	5.10	4.73	5.04	4.62	4.95	4.63	5.28	5.08	5.57	5.24	5.16	–0.21
3 (Lavender oil)	5.94	6.44	5.86	4.71	4.38	5.47	6.08	6.76	3.95	3.49	4.48	4.95	+0.52
4 (Musk lactone)	5.45	3.81	4.76	4.02	3.36	4.28	5.68	4.75	5.32	5.18	4.36	5.06	–0.78
5 (Vanillin)	3.39	3.64	3.43	4.53	4.65	3.93	4.59	3.62	4.82	4.34	4.08	4.29	–0.36
6 (Neroli oil)	7.20	6.31	6.89	5.51	5.52	6.29	7.83	7.89	7.29	6.23	4.66	6.78	–0.49
7 (Almond essence)	6.44	6.05	5.31	6.07	5.32	5.84	5.17	5.14	4.61	4.43	5.26	4.92	+0.92
8 (Naphthalene)	5.64	5.35	5.75	5.30	5.26	5.58	5.07	4.06	4.60	5.03	5.67	4.89	–0.69
9 (Rape oil)	5.66	7.25	7.46	7.80	7.98	7.23	5.91	7.59	7.46	7.79	7.95	7.34	–0.11
10 (Chlorophyll oil, soluble)	8.52	9.15	8.90	9.14	9.80	8.98	8.57	8.61	9.51	9.39	9.34	9.08	–0.10

Berglund, Berglund, Ekman, and Engen[14] (at the University of Stockholm, Sweden) obtained individual scales of odor intensity for 28 different chemical compounds. One hundred ninety-six olfactory stimuli, which differed in quality and intensity, were administered to five male and six female subjects. The results were in agreement with the above, in that they were varied. This variation was attributed to perceptual differences rather than to response bias. "Perceptual differences"–after Cassel's[7] definition–are caused by varied backgrounds and environmental conditions and result in different mental images produced by the odor stimuli. The female or the male mind is conditioned to certain odors and can therefore respond to those stimuli faster than the opposite sex.

However, Wouters and Koster[11] observed that the sensitivity of women is greatly influenced by the menstrual cycle. Studies[4] have shown that hormones, androgens, and estrogens control olfactory sensitivity. And, in agreement with Wouters and Koster,[11] these studies show that the removal of the ovaries (thus preventing menstruation) may raise the threshold limit (*i.e.*, lower acuity) for certain odorants by a hundred- or even a thousand-fold. The use of estrogen may restore normal sensitivity. This linkage of odor sensitivity to the sexual system does not seem accidental. As noted previously, odors play a major role in sexual relations. Many odors, including body odors, are nervous stimulants which can lead to greater potency. Ozolagmy is the sexual gratification derived through the olfactory system,[10] and–surprisingly or not–much work has been done in this area.

Ellis[10] explains personal odors as mainly of an intimate, emotional, and imaginative character. Thus our brain's cognitive powers are extremely taxed as our entire nervous system is alerted. Reuben[15] describes smell as a segment of the whole sexual experience. Smells, in cooperation with the other senses, contribute to a "rising sexual excitement."

The sense of smell has an enormous effect on humans, and our responses to them are more than casual. We have tried to show this by discussing areas of olfaction familiar to readers and allowing their past experiences (because odors and experiences are interrelated) to provide the proof.

ODORS ARE ESSENTIAL

Odors are essential as warning signals. Most mammals use odors to outline an area to "mask off" the limits of their territories. Some animals, such as the buck, have odor-producing glands which they use to indicate their domains. Others use urine to imprint on their regions an odor which broadcasts "Stay clear of my turf."[16]

Humans use odors in a similar manner. They warn us of impending dangers, since certain odors are synonymous with perils. There are many instances where our lives have been saved only because of the odor of a hazard.[5] We smell smoke and know there is a fire somewhere; we smell gas and realize there is a leak in the vicinity. We fail to realize the importance of this sense until we come in contact with an odorless substance like carbon monoxide.

Bienfang[5] quotes the order adopted by the Railroad Commission of Texas in 1941 pertaining to odorizing gases:

> . . . From and after the effective date of this order, each person, firm, or corporation, including municipal corporations, engaged in the business of handling, storing, selling, or distributing liquified petroleum gas for private or commercial uses, or supplying the same by pipe lines or otherwise to any public building or buildings or to the general public, shall continuously odorize such gas by the use of a malodorant agent to be introduced into the gas in either its gaseous or liquid state so as to indicate by a distinctive odor the presence of gas when it is present in the air in concentrations of as much as one-half of one percent by volume. Such odor must be readily perceptible to normal or average olfactory senses of a person coming from fresh ungasified air.

The commission's order brings up two important points. First, the gas must be perceptible to average olfactory senses. (This is an unknown which will be discussed later in this chapter with odors' classification and interaction with one another.) Second, a person must be able to smell the gas "coming from fresh ungasified air." Here the commission notes the olfactory sense's ability to adapt to an odor and/or become fatigued.

Before we discuss the average olfactory sensory system, let us investigate those who are not "normal persons." It has long been assumed that primitive man or those who live in a primitive society have an extraordinary olfactory acuity. Not many studies have been done in this area; however, McCartney[4] cites a few. This notion has come forward probably because of the primitive's innate ability to track animals for food, or detect slight but perilous odors. We often neglect to note that, in tracking, vision is most important and that through use and practice we can sharpen any of our senses to its highest degree.

The cited studies produced contradictory conclusions. Olfactory acuity, odor strengths, discriminatory ability, and the power to remember odors were tested in a primitive society. Problems arose in language, native dislike for instruments, and an insufficient number of individuals. The studies failed to prove conclusively that the senses of the primitive person were superior to those of the "civilized" individual or one who lives in a more modern society.

Studies cited by McCartney[4] also proved contradictory with reference to the blind. However, this is another area of olfaction where very little work has been done. Some experiments showed the blind have a greater ability to smell substances than the seeing, and other reports showed the opposite results.

The blind are forced to utilize their remaining senses to the utmost. Therefore, through use and necessity, the blind are more likely to develop their sense of smell. A blind person will learn to distinguish between odors, to develop his knowledge of odors, and to characterize each one so as to have a memory bank and reference points for new odors. The ability to do this does not necessarily mean such an individual's olfactory senses are more acute.

PLEASANT AND/OR UNPLEASANT ODORS

Classifying odors and distinguishing between them is very important to the way we respond to odors and their effects on us. Our response to odors is sometimes thought of as very personal, as when we say "I like this smell," possibly referring to a flower. We usually assume that most everyone's olfactory preferences are the same, and this is essentially true. However, as was previously noted, the nomenclature of the classification of odors presents a fuzzy picture. We note Zwaadermaker's classification system, as well as those of the other men previously mentioned, and their difficulties in trying to define a particular odor.

Evans, Moser, and List[17] explained that when it comes to tastes and odors, we are distinct individualists. Thus (although we have seen ourselves differentiated in age and sexual categories) our personal responses to odors are unique. The knowledge of smell is acquired through experiences with food and environment, and, according to the individual's olfactory acuity and memory skills, he is able to classify and describe an odor in his own terms. Evans *et al.* found that most people can identify a variety of aromas because of their acute ability in odor perception.

A rather broad classification system would be to categorize odors as pleasant or unpleasant. We easily know whether we like or dislike an odor—rather than try to classify it as floral or pepperminty.[18] We must realize that the pleasantness or unpleasantness is highly psychological in some instances. A particular food's smell might be highly pleasant at a time of extreme hunger, but not so attractive after the palate has been satisfied.

Moncrieff[1] explains that pleasantness or unpleasantness is a result of emotions in the individual. Things we enjoy and desire are those things that smell pleasant. Malodorous objects make us shy away. The smell of a flower might ease a state of depression and nostalgic odors might bring on a smile.

The following incorporates the idea of unpleasantness and pleasantness and the human response to odors.

1. Our reactions to odors are similar to our reactions to other sense stimuli: involuntary and spontaneous, either liking or disliking, or indifference.
2. Reasons for the above cannot be interpreted; that is, usually the reasons, if there are any, show no trends or give no explanations.
3. Previous experience with an odor or with similar odors sometimes determines if an odor is liked or disliked.
4. According to bodily needs, food smells are pleasant or unpleasant.
5. Odors which are pleasant tend to "feed" those emotions which are affected by "beautiful" things in the environment.[1]

Myddleton[19] thoroughly investigated different facets of olfaction and also concluded that pleasantness is directly related to the individual's olfactory experiences. He made 200,000 to 300,000 observations of several hundred subjects. Fifteen odorants were used as representative of different qualities of odors. Identification of the odor was not required by the subject. Rather, as systems of odors were passed by the observer, he was asked if a new odor was introduced to a particular system. Using single odorants, it was found that (1) pleasantness and intensity are independent factors of olfactory experiences, (2) familiarity and intensity are independent characteristics, (3) familiarity and pleasantness are directly related to the extent of the odor's being attractive, (4) a pleasant odor such as vanillin is rated pleasant, but is not rated more pleasant after smelling two unpleasant odors, such as butylamine and butryric acid, than after smelling two pleasant odors such as eugenol and sutrole.

Karl E. Henion[20] made a direct psychophysical scaling of the olfactory pleasantness of diluted n-amyl acetate. Observing 12 undergraduate students, he constructed a category and two magnitude scales. The subjects observed the pleasantness of a geometric dilution series of n-amyl acetate. The results showed pleasantness varied inversely with concentration.

Henion[21] did a follow-up study to note the relationship between pleasantness and intensity. He observed eight male and two female undergraduates. He constructed a category and three magnitude scales of these subjects, rating a geometric dilution series of n-amyl acetate for pleasantness and intensity. It was shown that pleasantness was inversely proportional to the concentration, while intensity was directly related to concentration.

Thus we can conclude that the pleasantness of an odor is not only related to familiar or unfamiliar and/or desirable or undesirable substances, but the pleasantness of an odor changes as its concentration or intensity changes. Table 1-3 shows the results of 5,200 observations of 104 subjects to the pleasantness and/or unpleasantness of the selected odorants.[3] In this system there was no relationship between odor intensity and the degree of pleasantness. Harper, Bate Smith, and Land[3] found that when similar expressions of perceived strength and pleasantness and unpleasantness were made, the range of the latter varied far more than the range of the former. This seems to suggest that our olfactory powers can be very discriminating about what they like and dislike, but tolerant of the differences in odor strength.

Table 1-3. Response to Selected Odorants

Odor	Average Intensity 0: None 4: Extreme	Average Pleasant/Unpleasant 1: Extremely Pleasant 9: Extremely Unpleasant
Methyl salicylate	1.5	7.0
Benzaldehyde	2.3	6.5
Eugenol	2.0	6.5
Safrole	1.6	6.4
Vanillin	1.7	6.3
Orange	1.7	5.8
Amyl acetate	2.2	5.0
Phenylethyl alcohol	1.2	5.0
Terpineol	1.9	4.7
Geraniol	1.7	4.7
Coffee (extract)	1.0	4.4
Ionone	1.8	4.1
Indole	1.4	3.7
Butyric acid (dilute)	2.9	2.1
Butylamine (dilute)	2.7	2.0

Reactions to odors are various but not without extreme responses. Unpleasant odors can cause violent facial contortions, whereas pleasant odors effect smiles and a happier and more colorful outlook. These are reactions that cannot be measured. How does one go about measuring a cringe or the degree of elation?—we find it's just not possible. However, it seems as if everything and anything can be made analogous to mathematical formula, so why not the pleasantness of an odor? Therefore a means was devised by Beebe-Center[22] to determine the percentage of pleasantness:

$$P = \frac{p + \frac{1}{2}i}{p + i + u} \times 100$$

where:

P = % of pleasantness
p = no. of pleasant reactions (1 point)
i = no. of indifferent reactions (½ point)
u = no. of unpleasant reactions (0 point)

The percentage of pleasantness for a set of stimuli is for that odor in this system. In other words, percentages from one group of odors cannot be compared with percentages from another group.

Moncrieff cites work done with 100 subjects (50 men and 50 women) as 15 odors were examined (Table 1-4).[1] The fact remains that certain odors are preferred by a large percentage of people, and that some odors—no matter how unappealing to the majority—are pleasant to some people.

Table 1-4. Rating Various Odorants to Pleasantness

Odorant	Pleasantness (group of 100 subjects)
Vanillin	93
Wintergreen oil (methyl salicylate)	90
p-dichlorbenzene	83
Geraniol	81
Camphor	70
Menthol	62
Empty bottle*	61*
Acetophenone	56
Diphenyl ether	44
Phenol	31
Ethyl cinnamate	31
o-bromto	20
Nitrobenzene	15
Quioline	15
n-caproic acid	9
Heptyl aldehyde	7

*Control—ideally should have been 50.

This is further proof of the individuality of the olfactory system in people. Although we generally agree on what is pleasant and unpleasant, we all have our own idiosyncrasies which destroy the theories. Along with different environments and personal experiences, it also might be possible for different persons to have varied olfactory systems, that is different receptors for different people. We know that the size and shape of the odor molecule affects its ability to be smelled (see "Primary Odors"). Therefore, if the receptors of each person are varied, each person might smell or not smell a different odor, resulting in different responses from each individual–to a large extent, if the differences are great; to practically nil, if the differences are slight.

PRIMARY ODORS

In order to classify an odor in a particular category, or to make a particular category, we must define the odor stimuli being smelled, that is the odor producing a certain sensation in the body. Before we can categorize it as pleasant or unpleasant, we must ascertain the sensation to be an odorous stimulus. Thus to perceive a sensation such as an odor we must first interpret it as an odor.

Perception = sensation + interpretation

Harper, Bate Smith, and Land[3] cite Peter Caws for the above equation and give us Figure 1-2, which is directly related to odor classification.

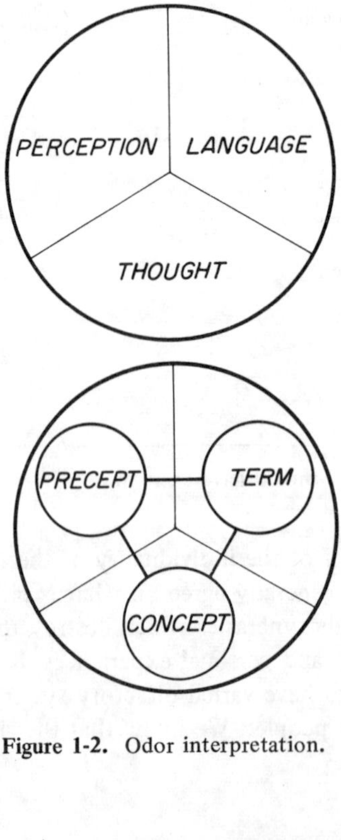

Figure 1-2. Odor interpretation.

Once we have perceived an odor, it is recognized by the brain. The brain then searches its memory banks for the experience or idea that surrounds this particular odor. When this is done, we know the odor to the best of our knowledge and a concept of the odor has been developed; it is pleasant or unpleasant, desirable or undesirable. Now we must describe the odor in terms that can be understood, in a language that will have meaning to other persons. The problem that has plagued us all along is the language or term portion of the triangle. As soon as we describe an odor, we classify it. Since everyone interchanges the adjectives of odors–an odor might be foul, but is it putrid, rancid, or disgusting?–the identity and description of each category are different for each individual.

Woskow, in Tanyolac,[23] notes the futility of attempts by many to arrange odors in categories which show psychological meaning. Anyone can devise a category for odors, but the object is to categorize the odors simply enough to be understood by the observers yet complex enough to incorporate the various aromas available.

Thus our olfactory discriminatory powers have been analyzed by many, but few of these latter have been acclaimed. Alexander Bain made his classification of odors[1] and described the categories with relation to the initial human response to them:

1. Fresh (odors of the outdoors)
2. Close or suffocating (crowds)
3. Nauseous (leather tannery)
4. Stinking (garbage wastes)
5. Pungent (ammonia)
6. Sweet (flowers)
7. Ethereal (alcohol)
8. Appetizing (food)

This is a good idea for a classification system, but many people find garbage wastes, as well as crowds, nauseating. Although there does not seem to be any real scientific basis for Bain's categories, they fit neat ideas and experiences that can be used by a majority of people. However, odors do not fit into categories like pegs into holes. A sweet, flowery odor like preserves cooking, or an odor that stinks like boiling cabbage, are all appetizing or unappetizing, as the case may be. But the point is that most odor categories are in some way related to other categories, and for one to list steadfast rules seems unreasonable.

It seems the best systems are those which try to pinpoint the primary odors–those that cannot be broken down any further–and allow all other odors, which are mixtures of the primary odors, to find their own places by equilibrium.

Hennings' classification system is a bit more complex; it includes six odors, all of which are interrelated. Figure 1-3 shows Hennings' odor prism.[1,3,8,24] His odors, when one classifies them, are allowed to stray among the categories. Hennings' prism takes into account a fragrant odor that may be putrid.

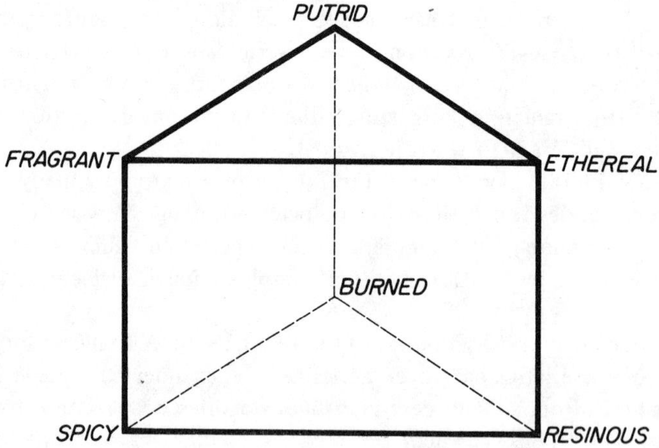

Figure 1-3. Hennings' odor prism.

Amoore, Johnston, and Rubin[18] give their odor classifications in Table 1-5. They investigated the olfactory systems, and the function of smelling, through an olfactometer developed by Johnston. In the nose are series of receptors, and each receptor is responsible for a different primary odor; hence Table 1-5. Each primary odor molecule seems to have a certain shape and size, thereby enabling it to react with its particular receptor.

Table 1-5. Odor Classification

Primary Odor	Chemical Example	Familiar Substance
Camphoraceous	Camphor	Moth repellent
Musky	Pentapecanolactone	Angelica root oil
Floral	Phenylethyl menthyl Ethyl carbinol	Roses
Pepperminty	Menthone	Mint candy
Ethereal	Ethylene dichloride	Dry cleaning fluid
Pungent	Formic acid	Vinegar
Putrid	Butyl mercaptan	Bad egg

Karl E. Henion[25] investigated the pleasantness of a homogeneous series of n-aliphatic alcohols. The above receptor principle seems to be correct, since he found magnitude scaling showing the degree of pleasantness to vary inversely with the molecules' chain length. This suggests that the size and shape of the molecule of a specific alcohol might fit a certain receptor or series of receptors, making the odor of one alcohol less pleasant than an alcohol of different molecular standing.

Hennings' prism is more appropriate under these terms. The lines interconnecting each category allow all odors to be represented by all the classifications, if necessary. His six primary odors are all interrelated; therefore a more flexible system is encountered. When we discuss colors and note a color blue, although there are various shades of blue we know what is on each other's mind. However, a putrid odor can be one of any number of the disagreeable scents known to us, calling for the description to be more explicit. Flexibility is also encountered with the theory of individual receptors for each primary odor.[25] Since all odors are mixtures of the primary odors, it is easier to fit the indescribable odors into their slots by noting the receptor or receptor series which is stimulated by the odor. Theoretically, then, a particular odor could eventually be identified by its receptor region and have a characteristic property, not unlike the wave length of a beam of light identifying a particular color in the spectrum.

ODOR INTERACTION

The human response to two or more odors can be very complex. When we are subjected to an odorous stimulus, it is usually in combinations of many aromas. We might be riding a bicycle behind a bus while passing a pizza parlor. Obviously, our responses to these two distinct and rather contrasting odors will be diverse. Bartley[8] explains that when two different stimuli are effected together to be presented to the olfactory system, any number of occurrences may result.

(1) Blending occurs when the two odors are brought together and the resulting mixture has properties of both the original odorants. New characteristics may develop; usually, however, a blend produces an odor resembling each of the two original odors independently. (2) When two distinct and contrasting odors are presented simultaneously (the situation described above), both odors result, with one odor being perceived and then the next, in a flip-flop manner. (3) Dichlorhinic stimulation, or the stimulation of one nostril with one odor and the other nostril with another odor, results in two odors being smelled alternately. (4) Bartley also cites some literature to the effect that two odors may be smelled concurrently and still be smelled as individual odors. It seems

this idea is hard to explain, as the literature is contradictory. (5) The masking of one odor with another (the chapter on odor modification covers this in detail), so that one odor covers up another entirely. (6) The neutralization of one odor by another–in which our response would be the same as in (5), as if we were smelling only one odor, or as if the original odors were not present.

The preceding effects of one odor on another with respect to sensitivity are briefly discussed by Wouters and Koster.[11] However, the reasons for the different responses are sometimes unaccountable. Corbit and Engen[26] investigated the interrelation of odors by determining human sensitivity to n-aliphatic alcohols. The subjects were three nonsmoking females who were required to give yes or no answers upon request. The study showed that the effects of one odor on the detection of another may be related to the differences in the water solubility of the two odors. This result seems in accord with the requirement that odors be soluble in water, even to a minute degree.

The receptors and nerve endings in the olfactory complex are coated with a watery film, which will act as a barrier if the odors cannot be absorbed by it. Odorous substances also should be soluble in lipids, where they must be able to penetrate the lipide surface layer of the cell membrane to get to their appropriate nerve endings.[18]

Mixing two organic sulfides to obtain different levels of olfactory intensity, Berglund, Berglund, and Lindvall[27] explored the principle of odor interaction. Eight untrained observers between the ages of 19 and 31 were used. The two odors were qualitatively alike and produced a homogeneous end product. Each component odor's perceived intensity was measured, as well as the mixture's perceived intensity. The results showed the intensities of the mixtures to be directly proportional to the arithmetic sum of the intensities of the components of the mixtures. The concept here is that the individual odors are independent and odor intensities are additive similar to light or sound intensities.

Myddleton[19] did further work in mixed odorants, and his results concur with the above study.

1. Intensity of a mixture is about the same as the average of the component odors.
2. Pleasantness of a mixture is about the same as the average of the component odors.
3. Mixture nomenclature or response is the same as for single odorants.
4. No new qualities develop as a result of the mixture; that is, the qualities of the mixture are defined by the qualities of the component odors. Note that this is a contradiction of Bartley,[8] where he finds that new characteristics may develop.

5. Berglund, Berglund, and Lindvall's work[27] is confirmed by Myddleton, since he also found the characteristics of the individual odors to be additive in some way, as to develop the new characteristics for the mixture.
6. Mixtures that involve three or four component odors do not become more complex olfactory experiences. We respond to mixtures in the same manner that we respond to single odors.

ADAPTATION

Small variations in the environment can affect our lives dramatically. These variations can be of comfort or discomfort; and in order to accommodate to minor changes in the environment we change our behavior unconsciously, which is called adaptation.[6]

If an odor produces certain stimuli in the nose and the sensitivity of the nose is reduced during the next stimulus of the same odor, adaptation has occurred. Usually this reduction occurs for the smelling of one odor. Where two different odors are introduced, the reduction in sensitivity, if it occurs, is not so great as if one were smelling one odor.[1]

Adaptation to a particular odor can be measured by noting the increase in threshold limit.[1,24] During the first several minutes of stimulus and adaptation, the threshold limit increases with the adaptation time, but after a certain time the rate of adaptation levels off as the saturation is reached. Figure 1-4 shows two concentrations of benzoin and two concentrations of rubber. It identifies two basic phenomena pertaining to adaptation: (1) the threshold limit changes faster when we are presented with an odor of high intensity than when we are presented with one of low intensity, and (2) the adaptation rate is different for different odors.[24]

The conclusion to be drawn is that the maximum olfactory response during normal breathing occurs upon the first breath, if the odor concentrations are to be rated medium to high. At high concentrations, the olfactory response becomes limited after a certain exposure time, due to adaptation. Adaptation reduces the response until a state of equilibrium or a steady level of response is attained. This steady level of response can be zero (total fatigue) when a high concentration of odor is present. At low concentrations, adaptation occurs more readily when the stimulus is presented continuously rather than haphazardly, as during heavy breathing.

A further complication is that the rate of adaptation can be controlled by the manner of application of the stimulus. The flow rate across the olfactory region may vary in a number of ways and for a number of reasons. Thus the adaptation rate is not only dependent upon the odor and odor intensity but also on how the odor is perceived. We must also note that the response is affected by the sympathetic fibers in the ethmoidal nerve in the nasal area.[28] (In the past, air-freshener manufacturers have added ingredients to their products not to mask or neutralize an odor but to numb the nerve endings in the olfactory area to effect masking, neutralization, and/or adaptation.) The saturation point is reached in the nerve endings in the olfactory region and adaptation begins.

Total adaptation to a specific odor occurs as the sense of smell becomes fatigued. Workers in a leather tannery, during the first few days on the job, may become nauseated by the pungent odors of hydrogen sulfide. However, after a prolonged exposure to the odor, it disappears from their consciousness, as they become fatigued. This may seem good, that they are allowed to work in relative comfort, but hydrogen sulfide can cause physical damage; so they are neglecting one of the body's warning and defense mechanisms.

An example of adaptation is clearly shown by the stimulus produced by acetone in water. The threshold limit for acetone in water is 0.03%. However, upon the inhalation of one breath, the next breath must contain an acetone-in-water concentration of at least 5% to be perceived. Similar results were obtained when experiments were made with n-butyl alcohol. The threshold limit for n-butyl alcohol in water rose from a 0.005% concentration to a 1% concentration between the first and second breaths.[1] We can also see from the data the adaptation rate varying from substance to substance. The threshold limit changes are different for different odorants (Figure 1-4).[24]

Adaptation is a selective process that may or may not be affected by other stimuli. If we adapt to one odor, it does not necessarily follow that we will adapt to other odors. However, adaptation to one odor affects adaptation to other odors. Each initial odor affects the olfactory system differently, so that the responses to accompany or follow odors are affected differently.

Berglund, Berglund, Engen, and Lindvall[29] studied the effect of adaptation on odor detection and found the sense of smell to be fairly stable. The sensitivity to a weak odor of hydrogen sulfide following prior exposure to it was examined. The subjects were all female, ranging in age from 2 to 20. The exposure time was varied from 0 to 5 minutes, and the concentration from 4.05×10^{-7} mg/l to 24.4×10^{-7} mg/l. A yes-no detection method was used. The effect of adaptation was demonstrated by the results. It was shown that an odor, to be recognized, is independent of exposure time. In other words, no matter what the adaptation concentration, the length of exposure time will not increase sensitivity by allowing us to smell something originally unperceivable.

Much work has been done on the adaptation of odors and threshold limits. Pryor, Steinmetz, and Stone[30] worked with methyl isobutyl ketone

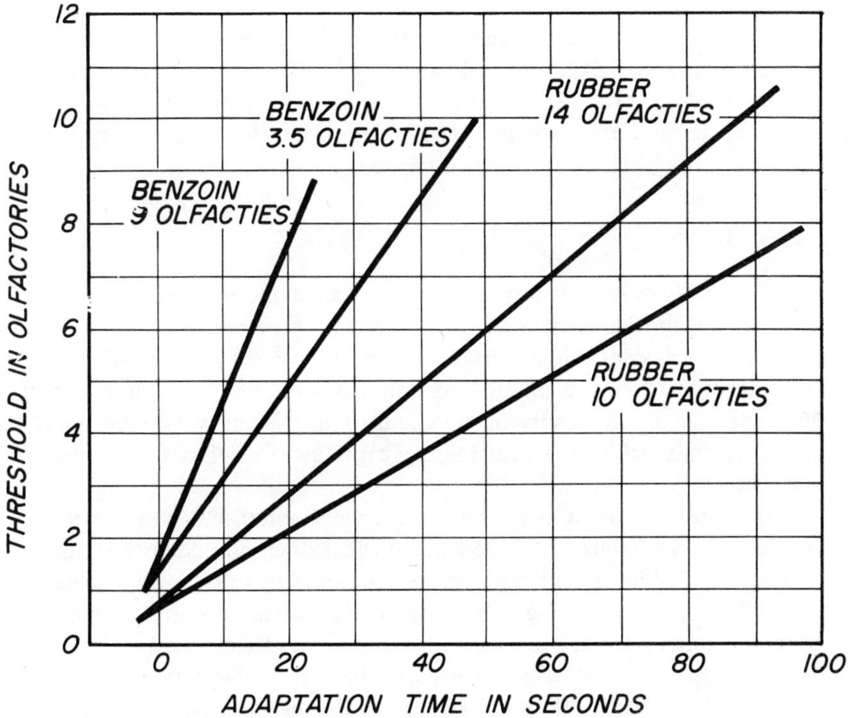

Figure 1-4. Threshold limit changes.

and compared changes in threshold limits with changes in intensity and stimuli.

Stoll[31] examined the effects of odor stimulation produced by slight chemical differences. The major odorant compounds of musk with molecule weights around 280 were tested upon the upper limit of perception. The difference in the compounds differed only in the configuration of the functional groups. That is, α and β-epimers of the C_{18} series produced a strong odor, while ambreinolide, scharoel lactone, and scharoel oxide were odorless. Adaptation and the disappearance of the odors were attributed to the nerve endings of the olfactory system becoming unapproachable through overadsorption (saturation). Adsorption at the olfactive receptors is different for different molecular weights and configurations. Thus changes in olfactory perception occur for complex compounds and their derivatives.

Recovery from adaptation, or the time it takes the olfactory system to reach its original state of sensitivity, has not drawn much attention.

Work[23,32] done in the area of recovery from adaptation seem to draw no conclusions for this phenomenon. Different hypotheses, for different recovery rates and for different odors, are not solidly backed, or the discussions have loopholes. The agreement that occurs is based on the receptor sites having some responsibility for recovery time through their sensitivity or locations within the olfactory complex.

ANOSMIA

The total loss of olfactory sensitivity is called anosmia. Hyposmia is a partial loss, and parosmia is a distorted olfactory perception. If the distortions are unpleasant, the condition is called cacosmia. It is possible for our minds to become involved in the olfactory response to such an extent that they distort the reality of the stimulation. Normally pleasant odors might be intolerable, as a situation can cause us to react in an abnormal manner.

Hallucinations occur as a result of disorders, whether they be physical or mental. Sometimes substances are smelled when no olfactory stimulus is present. Hallucinations can also be a response to an odor that is mentally pleasing or unpleasing. (We have already discussed nostalgia and the response to an odor in bringing back memories and old situations.)

Where most people have experienced olfactory hallucinations at the same time, anosmic persons are very rare. However, specific anosmia, or the ability to perceive odors, is widespread. John E. Amoore in *Theories of Odor and Odor Measurement*[23] notes that 2% of the population are unable to detect the odor of isobutyric acid, while they have a normal olfactory sense when it comes to most odors.

Amoore tested 10 anosmics and compared his results with 97 other subjects. Eighteen compounds related to isobutyric acid were used and the first 10 aliphatic acids were plotted against odor threshold concentrations. The aliphatic acids threshold limit was greater in all cases for the specific anosmics than for the normal observers (Figure 1-5).[23] Note again: the size and shape of the odor molecular structure is extremely important. The olfactory receptors seem to be very selective in their reception of odorous substances. The variance in the curve with respect to the addition of a carbon atom proves the above statement, and identifies olfaction as a complex science.

Obviously, the value in the use of the anosmic can be enormous for the researcher. By using a specific anosmic for a certain odor, and comparing and matching results, the primary odors could eventually be defined. Or the converse might be true: a specific anosmic or an individual's inability to smell certain odors could be due to one of his primary odor receptors being defective.

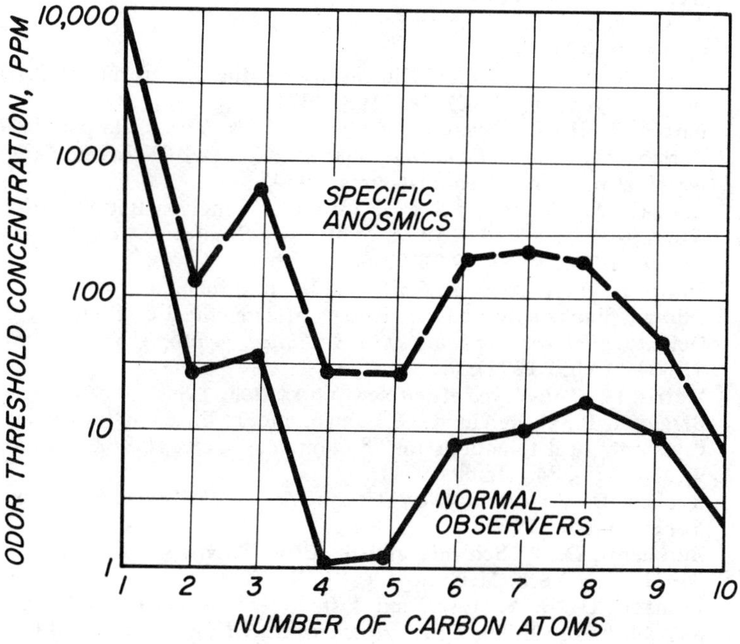

Figure 1-5. Odor thresholds for normal fatty acids.

The loss of odor sensitivity can be attributed to (1) a lack of receptor protein; protein defects or deficiencies in the receptor cells can cause total anosmia; (2) missing olfactory cells, whether through a birth defect, or disease, or accident, which results in a loss of olfactory sensitivity; as well as (3) broken nerve connections; (4) delayed recovery from adaptation; or (5) psychological block.[23]

REFERENCES

1. Moncrieff, R. W. *Odour Preferences* (London: Leonard Hill, 1966).
2. Vlastos, G. "Plato's Supposed Theory of Irregular Atomic Figures," *ISIS* 58 (191), 204-209 (1967).
3. Harper, R., E. C. Bate Smith, and D. G. Land. *Odour Description and Odour Classification* (New York: American Elsevier Publishing Co., 1968).

4. McCartney, W. *Olfaction and Odors* (New York: Springer-Verlag, 1968).
5. Bienfang, R. *The Subtle Sense* (Norman, Okla.: University of Oklahoma Press, 1946).
6. Alland, Jr., A. *The Human Imperative* (New York: Columbia University Press, 1972).
7. Cassel, R. N. "Functional Dimensions for the Structuring of Perceptions," *Psychology* **8** (2), 1-5 (May 1971).
8. Bartley, S. H. *Principles of Perception* (New York: Harper, 1958).
9. Hartshorne, C. *The Philosophy and Psychology of Sensation* (Port Washington, N.Y.: Kennikat Press, 1934).
10. Ellis, H. *Psychology of Sex* (New York: Emerson Books, 1964).
11. Wouters, O. and E. P. Koster. "Olfactory Studies," *Arch. Biochim. Cosmetol.* **9** (87-88), 62-72 (1966).
12. Smith, D. B., T. Allison, Goff, and J. J. Principato. "Human Odorant Evoked Responses: Effects of Trigeminal or Olfactory Deficit," *Electroencephalography & Clinical Neurophysiology* **30** (4), 313-317 (April 1971).
13. Morris, D. *The Naked Ape* (New York: Dell, 1967).
14. Berglund, B., U. Berglund, G. Ekman, and T. Engen. "Individual Psychophysical Functions for 28 Odorants," *Perception & Psychophysics* **9** (3B), 379-384 (March 1971).
15. Reuben, D. *Everything You Always Wanted to Know about Sex* (New York: Bantam, 1969).
16. Burkhardt, D., W. Schleidt, and H. Altner. *Signals in the Animal World* (New York: McGraw-Hill, 1967).
17. Evans, C. D., H. A. Moser, and List. "Odor and Flavor Responses to Additives in Edible Oils," *J. American Oil Chemists Soc.* **48**, 495-498 (Sept. 1971).
18. Amoore, J. E., J. W. Johnston, Jr. and M. Rubin. *Perception: Mechanisms and Models*, R. Held and W. Richards, eds. (San Francisco: W. H. Freeman and Co., 1972).
19. Myddleton, W. W. "Cosmetic and Toilet Preparations," *Manufacturing Chemist* **35**, 71-72 (April 1964).
20. Henion, K. E. "Direct Psychophysical Scaling of the Olfactory Pleasantness of Diluted n-amyl Acetate," *Perception & Psychophysics* **10** (3), 158-160 (Sept. 1971).
21. Henion, K. E. "Odor Pleasantness and Intensity: A Single Dimension?" *Journal of Experimental Psychology* **90** (2), 275-279 (Oct. 1971).
22. Beebe-Center, J. G. "Law of Affective Equilibrium," *American Journal of Psychology* **41** (1), 54-69 (Jan. 1929).
23. Tanyolac, N., Ed. *Theories of Odor and Odor Measurement* (Pelham, N.Y.: Circa Publications, 1968).
24. Mueller, C. G. *Sensory Psychology* (Englewood Cliffs, N.J.: Prentice-Hall, 1965).
25. Henion, K. E. "Psychophysical Scales of the Olfactory Pleasantness of Homologous Alcohols," *Perception & Psychophysics* **9** (2B), 234-236 (Feb. 1971).
26. Corbit, T. E. and T. Engen. "Facilitation of Olfaction Detection," *Perception & Psychology* **10** (6), 433-436 (Dec. 1971).

27. Berglund, B., U. Berglund, and T. Lindvall. "On the Principle of Odor Interaction," *Reports from the Psychological Laboratories, University of Stockholm, Sweden* **318**, 12 (Feb. 1971).
28. Rosenblith, W. A., Ed. *Sensory Communication* (New York: John Wiley & Sons, 1961).
29. Berglund, B., U. Berglund, T. Engen, and T. Lindvall. "The Effects of Adaptation on Odor Detection," *Reports from the Psychological Laboratories, University of Stockholm* **300**, 10 (June 1970).
30. Pryor, G. T., G. Steinmetz, H. Stone. "Changes in Absolute Detection Threshold and in Subjective Intensity of Suprathreshold Stimuli During Olfactory Adaptation and Recovery," *Perception & Psychophysics* 8 (5B), 331-335 (Nov. 1970).
31. Stoll, M. "Important Effects on Odor Produced by Slight Chemical Differences," *France Parfums* 8 (44), 227-232 (1965).
32. Schneider, D., Ed. *Olfaction and Taste IV* (Stuttgart: Wissenschafliche Verlagsgesellschaft MBH, 1972).

CHAPTER 2

THRESHOLD OF SMELL AND MEASUREMENT

Andrew Dravnieks

Odor Sciences Center
IIT Research Institute
Chicago, Ill. 60616

Air pollution by odorous effluents from various industrial processes tends to be one of the most frequent sources of pollution complaints. The odor does not have to be definitely unpleasant to cause complaints. Odors that are pleasant under certain conditions, *e.g.*, fragrances, or the aroma of freshly baked bread, become the source of complaints by individuals who are exposed to these almost constantly.

To reduce odor complaints, various control measures are taken. In all cases, however, some form of odor measurement is needed to establish the extent of the problem and to assess the degree of improvement achieved.

Sensation of odor is an effect caused by odorants when they reach chemoreceptors in the nose. Thus odor measurements may utilize two different approaches. The identity and the concentration of odorants in air can be established by appropriate analytical methods. The odor sensation, which is a physiological and psychological effect, can be evaluated using sensory methods which evaluate human responses to the odorous air. In addition, procedures such as the gas chromatogram/odorogram technique may combine both approaches (analytical and sensory) and help in solving some problems.

ANALYTICAL APPROACH

In a purely analytical approach, the concentrations of known odorous substances in air or an emission are measured. Since many odorants evoke an odor sensation at concentrations as low as 10^{-9} ppb (vol.), such measurements may require very specialized techniques.

However, analytical data must be interpreted by reference to some target level. Usually the target level is related to the detectability of the odorant by the nose. For example, if a substance begins to smell at a concentration of 10 ppb, the analytically determined concentration of 15 ppb would possibly be considered odor evoking. In reality, such interpretations are rarely clean cut. An analytical method may focus on some single substance if it is known that its concentration is the prime cause of odor, or that its concentration varies proportionally to other odorants, is found to correlate with the odor of the entire emission, and can be used as an odor index.

Other analytical methods may measure the combined concentration of compounds of chemically related substances. Total reducible sulfur or total aldehyde content are two examples. In principle, this approach is subject to pitfalls. Odors of a series of related compounds, *e.g.*, sulfides, are strongly dependent on the molecular size of the compound. Odorosity usually reaches a maximum value of 3 to 6 carbon atoms in the molecule, then subsides for larger molecules. Measuring the total concentration of sulfides would give useful correlation to the odorosity only if all the sulfides would vary in concentration reasonably proportionally, as in the case where only one component's concentration is measured. Basically, it is much more desirable to measure the concentration of each odorant separately.

The most adequate approach involves scanning the composition for all possible odorants and measuring the concentrations of those which are found to be present. This is approximated by gas-chromatographic techniques. However, because of the necessity to measure very low concentrations, such analysis must be preceded by a preconcentration step. Odorants must be separated from bulk air and excess moisture, and only then be analyzed.

The classical method of concentrating by cryogenic condensation produces a mixture of odorants and condensed moisture, and requires separation from water before analysis. Modern techniques have abandoned this approach and utilize nonpolar sorbents to collect organic odorants (most odorants are organic substances) while adsorbing negligible amounts of water. Figure 2-1 shows one type of collector. Five grams of a highly purified porous organic polymer (Chromosorb 102), with a total surface area of about 1,500 m^2, placed in a stainless steel cartridge, serve as the collector. An air or emission sample is taken at a rate of up to several liters per minute. In the case of ambient air, as much as 50 to 100 liters of air may need to be passed through the collector for sampling.

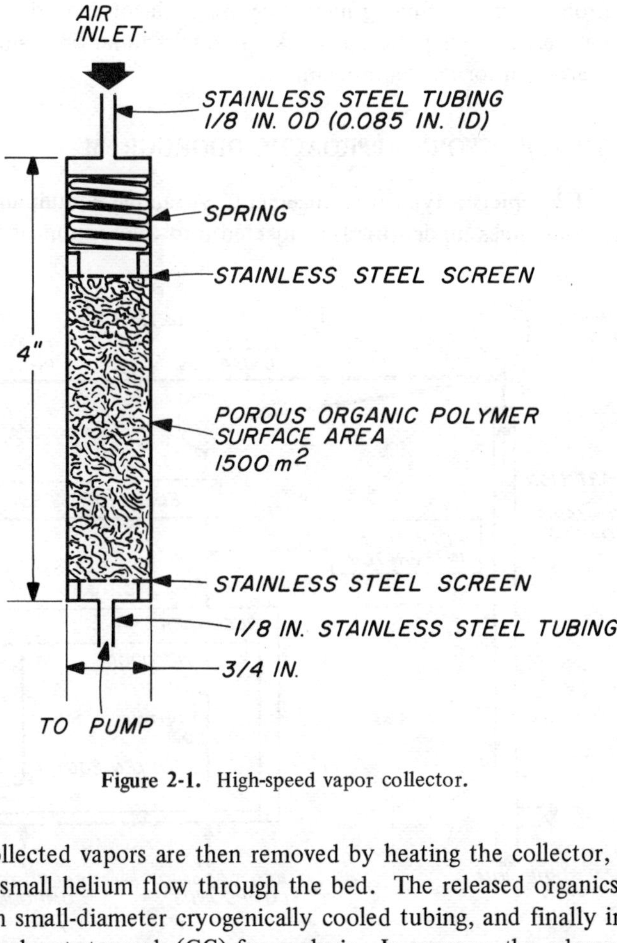

Figure 2-1. High-speed vapor collector.

The collected vapors are then removed by heating the collector, while passing a small helium flow through the bed. The released organics are trapped in small-diameter cryogenically cooled tubing, and finally injected into a gas chromatograph (GC) for analysis. In essence, the odorant content of many liters of air is analyzed in one analysis, and compounds at concentration of 0.1 ppb in air produce measurable GC peaks. The chromatographic stationary phase used for the analysis is selected to suit the particular analytical task.

The GC technique permits simultaneous analysis for many odorants.

In many cases, the identity of odorants may not be known. Identification using mass spectrometry is possible, but very expensive. Frequently it does not succeed because of inadequate sample size, ambiguities in the mass spectrogram interpretation, etc.

A technique where the analyst's nose is used to aid in the interpretation of the chromatogram is finding increasing use in flavor and odor studies. It is also effective in air pollution work. It results in an odor-annotated gas chromatogram, or an "odorogram."

ANALYTICAL/SENSORY APPROACH: ODOROGRAM

Figure 2-2 depicts a typical arrangement. A sample containing many odorants, sometimes unidentified, is injected into the septum of the GC.

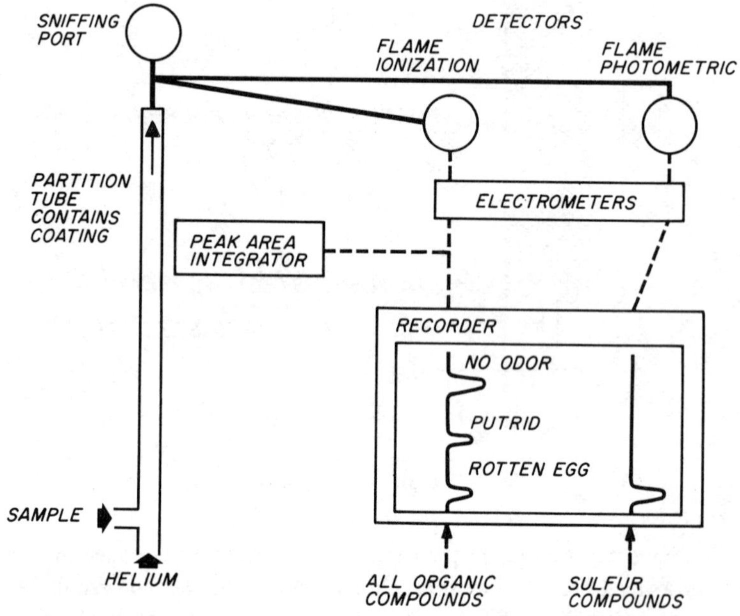

Figure 2-2. Obtaining an odorogram (odor-annotated gas chromatogram).

Components separate in the GC partition column (which actually is a coiled tube) and exit spaced in time. Those more volatile or less soluble in the stationary phase coating in the tube exit first, etc. Detectors monitor their exit and produce a readout on the recorder. A sniffing port is also attached and permits smelling the components as they emerge from the column. Odor notes are marked on the chromatogram.

The same compound always produces a peak in the same spot (same value of the retention time) in the chromatogram. The area under the peak measures the mass concentration of the component in the sample (on that part of the record which utilizes the output of the hydrogen flame ionization detector). Another detector is sensitive only to sulfur compounds. Thus an odorogram contains information on the odor of the components. The peak size gives objective information on the amount of this component.

If the air sample preconcentration is conducted by a device, as depicted in Figure 2-1, the components emerge from the sniffing port at concentrations many times higher than in the original air sample. Hence if a peak is not accompanied by an odor, the corresponding component was present in air much below its odor threshold concentration and is most probably not odor relevant.

Analysis is conducted with programmed temperature rise, to include compounds of high as well as low volatility. One suitable column is Teflon-coated (inside) aluminum tubing with a packed coating of SP 1000 as the stationary phase. This phase produces good GC peaks for most every compound, including free fatty acids, which in most phases produce poor peaks. If the analysis with SP 1000 proceeds to the C_{22} alkane position, it covers the entire range of possible odorants since, according to existing knowledge, compounds beyond this GC position do not exhibit odors.

Comparison of odorogram samples before and after a scrubber indicates which odorants seem to be reduced in concentration, or whether new odorants were generated. Peak-size reduction gives a quantitative measure of the reduction.

If the odorogram "signatures" of various odor sources are known, this technique permits examining ambient odor samples for the presence of odorants from specific sources.

In the odorogram technique, the chemical identity of the odorants may remain unknown, since one simply compares peak sizes of peaks occurring in different chromatograms at the same retention time and accompanied by a similar odor. Interpretations do not need to be based on one single peak; several peaks may be useful.

SENSORY APPROACH

In the sensory approach, individuals called "panelists," "judges," "observers," "participants," etc. are presented with odor samples and asked to indicate their responses. The panelists serve as sensory instruments.

Odor sensation has several sensory dimensions: intensity, detectability, acceptability, and quality (character). Intensity and detectability are interrelated. As the sample is diluted more and more, the odor becomes weaker, and finally reaches an odor intensity that is too weak to be consistently noticed. Of these four dimensions, most regulations deal with detectability, and some with acceptability.

ODOR THRESHOLD, DETECTABILITY

At some low concentration (much dilution), the consistent detection of the odor of a sample becomes impossible: panelists cannot discriminate, except by chance, between pure air and a diluted odorous sample. In between is a range where better-than-chance detection is more or less possible.

Presumably, a dilution level can be defined for the particular sample, at which the success of a correctly identified odorous sample reaches some statistical degree of reliability. This is the threshold concept. Modern psychologists consider such a concept a great oversimplification: judgments depend on the criteria which are used internally by the panelist, and these in turn are influenced by expectations, motives, possible consequences of correct and incorrect judgments, etc.

Despite this fundamental objection, various procedures exist in which reasonably reproducible values of thresholds can be obtained for the same odorous sample. Other techniques may yield equally reproducible, but not necessarily the same, values. Consequently, thresholds are not constants, such as boiling points. Rather they are values within some broader range and dependent on the measurement technique.

Instrumentally, odorous sample dilution techniques subdivide in two groups. In static dilution measurements, a known volume of odorous sample is diluted by the addition of a known amount of nonodorous air. The ASTM D1391 syringe dilution technique is in this category. In dynamic dilution measurements, the sample flows into some mixing volume at a constant rate, mixes with dilution air, which also flows at some constant rate, and is continuously available for smelling. Provisions are made for changing the mixing ratio to produce the desired dilutions. Devices for such manipulations are called dynamic olfactometers.

Ideally, odor threshold measurement should be conducted by a panel directly at the odor source, where the sample can be continuously obtained without need for storage. This requires special vehicles in which panelists can be maintained in an odorless atmosphere. Odorous samples, dynamically diluted, can then be presented to panelists from the dynamic olfactometer.

Such a procedure is cumbersome and expensive, since much of the panelists' time is wasted in transport, waiting, etc. Therefore other techniques have been developed to sample the odorous air or emission and then to evaluate the odor in another location which is well equipped for odor work and is accessible to panelists.

Figure 2-3 illustrates some sampling techniques. In the ASTM syringe method, the sample is taken into a 100-ml syringe or in a 250-ml glass bulb. Most experts believe that the relatively small volume surrounded by a large glass surface results in difficult-to-control and species-specific

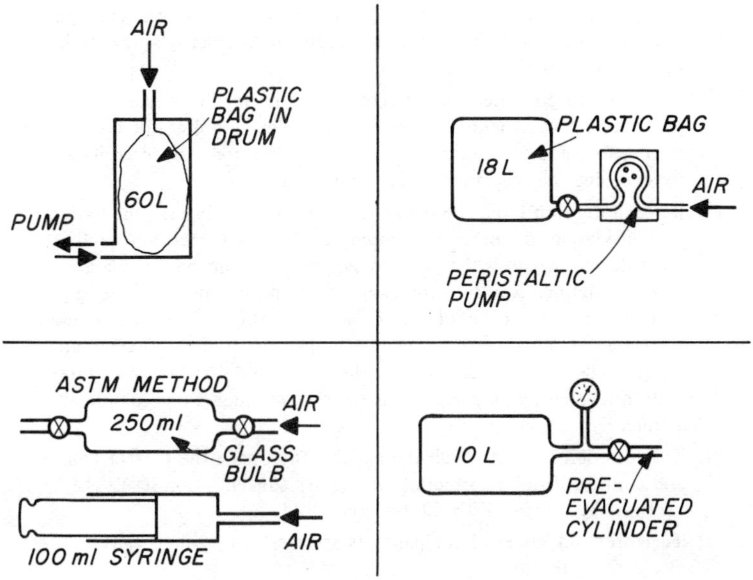

Figure 2-3. Sampling techniques for sensory evaluation of odorous emissions.

adsorption, and sometimes in catalytic degradation losses. Sampling into preevacuated stainless steel cylinders is convenient, but the cleaning of cylinders requires special provisions and some question on adsorption and chemical losses may be raised.

The most often used sampling method utilizes plastic bags. These can be filled either by a drum technique, where the air between the bag and the drum is manipulated to fill the bag with a sample, or by a peristaltic pump technique where the sample is "milked" through disposable odorless plastic tubing into a plastic bag. To reduce sample losses by adsorption and penetration through the bag walls, proper selection of the bag material is made. In the peristaltic pump technique, the tubing and the bag are first preflushed with the sample and then the actual sample is taken. With adequate bags and preflush, samples are useful for up to 48 hours after collection.

The ASTM D1391 test consists of presenting various dilutions of the sample in the form of a few-seconds blast from the 100-ml syringe into the nose. In 1973 the test procedure was reviewed at length by the ASTM E18 Sensory Evaluation Committee and changes were recommended to bring it in better accord with the best sensory evaluation principles. It is reasonably anticipated that the revised version will appear in the 1976 ASTM standards book. Briefly, the following changes are anticipated.

1. Panelist prescreening, if conducted, should use odorants pertinent to the problem; sensitivity of individuals to vanillin-methylsalicylate has been found not to correlate with sensitivity to sulfides or fatty acids.
2. Dilutions are presented, beginning with the most diluted sample and proceeding toward more concentrated samples, doubling the concentration for each subsequent presentation. This prevents desensitization by exposing the nose to stronger odors first.
3. At each dilution level, the panelist receives two syringes: one with a diluted sample, the other with room air. He must smell both and *must* decide which is the odorous syringe. This introduces blanks in a well-defined way into the experiment and permits proceeding systematically from the diluted to more concentrated samples without risking an anticipation effect where the panelist begins to expect and imagine odors. Also, this reduces the possibility that the panel leader may influence the judgment, as is the case in judging one sample at a time.
4. That dilution level at which the panelist first made the correct choice and continued to be correct at two higher concentration levels is taken to be the panelist's dilution threshold.
5. The threshold values of the panelists are averaged geometrically to give a panel threshold.

Several dynamic olfactometric devices are presently being used. TRC (the Research Corporation of New England) utilizes a long glass manifold with eight valved glass ports. Each of the eight panelists sits at his own port. On signal, the ports are opened, the panelists smell, and indicate by electric switch if there was "an odor" or "no odor." Dilution of the sample is controlled by adjusting flows, and the dilution ratios are read from flow meters.

The California PHS (Sanders) portable olfactometer utilizes a mask to which a mixture of a sample and air deodorized by active carbon is delivered. The test begins with odorless air, and the sample addition rate is gradually increased until the panelist detects the characteristic odor of the sample.

Hellman's Odor Fountain uses three ports: one with a jet flow of diluted sample, two others with odorless air jets. The jets serve to carry odorous samples into a hood exhaust, preventing the odorant from accumulating in the room.

The Hemeon olfactometer utilizes three ports, so that three panelists can judge simultaneously.

A. D. Little Company uses an odor test-chamber technique. An odorant/air mixture is generated in a special room and rapidly mixed by fans. Panelists enter the room and judge if there is an odor.

In most such techniques, a parasitic effect of "background" odor vexes the tests to some degree. It is very difficult to reduce odor to zero in what are meant to be blank samples. The practical solution is that the panelists are either taught or learn by themselves that this background does not count as an odor, and what is to be judged is the possible presence of additional odor effects.

These and other considerations (size, convenience, speed) prompted Fats and Proteins Research Foundation Inc. to initiate development of an olfactometer that reduces such problems.

Figure 2-4 shows a device that utilizes a forced-choice triangle principle. A dilute sample is presented with two odorless (room air) samples. The panelist must choose which is the odorous, and signals his choice by depressing a push button placed at the selected port. By guess alone, there is only a 1:3 chance to signal the correct port. There are six dilution levels, typically from 4,500x to 15x, or from 80,000x to 450x. Adjoining levels differ in concentration by a factor of 3. All of the required dilutions and blanks continuously emerge from the appropriate smelling ports at 500 ml/min. Such flow rates do not produce an odor buildup in a normally ventilated room, so that special room-air treatment is not needed.

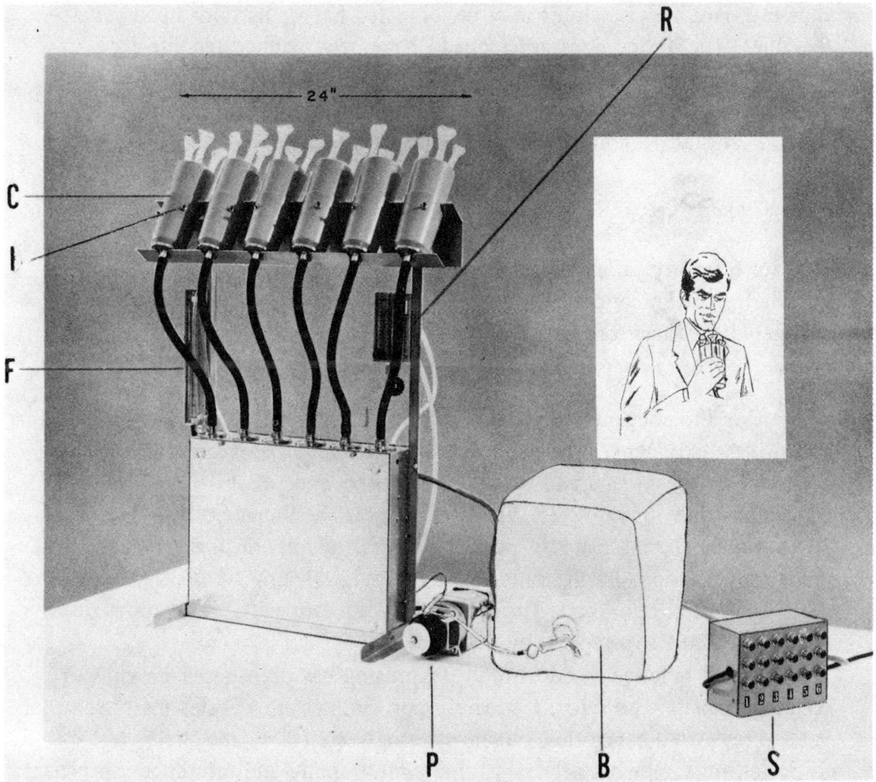

Figure 2-4. Six-level dynamic triangle olfactometer.
B = bag with odorous air sample; P = pump; C = cups with three sniffing nozzles each; I = pushbuttons for indicating panelist's choice through light bulb system on box S; F and R = flowmeters.

Panelists begin with the most diluted level and proceed toward higher concentrations. Nine panelists complete one sample within 12 to 15 minutes. The electrical signaling system and the circular arrangement of ports at each dilution level eliminate bias and the influence of the panel leader on the judgments. Simple statistics and a plot are used to calculate the panel threshold.

Dilution thresholds of odorous ambient air are usually quite high. Dilution by a factor from 2 to perhaps as high as 30 suffices to reduce the odor intensity to a nondistinguishable level. Such weak odors cannot be reliably diluted in syringes, and the techniques of sampling for off-site evaluation are not too well developed. An instrument called a Scentometer is sometimes used. The panelist, situated in the odorous ambient air, breathes through the device, in which the outside air is deodorized by passage through an active carbon layer. He can then admit, by opening various size holes, various proportions of ambient air to the deodorized air and thus estimate how much air may be admitted before its odor becomes detectable. The device does not seem to have been studied technically. At present, the Fats and Proteins Research Foundation has concluded an effort at I.I.T. Research Institute to devise a dynamic triangle olfactometer for ambient air and to compare its performance with the Scentometer.

ODOR INTENSITY

Odor intensity refers to the intensity as perceived, regardless of the knowledge of the concentration or dilution. The intensity is known to change with concentration in accord with a power law:

$$\text{Intensity of sensation} = K\,C^n$$

where C is the concentration of the odorant in the air, K is a coefficient, and n is an exponent. The value of n for different odors generally falls in the range between 0.20 and 0.80. The consequence of this fractional coefficient is that for n = 0.25, the dilution must be increased by a factor of 16 to reduce the intensity sufficiently so that people, on the average, would judge that the odor has become twice as weak. Figure 2-5 shows some plots for several odorants. The odor of an odorant with the shallow slope is more difficult to suppress by dilution.

The usual practice in odor intensity judging has been based on category scales such as 0 = no odor, 1 = barely perceptible, etc. Scales may use from three to 10 categories, but most usually six (0, 1, 2, 3, 4, 5).

Scales are somewhat arbitrary, and some training and reference samples are needed to arrive at a more uniform usage of terms. The ASTM E18 Sensory Evaluation Committee has studied this problem for several years and arrived at the conclusion that a reference odorant scale may as well

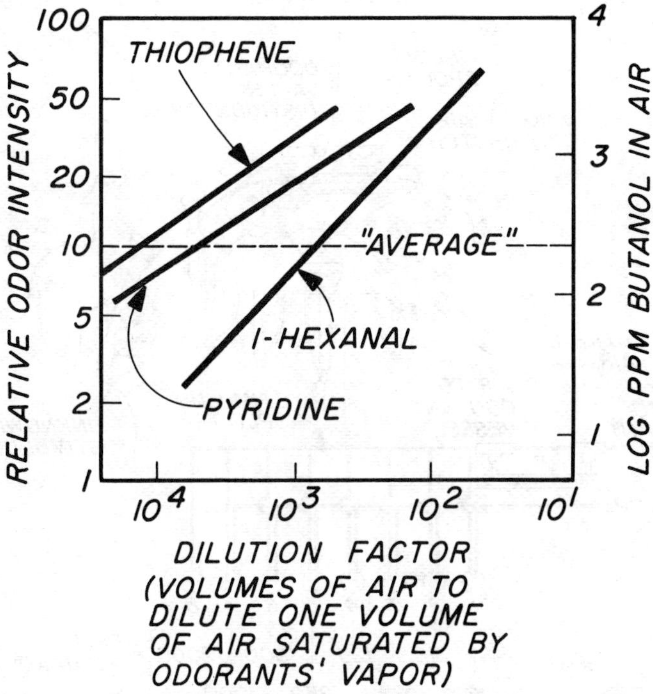

Figure 2-5. Influence of vapor dilution on perceived intensity of odor. Scale at left indicates numbers in proportion to the odor intensities; numbers do not represent categories. Scale at right is in terms of proposed ASTM reference scale, based on 1-butanol as a reference substance.

be used directly, without involving categories. The proposed scale consists of a series of 1-butanol concentrations. The odorous sample is compared for odor intensity to the 1-butanol scale, the panelists' judgments are averaged geometrically, and the odor is reported to be "equivalent in intensity to x ppm (v/v) of 1-butanol in air." The typical scale is depicted in Figure 2-6. The 1-butanol vapor is generated by air flow over its surface in a horizontal tubular vessel and is delivered at calibrated flow rates to each of eight ports. Air is added to the ports as needed to assure that the total flow rate from each port is 160 ml/min.

This procedure requires very little training for the panelists, since a comparison of different odors for intensities is surprisingly easy to learn. The scale is used to compare odors at their as-is intensities, to record intensities for a later comparison, for communicating data to others, etc.

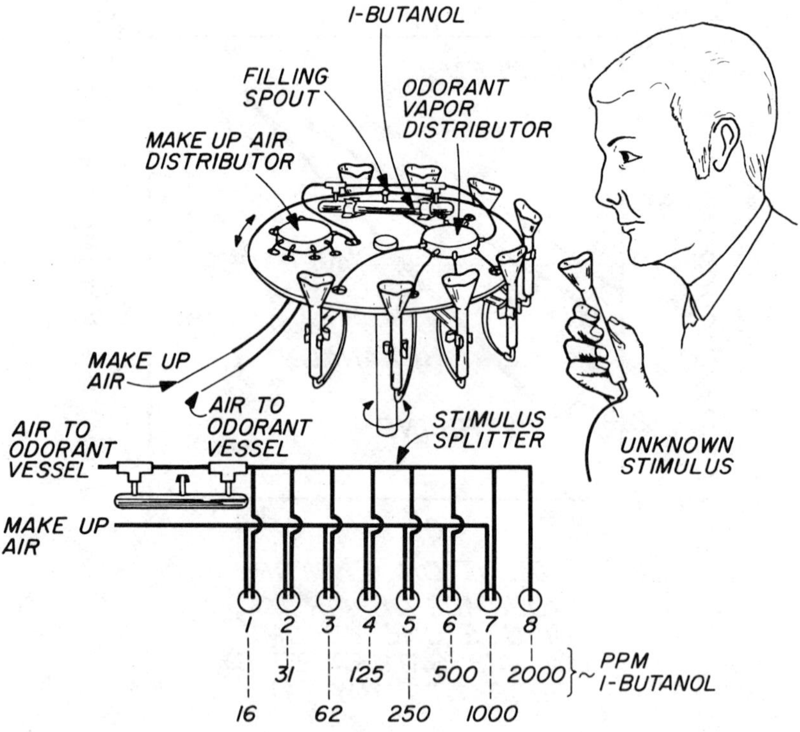

Figure 2-6. Odor intensity reference scale based on l-butanol.

Such reference scales permit characterization of the odor intensity/sample dilution relationship for an industrial odorous emission. Figure 2-7 illustrates the properties of two emissions. The numbers at the right are proportional to the perceived intensity of the respective samples. Thus the odor exhibiting an intensity of 5 on this scale is twice as weak as the odor that corresponds to an intensity of 10.

> The right-hand scale is set by defining the odor intensity of 250 ppm (v/v) of l-butanol in air as equal to 10. Other numbers are spaced so as to correspond to n = 0.66, which is tentatively considered a mean value for l-butanol (see the equation above). The numbers for the threshold region are of the order 0.5.

Such plots are useful in estimating the impact of the odorous emissions on the surrounding area. Emissions A and B both have the same dilution threshold for odors; thus one would think they are equivalent with respect

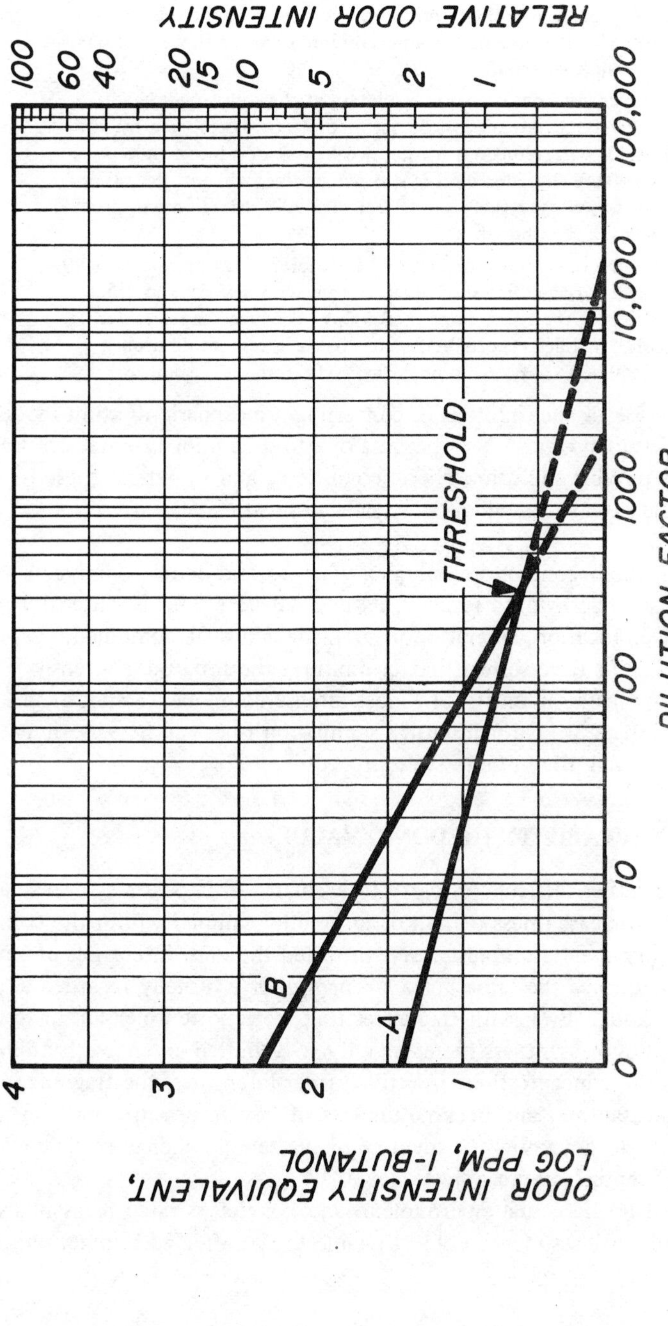

Figure 2-7. Two emissions with the same odor threshold but different slopes of intensity versus dilution function.

to the impact on the people in the vicinity. However, emission A has a lesser slope. Therefore:

1. At the same atmospheric dilution, but as long as the odor is still noticeable, B will exhibit a more intense odor, at concentrations above the threshold.
2. However, odorous emission reduction at the source will have a greater effect on emission B. Reduction of this emission by 90% (10-fold) will result in a marked weakening of odor in those locations where the odor of B is still perceivable, while in the case of A, with the lesser slope, the effect of a 90% reduction will provide a barely noticeable effect.
3. Since the odor dilution threshold is different for different individuals, and for the same individual changes with activity and other conditions, the larger slope of B helps to reduce the relative incidence of odor reports below the formal threshold dilution, *c.f.*, dotted-line extrapolation, into the "subthreshold" region.

Another use of such plots is in converting atmospheric dilution models to odor-relevant profiles. This permits one to map odor iso-intensity lines for various terrains and atmospheric conditions, and to extend these to the subthreshold intensities, which will relate to less-than-50% detection frequencies.

Applications become more complex when several odorous emission sources exist. Each would result in its own separate odor iso-intensity map. For any given location, several odor intensities must be combined. Laboratory experiments have shown that in mixtures the total odor intensity is usually between the intensity of "most intense component" and, at most, twice that intensity. Odor intensity summation rules can be experimentally obtained by using the butanol scale procedure.

ODOR ACCEPTABILITY (HEDONIC VALUE)

This is a rather difficult property to evaluate. It is possible to evaluate the relative unpleasantness of two or more odor samples within the context of the sensory evaluation laboratory, provided the odor intensities are evaluated as well, since the same but a stronger odor is usually regarded as more unpleasant. Even with fragrances that are reported pleasant at low intensities, the pleasantness increases only to a limited extent with a limited increase in the concentration. At still further intensities, the fragrances decrease in pleasantness and become unpleasant. Such measurements, in which the panelists do not know the identity of the sample, indicate which emissions are potentially more objectionable.

However, in the actual environment, an odor that is rated pleasant by uninformed panelists may be objectionable to the affected population. If an odor "does not belong," it is objected to. Thus the aroma of some fragrant chemical is pleasant, but not if an individual is constantly subjected to it. Ambient air is not expected to continuously exhibit such an odor, unless the location is a flower garden.

The actual measurement of acceptability typically involves some category scales: numbers, words, cartoons with various facial expressiones, or poses to depict the type and the degree of response. Recently we have experimented with a combination of the category and a length-of-scale method in which the panelist first makes a category judgment: pleasant, neutral, or unpleasant. He then indicates his reaction by pulling out a flexible, unmarked (on the panelist's side) steel tape ruler to expose that much of the tape as corresponds to the degree of unpleasantness. Each panelist is also calibrated for his length concept of "moderately unpleasant." This length, individual for each panelist, is equal to 100, and the responses to the odor are referred to such individual calibration values.

The method appears workable. For the panel, mean values—the categories "slightly," "moderately," and "strongly"—plot in a straight line versus the log of the indicated tape length. Thus the hedonic magnitude scale is simply equivalent to a nonstructured category scale, which appears to give more resolution and freedom in response than a structured (numbered) category scale. The line in Figure 2-8 helps to indicate which odorous emissions are more unpleasant (above the line) than an industrial average for odors of similar intensity.

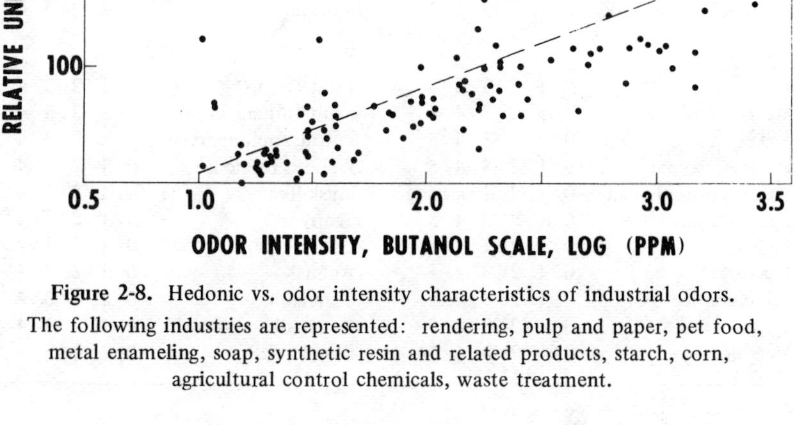

Figure 2-8. Hedonic vs. odor intensity characteristics of industrial odors. The following industries are represented: rendering, pulp and paper, pet food, metal enameling, soap, synthetic resin and related products, starch, corn, agricultural control chemicals, waste treatment.

ODOR QUALITY (CHARACTER)

At present, the most practical technique utilizes lists of odor descriptor terms, with a simple category scale for each descriptor. Table 2-1 depicts an Americanized Harper's scale. The odor is rated for applicability of each descriptor on a 0 to 5 scale. Recently we added more descriptors to produce more differentiation and to offer more choice to the panelists.

To use such scales in a more absolute fashion, it is important to train panelists in the usage of specific terms. This is a difficult and expensive procedure since initially there is much scatter in the meaning of terms. Training results in a more uniform but not necessarily less provincial use of the descriptors and categories.

Table 2-1. Harper's Odor Descriptor Scale (Americanized)

Quality Scale					
Absent	Slightly		Moderately		Extremely
0	1	2	3	4	5

Q		S	
Fragrant	0 1 2 3 4 5	Oily, fatty	0 1 2 3 4 5
Sweaty	0 1 2 3 4 5	Like mothballs	0 1 2 3 4 5
Almond-like	0 1 2 3 4 5	Like gasoline, solvent	0 1 2 3 4 5
Burnt, smoky	0 1 2 3 4 5	Cooked vegetables	0 1 2 3 4 5
Herbal, green, cut grass, etc.	0 1 2 3 4 5	Sweet	0 1 2 3 4 5
Etherish, anaesthetic	0 1 2 3 4 5	Fishy	0 1 2 3 4 5
Sour, acid, vinegar, etc.	0 1 2 3 4 5	Spicy	0 1 2 3 4 5
Like blood, raw meat	0 1 2 3 4 5	Paint-like	0 1 2 3 4 5
Dry, powdery	0 1 2 3 4 5	Rancid	0 1 2 3 4 5
Like ammonia	0 1 2 3 4 5	Minty, pepperminty	0 1 2 3 4 5
Disinfectant, carbolic	0 1 2 3 4 5	Sulfidic	0 1 2 3 4 5
P		R	
Aromatic	0 1 2 3 4 5	Fruit (citrus)	0 1 2 3 4 5
Meaty (cooked)	0 1 2 3 4 5	Fruit (other)	0 1 2 3 4 5
Sickening	0 1 2 3 4 5	Putrid, foul, decayed	0 1 2 3 4 5
Musty, earty, mouldy	0 1 2 3 4 5	Woody, resinous	0 1 2 3 4 5
Sharp, pungent, acid	0 1 2 3 4 5	Musk-like	0 1 2 3 4 5
Camphor-like	0 1 2 3 4 5	Soapy	0 1 2 3 4 5
Light	0 1 2 3 4 5	Garlic, onion	0 1 2 3 4 5
Heavy	0 1 2 3 4 5	Animal	0 1 2 3 4 5
Cool, cooling	0 1 2 3 4 5	Vanilla-like	0 1 2 3 4 5
Warm	0 1 2 3 4 5	Fecal (like manure)	0 1 2 3 4 5
Metallic	0 1 2 3 4 5	Floral	0 1 2 3 4 5

We are experimenting with another approach to the descriptor technique, which appears to be workable with most any panel. The approach is based on two assumptions.

1. Panelists differ in the interpretation of the meaning of the descriptors and in the degree of perception of different odor notes.
2. Each panelist, after a short experience with a descriptor list and 10 to 20 odors, becomes reasonably consistent in his usage of descriptors.

This approach still permits approximating the odor note profile of the odor by combining the responses of the panelists. However, it is most useful in comparing the odors among themselves, or with respect to some reference odors. This is done by using the χ^2 statistic with a 2 x 2 contingency table. In essence, it is assumed that there are 44 x 9 = 396 descriptor terms for a panel of nine. Both odors are rated for all 396 descriptors.

A tabulation is made of the usage of the descriptors separately, for each panelist and for each pair of odors, A and B. For example, from the 44 descriptors, a panelist may have used four of the descriptors for odors A and B; used seven in the description of A but not B; used five others in the description of B but not A; and did not use the remaining 28. The response patterns are added over the panel. The resultant contingency table (panel of nine) for anisole and toluene was developed:

		Toluene	
		+	–
Anisole	+	38	14
	–	8	350

The resulting chi-squared value is 220, and the coefficient of association 0.55, indicating a significant similarity between the odors of toluene and anisole. For odorants with very different odors, the coefficient may be as low as 0.01. For the same odor repeated at different sessions by the same panel of ordinary subjects, the coefficient usually exceeds 0.50.

This method should be useful, if developed further, for characterizing air pollution odors by a comparison of their descriptors to the descriptors of references. The only requirements are that the same panels are used for the sample and the references, and that the reference odors are reasonably pertinent.

PANELS

Eight or more panelists should be used for odor measurements in critical cases, wherever a costly argument may be expected. With a lesser number, the values obtained are still informative, but simple statistical tests for judging the significance of differences between samples become difficult, unless the differences are large.

Panelist selection depends on the rationale of the measurement. Selected, trained subjects are best able to discriminate small differences in odor character, but would not reflect the response of the broader population. The ability to detect odor cannot be substantially increased by training, but the ability to recognize odor character can be improved.

For the most usual task in industrial odorous pollution measurement—evaluation of the dilution threshold—the most suitable panel would be one that reflects the odor sensitivity of the broader population, both with respect to the mean value and to the distribution of sensitivities within the population. Thus panels consisting of subjects with average, lower-than-average, and higher-than-average sensitivities provide more useful information than a panel carefully selected to contain individuals with similar sensitivities. The latter would generate very homogeneous and consistent data, but would give no information on the sensitivity distribution. It would be impossible to estimate the dilution at which, for example, only 10% of the people would still be able to detect odor.

CHAPTER 3

MEASUREMENT OF ODORS

Thomas M. Hellman

Allied Chemical Corp.
Morristown, N.J. 07960

The development of a rational basis for attacking the problem of annoying odors in the atmosphere is one of the major tasks facing both industry and control agencies. Reasonably accurate and reproducible measurements are essential both in defining the magnitude of the problem and in ascertaining the effectiveness of control techniques. Since the ultimate goal is to eliminate community annoyance, odor measurements are needed to establish a relationship between the community's reaction to an odor and the dose to which it is exposed.

The objectives common to most odor evaluations have been summarized[1] as follows:

1. to study the relative importance of various odoriferous processes
2. to study the effectiveness of various odor-control systems
3. to measure source intensity by determining odorant levels in discharges
4. to forecast odor distribution and intensity at various distances from a source on the basis of source concentrations, stack height, effluent gas temperature, topography, and meteorology
5. to investigate the feasibility of predicting odor intensity from the chemical and physical characteristics of the discharge.

Of the common forms of air pollutants, odors are unique in their requirements for measurement because their sole effect is the subjective response of humans to various intensities. In general, programs to detect and identify odors by chemical analysis are not easily accomplished. Except for a few simple cases, the chemical composition of odors is so

Reprinted with permission from *Proceedings: State-of-the-Art of Odor Control Technology*, published by Air Pollution Control Association, 1974.

complex and varied that the difficulties of detailed analysis are seemingly insurmountable. Moreover, it has been found that the human nose can, in most cases, detect odorants in concentrations impossible to identify and monitor with available instrumentation. Certainly, at the present time, the ultimate sensor is the human olfactory system, and the data output must be limited to a subjective reaction. Consequently, the only practical way to measure odors is through the use of human odor judges. In order to add an objective component to these subjective odor measurements, odor standards (intensity and quality) have been designed to fit a number of specific applications, thus affording the odor judges a basis upon which to make odor measurements. In addition, a number of elaborate systems have been devised to introduce the odorant to the judge.

SELECTION OF ODOR JUDGES

Since at the present time there is no instrument available for measuring sensory characteristics of odors other than the human nose, a suitable selection of odor judges is essential in obtaining meaningful data. It is fundamentally important that odor judges be selected on careful consideration of the type of information or measurement desired. If an evaluation is to be made of a community nuisance odor, the subjects selected should be representative of the entire population. On the other hand, intensity or threshold determinations may require only a few carefully trained experts. A discussion of odor characteristics can be found in Chapter 2 written by Dr. Andrew Dravnieks.

Odor panels varying widely in size and composition have been employed in odor characterization studies. Hemeon[7] has found two or three subjects to be both convenient and adequate for all practical purposes. At the other extreme, Washington State University[11] has converted a 35-foot bus into an air-conditioned, mobile odor-perception laboratory to permit evaluation of responses from diverse groups of test subjects at fairs, expositions, elementary schools, and universities. Arthur D. Little Inc.[6] used four trained staff members, each with more than one year's experience in analytical work, to conduct an evaluation of the recognition threshold for 53 commercial chemicals for the Manufacturing Chemists Association. In an investigation where it was necessary to more closely approximate the total population, Southern California Gas Company[10] selected an untrained panel of 35 people to make threshold measurements of 18 sulfur compounds occurring as odorous constituents of natural gas.

The size of the panel can have a significant effect on the cost of an investigation, as well as on the complexity of data reduction. When large panels are used, the proportionately larger number of man-hours required is an obvious contributor to the cost of the measurement. Moreover, the relatively large amount of data collected may necessitate data processing by digital computer. With small panels, tests can be conducted quite conveniently, although the increased requirements of training may become a significant expense factor. In general, panel size must represent a compromise between measurement cost and desired statistical confidence in the results.

Wittes and Turk[12] have presented an in-depth discussion of potential methods of screening candidates for sensory odor panels and statistical procedures for evaluating test results. They suggest that panel members should be selected on their demonstrated abilities to (1) differentiate among odors at low levels of intensity, (2) focus their attention on more than odor quality in a given sensation, (3) understand test procedures, and (4) exhibit satisfactory behavior during tests.

One of the most common and convenient methods that can be used to screen odor panel candidates is the triangle test. In this test the subject is presented three samples, two identical and one different. He is asked to choose the "odd" sample. Dilute aqueous solutions of food flavors, which the subject can either smell or taste, are frequently used.

Candidates can be tested on their ability to distinguish among different intensities of the same odor by a somewhat more sophisticated procedure. A dilution series of an odorant in an odorless diluent is set up. One sample of the series is removed, and the candidate is asked to replace it in its proper position. Wittes and Turk[12] recommend a 20-sample series, with the odor of the nineteenth or twentieth being detectable only by a very sensitive person. They suggest the use of fruity or fragrant odors, such as amyl acetate, eucalyptol, oil of wintergreen, or heptaldehyde. These odorants can be diluted with mineral oil, propylene glycol, or benzyl benzoate.

If a panel is to be used to evaluate odor qualities, a multicomponent odor identification test may be required in the screening process. In this test the subject attempts to identify the components in mixtures containing two, three, and four odorants by a comparison with known pure-odorant standards. The subject is told how many components are in the sample and has the standards available for reference at all times during the test. Wittes and Turk recommend the following eight standards: oil of cade, cassia, eucalyptus, amyl acetate, clove oil, orange oil, almond oil, and vanillin.

Other qualities of an ideal odor judge that should be considered by the test director include speed, interest, independence, and honesty. These qualities cannot be quantitatively evaluated but can be observed and rated during the screening process. In assigning relative weights to the scores on different tests, the test director should consider the tasks the panel will ultimately perform, but must, in general, rely on his own judgment.

The three test procedures discussed above can also be used in the training of an odor panel.[3] As soon as a group of trainees completes a test, the correct answers are disclosed, and the test is repeated with attention focused on eliminating any errors. Another training procedure involves each panel member's learning to consistently identify a group of standards. The chosen standards should be tailored to the odorant the panel will eventually evaluate.

SAMPLING PROCEDURES

If the purpose of an odor evaluation is to study an ambient air malodor problem, it is usually preferable for the judges to expose themselves directly to the odorous atmosphere. This direct approach eliminates many possible sources of measurement error and allows the collection of large amounts of data that would be economically prohibitive by sampling methods. However, unfavorable local conditions and specific measurement objectives may make direct measurement impractical and necessitate sampling the odorous atmosphere. Turk[3] states that such instances occur when (1) the odorant must be diluted, concentrated, warmed, cooled, or otherwise modified before exposing the judges to it, (2) it is necessary to have a uniform sample large enough to be presented to a number of judges, or (3) the samples must be transported to another location where sensory evaluations can be made. Of course, if any chemical analysis is to be performed in conjunction with the sensory evaluation, sampling is unavoidable.

If conditions are such that the sample requires neither dilution nor concentration, simple grab sampling techniques may be adequate. The desired volume of odorous air is collected by allowing the gas to fill an evacuated container or to displace a liquid from the container. Grab sampling is subject to some severe limitations that must be considered for each specific application. If the sample is initially hot and later cooled, condensation may occur and alter the concentration of the odorant. Further limitations include the possibility of chemical reactions or sorption of the odorant on the container wall. These effects are minimized if the container is constructed of inert material and designed to have a small surface-to-volume ratio. Suitable materials may include stainless steel, glass, Mylar, Teflon, or Saran, but an advance test of the container with the particular odorant is desirable.

A recent study by Duffee and coworkers[19] has indicated considerable differences in the effect of sample container materials on off-site odor measurements of rendering process emissions. They found preconditioned Mylar, Tedlar (and Teflon), and polyetyhlene containers all will maintain rendering process emission sample integrity for 24 hours, after which there is a rapid decrease in odor level.

Hemeon[7] has found that odorant adsorption on the walls of a five-cubic-foot polyethylene sample bag is generally insignificant during a period of two to three hours. It should be noted, however, that other workers do not consider polyethylene sufficiently impervious for use in gas-sampling containers. Benforado and his coworkers[8] have experienced no odorant adsorption problems with 250-milliliter glass sample tubes even though the surface-to-volume ratio is relatively large.

The problems associated with sampling hot exhaust gases can be minimized by diluting the sample as it is collected at the source. The sample can be reduced to concentration and temperature levels suitable for human exposure without condensing any of the odorant.

Several techniques for sampling with dilution have been reported.[1] Sampling may be accomplished dynamically by ratioing the flow rate of odor-free air to that of the source stream. Alternately, a sample taken directly from the odorant source can be transferred to a larger container having enough odor-free air to satisfy the dilution requirements.

Very hot exhaust gases can be sampled in metal tanks and diluted sequentially by pressurizing the tank with odor-free air. If necessary, the tank can be vented to atmospheric pressure and repressurized until the desired concentration is reached. Another interesting technique, which yields an exponentially changing odorant concentration, employs a rigid container with constant mixing. The pressure of the container is kept constant by exhausting the gaseous contents at a rate equal to the input. The exhaust gas has a known concentration at any instant of time, and can be presented directly to judges for sensory evaluation.

An increase in the concentration of an odorous air sample is usually required only as a prelude to chemical analysis, and is seldom necessary to sensory evaluation. Chemical analysis in conjunction with sensory measurements may be appropriate if a community malodor is to be traced to one or more possible sources. Common methods of increasing sample concentration include freeze-out sampling, adsorptive sampling, and absorptive sampling.[1,3]

In *freeze-out sampling* the source stream is passed through a cold trap that freezes and holds the odorous components but does not retain the major odor-free components. Liquid air is a commonly used refrigerant, although unfortunately it traps both water and carbon dioxide. Use of a desiccant upstream of the trap to remove water is not advisable because of the possibility of selectively adsorbing or absorbing odorous components. A major advantage of freeze-out sampling is the minimal probability of chemical interaction among the concentrated odorants at low temperatures.

Adsorptive sampling involves the use of an activated carbon or a silica gel column to concentrate the odorous material. Activated carbon is an efficient medium, in that it selectively adsorbs organic matter in preference to water, but this strong attraction for odorous material makes recovery difficult. Silica gel is a less efficient adsorbing medium but offers the advantage of milder recovery conditions.

If the concentration factor need not be known precisely, *absorptive sampling* may be convenient. In this method the odorous air is bubbled through an odorless, high molecular weight solvent such as mineral oil or ethylene glycol. Alternately, a piece of some absorbent material may be suspended in an odorous space and subsequently removed for odor evaluation. Absorptive sampling is highly inefficient and generally is used only if the amount and complexity of equipment must be minimized.

EQUIPMENT TO AID SENSORY EVALUATIONS

All odor threshold measurements involve the determination of the number of dilutions of an odorous gas sample required to render it nonodorous. To a lesser degree, measurement of the other odor attributes also involves evaluation of the odor after dilution to varying levels of intensity. Numerous devices have been designed for both field and laboratory use to control and measure the required dilutions. These instruments, often termed olfactometers, are essential in studies of the contributions of odor to air pollution.

A number of devices have been designed specifically for making organoleptic measurements in the field. The Scentometer, manufactured by Barneby-Cheney in Columbus, Ohio, is among the most commonly used instruments in this portable category. This instrument is capable of producing dilution ratios up to 128 to 1 and can be conveniently hand carried. In principle, the Scentometer operates by blending two streams of ambient air, one of which is passed through an activated carbon chamber for deodorization. The observer simply breathes through the instrument and controls the dilution ratio by covering selected air inlet openings. The observer's olfactory sense can be defatigued by breathing through the Scentometer with all odorous air inlets closed.

Major limitations of portable olfactometers include a limited dilution range and the necessity of relying on a single odor observer. Although the Scentometer has one of the widest dilution ranges among the portable devices,[4] it cannot be used for measuring odor numbers larger than 128.

Test chambers may be used either to present the odor to a number of judges by surrounding them with the odorous gas or to provide an odor-free environment for sensory testing by other methods of sample presentation. In either case, the test chamber must be provided with some system for space odor control to minimize interference from extraneous background odors and to remove vapors generated by the tests themselves.

This odor-free environment provides an opportunity for the judges to defatigue their senses between tests. Space odor control can be accomplished by recirculating the test-chamber air through an activated carbon unit or by exhausting the test-chamber air to the outside while admitting fresh air through an activated carbon unit. If the available ambient air is sufficiently pure, activated carbon treatment of the incoming air may not be necessary.

The odor test chamber should be free from competing distractions, such as extraneous sounds relating to sample preparation. Color schemes should be neutral. Suitable materials of construction include hard-surface wall panels and asphalt tile flooring which have been found to be low in background odor. If highly odorous materials are to be tested, it may be wise to spray the walls of the chamber with odorless, strippable, replaceable coating.

Turk[13] has studied the quantitative design of test chambers. He states that the concentration of odorous vapor in a chamber will approach an equilibrium value at which the rates of vapor-reducing and vapor-increasing processes are equal.

A variety of techniques and devices has been successfully employed to present samples of odorous air to judges for their evaluation. The fundamental objective of any sample presentation technique is to allow the observer to sniff a reproducible odorant sample of known dilution without interference from extraneous odors.

As previously mentioned, the test chamber itself can be controlled at the desired odorant concentration so that all judges can enter the chamber to make a simultaneous evaluation. This technique was used by Arthur D. Little Inc.[6] in making threshold determinations for 53 odorant chemicals. The test room had two chambers, both with activated carbon treatment of the influent air. The judges conditioned their olfactory senses in the first odor-free chamber while the odorant concentration was adjusted to the desired level in the second chamber. After entering the second chamber and making the required evaluation, the judges returned to the first chamber for a defatiguing period. The air in the second chamber was then purified and readjusted for the next evaluation. This procedure allows rapid sample presentations and assures that all judges experience the same stimulus.

The "odor fountain" developed by Hellman and coworkers at Union Carbide's South Charleston Technical Center[20] has the advantages of low residence time, rapid turnover, and minimum effect on the observer. The odor fountain system does not have the inherent limitation of a contained system. For example, potential adsorption problems are minimized. Concentration changes are effected in less than a minute and the odor judges are able to make observations in an undistracted manner.

A valid criticism of the odor fountain is that it presents the odor stimulus in an artificial manner that may compound the difficulty of separating an odor response from other physical responses. The rushing flow of odorant into the nose may influence odor perception because of the unique aerodynamics of the nose. To minimize the difficulty in separating the responses, the odor judges are exposed to a reference odor fountain that delivers relatively odor-free air at a flow rate similar to that of the working fountain. It is actually the difference in response of the odor judge to the two fountains that is measured. The extra-odor physical responses appear to be canceled out by the reference fountain. The effect of air flow on the odor judge is additionally reduced by having the judges sniff the odor stream rather than gently inhaling it. The rapid inhalation rate of a sniff is closer to the flow rate out of the fountain than is the slower inhalation rate of a gentle breath.

A technique that has been frequently used for threshold measurements is the standard ASTM dilution method.[8,14] A diluted sample of odorous air is prepared in a 100-milliliter glass syringe and allowed to stand 15 minutes to assure uniformity. The observer suspends breathing for a few seconds and slowly expels the 100-milliliter sample into one nostril. He then makes and records his evaluation. As with other methods, the evaluation is made in an odor-free room, and a minimum of 15 minutes is allowed between tests to avoid olfactory fatigue.

A number of studies[7,10,11] have been conducted with devices that prepare continuous streams of diluted odorant by controlling the flow of both pure and odorous air. An operator adjusts the ratio of the two flow rates to produce the desired level of dilution. The observers may approach a sample port to evaluate the odor or may sit beneath an odor hood with the operator controlling the length of exposure.

COMPLICATIONS IN VALID ODOR MEASUREMENT

Sensory analysis can prove to be a convenient solution to the problem of odor measurement, but is far from being free from difficulties. It is well known that there are variations in the odor-perception characteristics of different persons and in the same person from day to day. These variations greatly complicate the task of obtaining consistent and reproducible odor evaluations by sensory methods. Hemeon[7] believes that the odor-perception differences among individuals have been overemphasized and reports that threshold values measured under identical conditions rarely vary by more than 3 to 1. On the other hand, Wilby[10] reports studies that show threshold values for 35 untrained observers to have a minimum variation of 10 to 1, with a maximum variation of 50 to 1.

The response of an observer is dependent not only on the nature and strength of the stimulus but also on the degree of personal adaptation, motivation, attitude, expectancy, previous experience, and variation in background. Lindvall[9] has presented both an excellent review of the literature and the results of some recent experiments of his own on the importance of these variables.

The existence of odor adaptation is demonstrated by an increasing threshold value with increasing time of exposure to a constant stimulus. Studies by Lindvall,[9] using hydrogen sulfide, indicate that adaptation commences quickly and generally levels off after two to five minutes. The relative change in response is independent of the absolute concentration level, and recovery is complete after about four minutes of exposure to odor-free air.

Experiments with auditory, visual, and olfactory behavior have shown that a subject's response to a given stimulus can vary significantly with changes in motivation. In particular, monetary rewards and punishments can produce up to a 20:1 variation in an observed odor threshold, and significantly influence the probability of a subject's reporting a sensation of odor when none is present.

It is generally accepted that, for a given subject, repeated olfactory testing produces a tendency to improved sensitivity until the subject has become familiar with the test conditions. Untrained subjects are commonly believed to exhibit higher thresholds than more experienced judges. Experiments by Lindvall,[9] however, show no evidence of any practical difference in the results of repeated threshold determinations. He attributes these results to the experimental conditions, but suggests that preliminary tests may be a worthwhile precaution.

Differences among observers with respect to age, health, sex, and smoking habits have been reported to influence olfactory sensitivity. However, investigations of these variables have generally produced contradictory and inconclusive results. In evaluating a community odor problem, it is a wise precaution to statistically match the test panel to the general population with respect to as many of these biological variables as possible.

In addition to differences among individuals there are many experimental variables that contribute to the total problem of obtaining accurate odor measurements. Although no generally accepted guidelines have yet appeared, there are indications that both temperature and humidity influence the human odor response to some degree. Until these effects can be quantified, odor evaluations or comparisons should be made at constant temperature and humidity whenever possible.

Other experimental factors that can invalidate odor measurements include background odors, sample-preparation errors, failure of judges to follow instructions, prior contamination of the olfactometer or test chamber, sample deterioration, and undetected equipment malfunctions. In spite of the uncontrollable human factor in sensory testing, rigorous attention to details can yield significant dividends in measurement accuracy, and any effort expended in perfecting reliable experimental procedures is clearly justified.

INSTRUMENTAL METHODS AND INSTRUMENTAL-SENSORY CORRELATIONS

Odorants, like other chemical substances, are amenable to chemical analysis. However, in most cases such an analysis is difficult and time-consuming as an odor may be a complex mixture of compounds at relatively low concentrations. Since, at the present time, the physical and chemical determinants of odor are not known, a chemical analysis by itself, no matter how complete, is not a measurement of odor.[15] One method for translating chemical analytical data of chemical mixtures to psychophysical descriptions of odor intensity and quality has involved the development of a complete inventory of all substances that can contribute significantly or be related to odor. Gas chromatography can provide a list of the detected species, which can then be arranged according to their physical and chemical properties. However, even if the components of the mixture were identified, conclusion on the odor properties which would occur in the subsequent dilution in the atmosphere are not yet routinely possible, on the basis of the present knowledge in this field.

A survey of the literature from 1958 to 1969 for examples of instrumental-sensory correlations was conducted under the sponsorship of the ASTM Committee E-18 on Sensory Evaluation of Materials and Products.[16] Of the several thousand articles reviewed in 65 major technical journals, only 45 were judged to have any direct value toward progress in establishing standards for instrumental-sensory correlations. Of these, 40 dealt with food, and the remainder with pure chemicals. None dealt with odoriferous air pollutants *per se.*

There have been significant advances in the analysis of various chemical substances that are known to be odoriferous. For example, the rapid analysis of malodorous sulfur-containing compounds can be carried out by an automated gas chromatographic process using a flame photometric detector, or by using a series of preselector filters with either a microcoulometric or flame photometric detector.[18] This work was prompted by the known importance of sulfur-containing substances from pulp and paper mills. For most odoriferous air pollutants, however, analytical methods that are demonstrably related to odor are not available.

REFERENCES

1. *Methods for Measuring and Evaluating Odorous Air Pollutants at the Source and in the Ambient Air.* Dept. of Environmental Hygiene, Karolinska Institute, Stockholm, Sweden (1970).
2. Turk, Amos, and Mehlmen Stanley. "Correlations between Instrumental and Sensory Characterizations of Atmospheric Odors," *Correlation of Subjective-Objective Methods in the Study of Odors and Taste,* ASTM STP 440, American Society for Testing and Materials (1968), pp. 27-35.
3. Turk, Amos. *Industrial Odor Control.* (Somerville, N.J.: Center for Professional Advancement).
4. Duffee, R. A. "Appraisal of Odor-Measurement Techniques," *J. Air Pollution Control Assoc.,* **18** (7), 474 (1968).
5. Turk, Amos. *Selection and Training of Judges for the Sensory Evaluation of Odor Intensity and Character of Diesel Exhaust.* Public Health Service Publication 999-AP-32 (1967).
6. Leonardos, G., D. Kendall, and N. Barnard. "Odor Threshold Determinations of 53 Odorant Chemicals," *J. Air Pollution Control Assoc.,* **19** (2), 91-95 (1969).
7. Hemeon, W. C. L. "Technique and Apparatus for Quantitative Measurement of Odor Emissions," *J. Air Pollution Control Assoc.,* **18** (3), 166-170 (1968).
8. Benforado, D. M., W. J. Rotella, and D. L. Horton. "Development of an Odor Panel for Evaluation of Odor Control Equipment," *J. Air Pollution Control Assoc.,* **19** (2), 101-105 (1969).
9. Lindvall, T. *On Sensory Evaluation of Odorous Air Pollutant Intensities.* Dept. of Environmental Hygiene, Karolinska Institute, Stockholm, Sweden (1970).
10. Wilby, F. V. "Variation in Recognition Odor Threshold of a Panel," *J. Air Pollution Control Assoc.,* **19** (2), 96-100 (1969).
11. Sullivan, D. C., D. F. Adams, and F. A. Young. "Design of an Odor Perception and Objectionability Threshold Test Facility," *Atmos. Environment,* 2, 121-133 (1968).
12. Wittes, Janet, and Amos Turk. "The Selection of Judges for Odor Discrimination Panels," *Correlation of Subjective-Objective Methods in the Study of Odors and Taste.* ASTM STP 440, American Soc. for Testing and Materials, pp. 49-70.
13. Turk, Amos. "Concentrations of Odorous Vapors in Test Chambers," *Basic Principles of Sensory Evaluation.* ASTM STP 433, American Society for Testing and Materials (1968), pp. 79-83.
14. "Standard Method for Measurement of Odor in Atmospheres (Dilution Method)," American Society for Testing and Materials, Des. D 1391-57 (1967).
15. "Air Pollution Aspects of Odor," unpublished EPA document.
16. *Reviews of Correlations of Objective-Subjective Methods in the Study of Odor and Taste.* ASTM Philadelphia, ASTM Spec. Tech. Publ. 451 (1969).
17. Stephen, R. K., *et al.* "Direct Chromatographic Analysis," presented at Am. Chem. Soc. Meeting, Minneapolis, April 1969.

18. Adams, D. F., *et al.* "Analysis of Sulfur Containing Gases in the Ambient Air Using Selective Pre-filters and a Microcoulometric Detector," *J. Air Pollution Control Assoc.,* **18**, 145 (1968).
19. Duffee, R. A., *et al. Evaluation of Odor Measurement Techniques, Vol. I: Animal Rendering Industry.* prepared for Environmental Protection Agency, Report no. EPA-650/2-74-008-a, Jan. 1974.
20. Hellman, T. M. and F. H. Small. "Characteristics of the Odor Properties of 101 Petrochemicals Using Sensory Methods," *J. Air Pollution Control Assoc.,* **24** (979) (1974).

CHAPTER 4

SOCIAL AND ECONOMIC EFFECTS OF ODORS

R. David Flesh

Copley International Corporation
La Jolla, California 92037

Amos Turk

The City College of
The City University of New York
New York, N.Y. 10031

The question of how to deal with community odor problems has been a subject of much concern among authorities throughout the United States and in other parts of the world. In recent years the number of complaints about disagreeable odors has approached the total of all other air pollution complaints received by authorities. We will cite data from California because some of its communities combine industry, population, and climate in patterns that generate many odor problems.

A federally sponsored survey, conducted in 1969, ranked the Los Angeles and San Francisco metropolitan areas among the 10 greatest potential odor-problem areas in the United States.[1] The relative importance of odor complaints to other air pollution complaints in these areas, as well as in Humboldt County, is illustrated in Table 4-1. Table 4-2 gives the number of odor complaints made in these metropolitan areas on a per capita basis.

The survey previously referred to found the national average to be 29 odor complaints per 100,000 persons, which is exactly that shown for the San Francisco metropolitan area. It is noted that the per capita figure for

*This chapter is based, in part, on studies performed under Contract Nos. CPA 22-69-50, CPA 70-116, and 68-02-0095 with the U.S. Environmental Protection Agency. The authors are grateful to R. Paul Weddell for his assistance in preparing the sections on economic impact.

Table 4-1. Comparison of Number (and Percentage) of Odor, Smoke, and Other Air Pollution Complaints Received by Los Angeles County and Bay Area Air Pollution Control Districts During 1971

Complaint	Los Angeles APCD	Bay Area APCD	Humboldt County APCD
Odor	3,812 (60%)	1,202 (33%)	746 (61%)
Smoke	1,423 (23%)	1,418 (39%)	388 (32%)
All other	1,074 (17%)	985 (28%)	88 (7%)
Total	6,309(100%)	3,605(100%)	1,222(100%)

Sources: Ralph E. George, Los Angeles County APCD; Tom F. Brennan, Bay Area APCD; and Jack L. Caufield, Humboldt County APCD.

Table 4-2. Number of Odor Complaints per 100,000 Persons Residing in Los Angeles Co., Bay Area, and Humboldt Co. Air Pollution Control Districts

District	District Population (1970)	No. Odor Complaints per 100,000 Persons
Los Angeles Co.	7,032,075	54
Bay Area[a]	4,174,233	29
Humboldt Co.	99,692	748

[a]Includes Alameda, Contra Costa, Marin, San Francisco, San Mateo, and Santa Clara counties only.

Source of population data: U.S. Bureau of the Census.

Humboldt County is more than 25 times greater than the national average. Virtually all of the complaints were directed against odors produced by two sulfate pulping plants operating near Eureka. The per capita figures for Los Angeles County and the Bay Area are much less, although still at or above the national average. The majority of complaints in these locations were also aimed at sulfide odors. However, these odors were produced by numerous oil refineries and chemical plants.

In California and in most state and local air pollution control jurisdictions across the nation, the legal weapons employed to attack odor problems have been the public nuisance laws.[2,3] The application of such laws is usually decided by the number of persons who have complained, since it must be shown that "any considerable number of persons" have been aggrieved. An example of such a regulation is Rule 51 of the Los Angeles County Air Pollution Control District:

> **Rule 51. Nuisance**
>
> A person shall not discharge from any source whatsoever such quantities of air contaminants or other material which cause injury, detriment, nuisance or annoyance to any considerable number of persons or to the public or which endanger the comfort, repose, health or safety of any such persons or the public or which cause or have a natural tendency to cause injury or damage to business or property.

It is generally acknowledged by air pollution control officers that public nuisance laws are helpful in widespread chronic-odor situations, but not in localized odor problems in which few persons are affected or in transient cases in which the complaints cannot be validated. In localized odor situations, the use of such laws is greatly hindered by the fact that it is difficult for the enforcement agency to convince enough complainants to attest to the existence of a problem. Without validation of complaint information by representatives of the enforcement agency, the words of complainants, even in courtroom testimony, are subject to the defense that they are biased and untruthful, or are not convincingly related to the alleged source, or are based on rare and transitory episodes.

To avoid the difficulties of using complaints, other than as indicators of where odor problems may exist, a few authorities have proposed to limit the emission of odorous subjects to levels related to the odor threshold. These approaches tend to equate the perception of odor with the existence of an odor problem. Rather, it is the personal evaluations of those who perceive odor, not the mere activation of the sense of smell, that may give rise to the problem and that determines its magnitude. This distinction must be made in order to explain why some odors are considered pleasant while other odors, or the same odors experienced under different conditions, are considered unpleasant, even to the extent of causing undesirable effects such as feelings of annoyance and the lowering of property values. Thus it seems more appropriate to base the existence of odor problems on investigations of the social and economic effects of odors. The methods for doing so, along with some of the results obtained, are discussed in the following sections.

SOCIAL EFFECTS

For convenience in definition, social effects can be said to differ from economic effects in that the former cannot be measured directly in monetary terms. The social effects of odors include interference with everyday activities, as well as feelings of annoyance on the part of persons exposed.

Numerous studies have described the types and extent of such interference. The basic tool used in these studies has been an attitude survey. Conclusive results have not been obtained, largely because people seem to find it difficult to speak with assurance that the presence of odors has, for example, reduced the utility of their yards or prevented them from entertaining their friends. However, people seem to find it easy to state their reactions to the odors they perceive. Attitudes expressing annoyance are typical in such cases.

History of Annoyance Measurements

Annoyance due to odors has been studied for many years, particularly in Sweden and the United States. But a series of Swedish studies, beginning in the early 1960s,[4,5,6] have been particularly helpful. They refined techniques for the measurement of annoyance and first suggested the use of such measurements in the establishment of legal standards.

Several of their results have also guided further inquiry. One of their findings,[6] for example, found that persons who volunteered their opinions tended to exaggerate. Twelve hundred persons signed a petition demanding that authorities eliminate the nuisance of offensive flue gases caused by a sulfate cellulose factory. In response to a later survey sponsored by public health authorities, almost half of these 1,200 persons stated they had not experienced annoyance during the period specified in the petition. The contradiction of statements could not be accounted for by a lessening of exposure since the time the petition was submitted. The petition was rejected as an unreliable measure of the extent of concern.

In the United States, three major studies of community odor problems have been completed since 1969. Two of these, conducted in Eureka[7] and Anderson,[8] California, focused on both annoyance and health aspects, and confirmed the existence of dose-response relationships. The third study, conducted in Los Angeles under the sponsorship of the U.S. Environmental Protection Agency, was concerned largely with annoyance and the effects of odor pollution on property values.[7] It is the latter study from which much of the information has been drawn for this chapter.

Methods of Odor-Problem Identification

The most direct method of measuring social effects and, at the same time, identifying community odor problems is an attitude survey. This recommendation is based on the effective nature of odor perception as manifested by expressed attitudes.

An attitude survey is designed to compare the attitudes of people residing in a community in which an odor problem is suspected with the attitudes of similar people residing in a community known to be relatively odor free. A test area is delineated within the suspected odor-problem community by driving an automobile in a grid pattern throughout the community while noting the geographical boundaries within which odors are perceived. Knowledge of local meteorological and complaint patterns is also helpful in establishing the test area.

Next, the socioeconomic characteristics of the test area must be determined. This is done by reference to the latest Censuses of Population and Housing tract reports if the test area is approximately coincident with a census tract, or by employing the temporary services of a professional appraiser. Five characteristics are noted: median family income, median value of owner-occupied housing units, median gross rent of renter-occupied housing units, median number of rooms per house, and percent of homes built in the same span of years.

An odor-free community, or control area, is located that matches the geographical and socioeconomic characteristics of the test area as closely as possible. Again, this is done by reference to census tract reports or with the assistance of a professional appraiser.

A randomly selected sample of residents in both areas is then interviewed by telephone or in person, using a questionnaire that does not initially reveal the true purpose of the inquiry. The degree of annoyance to odors in each area is measured by the number of responses to key questions.

The decisive step in odor-problem identification is to determine whether a significant difference exists between the proportion of residents interviewed in the test area and the proportion of residents interviewed in the control area who state they have been bothered by odors. For example, in the Los Angeles study an odor problem was said to be identified if a significant difference existed at or above a 95% level of confidence ($z = 1.65$, one-tail test). The true percentage of residents bothered by odors in the test area can be easily estimated by application of a Bayesian type correction for chance success commonly associated with signal detection theory.

ECONOMIC EFFECTS

Experimental results of the studies discussed above confirm that odors have been at least a source of annoyance to people. But do they cause economic loss? If so, how can this loss be determined? Again, the most direct approach is to survey the affected area. In doing so, a representative sample of persons in the community is asked how much they would be willing to pay to get rid of the odors. While theoretically valid, this approach runs into the practical difficulty of separating what people say they would pay from what they actually would pay if they had to comply with their own responses.

A more fruitful approach lies in determining what people actually *have* paid to obtain an odor-free environment. This requires recognition that the economic impact of odor pollution will most likely manifest itself in reduced property values, reduced productivity of industrial companies, and reduced sales in commercial areas.

Economic theory states that if odors are bothersome, people should be willing to pay more to live in an odor-free area. Thus two similar properties in similar neighborhoods should sell for different prices if one is affected by odors and the other is not. What is assumed here is that all economic losses due to the presence of odors are capitalized negatively into property values. It is also assumed that buyers regard land and structure as a bundle, not as separate entities, and that buyers need only know that they prefer some properties to others and are willing to pay more for them.

Odors may also affect industrial property values in somewhat the same way. However, another form of loss to commercial and industrial establishments is possible, namely, that odors may reduce the productivity of employees because of induced illness or distraction from work assignments. Productivity losses are likely to be particularly noticeable if such odors are intermittent as well as strong. Such a situation would be offensive, while also tending to inhibit persons from becoming adapted to their work environment.

Commercial areas may suffer economic losses from odors in the form of a general loss of customers and reduced sales per customer.

History of Attempts at Measurement

At the outset of the Los Angeles study it was learned that all previous studies had dealt only with the effects of dustfall, sulfation rates, and suspended particle concentration on property values. Yet there appeared to be no reason why an odor variable could not be added, if sufficient data were available.

By far the favorite technique that has been applied to economic effects research is multiple regression. Ridker and Henning[10] used this technique in a cross-section study of St. Louis. Property values and other demographic data were drawn from the 1960 Censuses of Population and Housing and used on a census tract basis. Pollution data were drawn from air quality agency sampling stations and interpolated to obtain values for census tracts without stations. Anderson and Crocker, in their 1969 study,[11] eliminated most of the specification problems incurred by Ridker and Henning and reworked the St. Louis data through an improved model. They also applied Kansas City and Washington, D. C., data to their model with good results.

Crocker's 1970 comprehensive study of Chicago[12] embodied another major advance, namely, switching from *ex ante* (offer-price) data to *ex post* (actual sales) data. In addition, he was able to avoid making the untenable assumption that response error to the census question on property value was both random and, on average, offsetting. Detailed information collected on two census tracts for the Chicago study did not offer much support to this often used assumption.

Another technique for assessing the impact of air pollution on property values is the use of professional appraiser judgments. While this method, employed by Stradbeck,[13] is undoubtedly less sensitive than regression analysis, it can be used to estimate property value differentials if they are large enough.

METHOD OF ODOR-PROBLEM ASSESSMENT

As the focus of the Los Angeles study was on the effects of odors rather than air pollution in general, modifications in the approaches of previous studies had to be made. A cross-section technique was judged unfeasible since it required the existence of widespread odor problems over an extensive area for which various intensities of odors could be measured. In reality, most odor problems tend to be localized over an area of only one to five census tracts.

Time series analysis appeared to be feasible, except that it depended on the availability of aerometric or comprehensive complaint data over previous years to serve as a proxy for odor intensities in the test area. Because it was found that such data would not be generally available, this technique was not employed.

Further considerations suggested that if the presence of odors in an area has a negative impact on home values, the impact should manifest itself in a slower growth in home prices, at least at the outset of the problem. Thus an examination of property value growth rates provided another method of assessment. If a control area that is homogeneous with the test area, except for odors, can be located, the growth in home values can be calculated and compared for each area during a particular period to isolate the influence of odors.

DESCRIPTION AND EVALUATION OF DATA SERVICES

Once test and control areas have been properly established, the next problem is finding accurate average-home-price data for each area. This problem is particularly difficult because these average price data must be collected over several years. During the Los Angeles study, three major sources of *ex post* data were located for single-family housing units. Although not all of these sources will be found in every metropolitan area, various workable alternatives may exist.

Individual sales data can be obtained from local boards of realtors as part of their multiple listing services. Use of this source is far more efficient than extracting data from the files of numerous real estate offices. However, three difficulties may be encountered: (1) the data may not be tabulated regularly, (2) some sales may not be final, and (3) the extent of sales coverage may vary over time, depending upon the interest of the boards. Usually the services are accurate in reporting dates of sales and sales prices.

In 1967 the Society of Real Estate Appraisers (SREA) began publishing data supplied by firms regularly engaged in making first-mortgage loans. As a part of SREA requirements, subscribers must provide complete sales and property description data on all loans they transact. Sales data coverage is generally better than that provided by multiple listing services.

During the same year a Los Angeles area firm began tabulating, on a monthly basis, home-sales data obtained from county recorders' offices. From this source, sales data can be estimated by tabulating transfer taxes. Coverage approaches 100%. However, several difficulties also exist in using this source: (1) sales prices can be determined only to the nearest $500, (2) sales prices of homes involving loan assumptions cannot be determined, and (3) a small percentage of the data represents sales to investors and indicates the addresses of the investors rather than the location of the properties.

While the property-value growth rate method is adequate for single-family homes, it is not likely to work well for commercial establishments or apartments, due to the lack of comparable facilities in odor-free areas. A similar difficulty is involved in applying the method to industrial developments. Fortunately, however, the economic effects on industrial developments are more easily measured in terms of loss of productivity.

USE OF THE METHODS

The method of identifying a community odor problem by measuring social effects was subsequently applied to four communities that lay within the jurisdictions of air pollution control agencies. The agencies were the

City of Houston (Texas) Department of Public Health
Hillsborough County (Florida) Environmental Protection Commission
Columbia-Willamette (Oregon) Air Pollution Authority
State of Maryland Bureau of Air Quality Control

Related field activities were conducted over a period of eight months. During that time, two of the agencies were required to conduct eight sets of attitude surveys (*i.e.*, eight surveys in test areas and eight in matching control areas) in accordance with a detailed procedure supplied to them. The other two agencies were required to conduct six sets of attitude surveys in accordance with the same procedure.

Each of the agencies investigated many communities in search of possible odor problems and delineated test areas in communities where the possibilities seemed greatest. Matching control areas were then located. A total of 28 sets of public attitude surveys were conducted in these areas. Secretaries and technicians in the employ of the agencies were used as interviewers. These employees had no previous experience with survey work. The field activities, including interviewing and subsequent determinations of whether or not community odor problems existed, were directed by agency officials alone.

The results obtained by the agencies are summarized in Table 4-3, which compares the socioeconomic characteristics of the test and control areas surveyed, in Table 4-4, which shows the survey information obtained for problem identification purposes, and in Table 4-5, which shows the number of odor complaints made against the suspected sources during the field activities.

As can be seen in Table 4-4, 17 community odor problems were identified from the results of the surveys (six by the city of Houston, six by Hillsborough County, one by Columbia-Willamette, and four by the state of Maryland). As can also be seen, four replicate sets of surveys were conducted. While the same decision was reached in three of the four cases, inadequate consideration was given to source production and emission control factors at the times the surveys were conducted to permit a reasonable evaluation of this method for judging odor control effectiveness.

An interesting side to this study was the apparent unreliability of the volume of odor complaints as indicators of where odor problems might exist. A comparison of the odor-problem index numbers of identified community odor problems (Table 4-4) with the number of odor complaints against the suspected sources from residents of the test areas (Table 4-5) supports this conclusion. For example, the Hillsborough County agency calculated an odor-problem index of 7 (10 is maximum) from one set of public attitude surveys. Yet no odor complaints were received by this agency from residents of the test area during the entire period of the field activities.

Upon the conclusion of the field activities, agency officials were asked to evaluate this method of measurement from the standpoint of overall utility to their agencies. Their responses were generally favorable and are

Table 4-3. Comparison of Socioeconomic Characteristics of Test and Control Areas Surveyed by Participating Agencies

Public Attitude Survey Set No.	No. of Homes in Area		Median Family Income		Median Home Value	
	T. Area	C. Area	T. Area	C. Area	T. Area	C. Area
1	1,802	803	$ 5,633	$ 8,100	$ 8,100	$ 9,200
2	2,467	3,319	11,658	14,624*	20,200	24,400*
3	2,025	1,462	7,809	7,297	9,600	11,500
4(a)	1,391	395	11,336	10,397	12,300	10,900
5	329	107	7,807	7,661	9,600	8,800
6(c)	650	233	10,168	8,971	12,900	13,100
7(b)	1,391	395	11,336	10,397	12,300	10,900
8(d)	650	233	10,168	8,971	12,900	13,100
Hillsborough Co. Environmental Protection Commission						
1	64	73	N.A.	N.A.	$13,000	$12,000
2	48	44	N.A.	N.A.	7,750	7,750
3	47	47	N.A.	N.A.	14,500	16,000
4	143	179	N.A.	N.A.	12,300	12,300
5	82	81	N.A.	N.A.	100,000	100,000
6	37	68	N.A.	N.A.	21,000	23,500
7	114	124	N.A.	N.A.	N.A.	N.A.
8	101	128	N.A.	N.A.	12,500	13,500
Columbia-Willamette Air Pollution Authority						
1	713	1,250	$ 6,596	$ 5,948	$15,392	$12,491
2(a)	1,259	1,481	5,877	5,963	12,766	12,670
3	1,578	1,481	6,172	5,963	13,228	12,670
4	1,334	1,210	6,833	6,827	14,922	13,450
5(b)	1,249	1,481	5,877	5,963	12,766	12,670
6	165	1,481	5,877	5,963	12,766	12,670
State of Maryland Bureau of Air Quality Control						
1(a)	350	933	$13,557	+13,001	$20,600	$22,100
2	218	1,550	8,839	11,298*	13,200	16,500*
3	114	803	14,827	15,293	28,000	27,200
4	310	971	13,014	14,730	18,600	21,200
5	141	170	11,855	13,075	37,300	34,815
6(b)	350	933	13,557	13,000	20,600	22,100

Note: Letters shown in parentheses after the survey set numbers refer to the initial and replicate sets of surveys conducted in the same test and control areas. For each of three agencies, set b is a replicate of set a and, for one agency, set d is a replicate of set c.

* Control area value differs from test area value in excess of recommended tolerance of ± 20%.

Table 4-3, continued

Public Attitude Survey Set No.	Median Gross Rent		Median No. of Rooms per Home		Percent Homes Built in Same Category of Years	
	T. Area	C. Area	T. Area	C. Area	T. Area	C. Area
1	$ 67/mo.	$ 65/mo.	4.6	4.6	46	32
2	145	166	5.5	6.3	35	30
3	71	71	4.3	4.3	42	25
4(a)	87	79	5.2	5.1	75	78
5	71	81	4.3	4.5	43	30
6(c)	116	86*	4.4	4.5	32	22
7(b)	87	79	5.2	5.1	75	78
8(d)	116	86*	4.4	4.5	32	22
1	N.A.	N.A.	5.5	5.5	N.A.	N.A.
2	N.A.	N.A.	N.A.	N.A.	N.A.	N.A.
3	N.A.	N.A.	6.5	7.0	N.A.	N.A.
4	N.A.	N.A.	4.9	4.9	N.A.	N.A.
5	N.A.	N.A.	N.A.	N.A.	N.A.	N.A.
6	N.A.	N.A.	7.0	6.5	N.A.	N.A.
7	$270/mo.	$245/mo	N.A.	N.A.	N.A.	N.A.
8	N.A.	N.A.	5.0	5.0	N.A.	N.A.
1	$100/mo	$102/mo	4.9	5.0	50	49
2(a)	84	102	4.6	4.9	68	67
3	84	102	5.0	4.9	53	67
4	85	87	5.1	5.5	51	85*
5(b)	84	102	4.6	4.9	68	67
6	84	102	4.6	4.9	68	67
1(a)	$ 89/mo.	$109/mo*	6.5	6.1	43	47
2	102	91	5.3	5.7	54	29*
3	189	203	6.3	6.4	59	71
4	169	211*	5.7	6.1	53	19*
5	159	183	4.5	6.7†	96	98
6(b)	89	109*	6.5	6.1	43	47

† Control area value differs from test area value in excess of recommended tolerance of ± 10%.

Source: U.S. Dept. of Commerce, Bureau of the Census, and estimates of professional real estate appraisers consulted by participating agencies.

Table 4-4. Public Attitude Survey Information Obtained by Participating Agencies for Problem Identification Purposes

Public Attitude Survey Set No.	No. of Interviews Completed		No. of Respondents Who Were Bothered by Odors		Value of Normal Deviate, z*	Odor-Problem Index
	T. Area	C. Area	T. Area	C. Area		
City of Houston Dept. of Public Health						
1	50	36	18	13	-0.01	--
2	58	44	27	12	1.98	3.8
3	51	37	25	12	1.56	--
4(a)	37	29	22	6	3.15	6.1
5	40	56	29	14	4.61	7.2
6(c)	31	36	16	7	2.79	5.7
7(b)	38	38	22	5	4.07	5.7
8(d)	29	33	15	3	3.69	6.9
Hillsborough Co. Environmental Protection Commission						
1	30	29	25	4	5.34	5.7
2	22	25	8	2	2.37	5.0
3	22	16	3	3	-0.43	--
4	20	30	5	3	1.42	--
5	27	27	21	3	4.75	4.1
6	21	29	9	1	3.44	8.3
7	22	23	10	0	3.66	7.0
8	30	30	10	2	2.58	3.6
Columbia-Willamette Air Pollution Authority						
1	31	30	9	5	1.15	--
2(a)	30	30	12	5	2.01	5.2
3	30	30	8	3	1.61	--
4	30	30	9	5	1.22	--
5(b)	30	30	9	6	0.89	--
6	31	30	10	6	1.08	--
State of Maryland Bureau of Air Quality Control						
1(a)	30	30	8	8	0.00	--
2	30	33	15	0	4.65	7.3
3	32	35	12	5	2.20	4.9
4	31	30	12	1	3.37†	7.9†
5	30	33	18	1	4.92	5.8
6(b)	30	34	10	4	1.56	--

Note: Letters shown in parentheses after the survey set numbers refer to the initial and replicate sets of surveys conducted in the same test and control areas. For each of three agencies, set b is a replicate of set a and, for one agency, set d is a replicate of set c.

* If the value of the normal deviate, z, equaled or exceeded 1.65, a community odor problem was said to exist.

† Values of the normal deviate and odor-problem index were adjusted to exclude control area responses about odors from a sewer line damaged in June 1972 by Hurricane Agnes.

Table 4-5. Number of Odor Complaints to Participating Agencies against Sources Suspected by Agencies of Causing Odor Problems in Test Areas (Jan. through Sept. 1972*)

Public Attitude Survey Set No.	Source(s) of Odor	No. of Odor Complaints against Source(s)	
		From All Persons	From Residents of Test Area
City of Houston Dept. of Public Health			
1	Sanitary landfill	2	2
2	Asphalt and pesticide manufacturing	3	2
3	Refineries and rubber manufacturing	88	35
4(a)	Pesticides and general chemical manufacturing	N.A.	N.A.
5	Refineries and rubber manufacturing	88	35
6(c)	Rubber manufacturing	88	32
7(b)	Pesticides and general chemical manufacturing	N.A.	N.A.
8(d)	Rubber manufacturing	88	32
Hillsborough Co. Environmental Protection Commission			
1	Sewage treatment plant	3	1
2	Fertilizer manufacturing	9	5
3	Barbecue restaurant	1	1
4	Citrus processing plants	N.A.	N.A.
5	Natural organic decomposition (Hillsborough Bay)	1	0
6	Asphalt manufacturing	N.A.	N.A.
7	Natural organic decomposition (Hillsborough Bay)	1	0
8	Natural organic decomposition (Hillsborough Bay)	1	0
Columbia-Willamette Air Pollution Authority			
1	Pulp mill	5	0
2(a)	Pulp mill	5	1
3	Animal rendering plant	4	0
4	Wood preserving plant	0	0
5(b)	Pulp mill	5	1
6	Pulp mill	5	1
State of Maryland Bureau of Air Quality Control†			
1(a)	Distillery	2	0
2	Sewage treatment plant	3	1
3	Asphalt manufacturing	2	2
4	Coffee manufacturing	9	0
5	Sewage treatment plant	0	0
6(b)	Distillery	2	0

Note: Letters shown in parentheses after the survey set numbers refer to the initial and replicate sets of public attitude surveys conducted in the same test and control areas. For each of three agencies, set b is a replicate of set a and, for one agency, set d is a replicate of set c.

* This was approximately the period during which the field activities were performed.

† The number of odor complaints shown for the State of Maryland include some complaints received by the Baltimore Co. and Price Georges Co. Health Departments.

documented in detail elsewhere.[14] Concern was expressed, however, that some of the survey tasks are too cumbersome for routine application.

During the Los Angeles study an attempt was made to apply the method of measuring the economic impact of odors to residential neighborhoods. This attempt was unsuccessful for any of three reasons: (1) the odor problem in the test area may have been too transitory to be capitalized into property values, (2) the technique adopted may not have been sensitive enough to detect the small differences in growth rates likely in this particular case, or (3) some other positive factor that opposed the odor effect may have been present. Nevertheless, it is felt that in extreme cases, where the results of attitude surveys show that more than half of those interviewed in test areas claim to be bothered very much by odors, further use of this method is warranted.

THE NEED FOR DOSE-RESPONSE RELATIONSHIPS

While the use of attitude surveys to identify community odor problems appears to be sound in theory, it leaves much to be desired in practice. The major difficulty with this technique is that it is cumbersome. Some success has been achieved in minimizing the effort through the design of a brief questionnaire and a plan requiring as few as 60 interviews to reach a decision. Yet, even with this design, at least a week and approximately $1,000 must be spent each time an odor problem is suspected.

After experiencing many failures in seeking solutions under existing laws, state and local authorities are frustrated with the use of odor complaints in dealing with odor problems. The requirements of attitude surveys are unattractive to them for enforcement purposes. Odor complaints and survey interviews provide different measures of the responses of persons to the presence of odors. They are subjective by nature and too often are biased. Because of these factors, air pollution control officers throughout the nation have voiced an urgent need for the development of more straightforward measures aimed at dosage instead of response. Such a "dosage approach" would take the form of a performance standard that could be evaluated at the stack. If some reasonable criteria could be established that would relate such performance to community odor levels, the dosage approach would offer practical advantages in effort, convenience reproducibility, and proof of violation or compliance with regulations. However, no criteria have yet been established that relate odorous emissions to community odor problems. In view of this lack, we must measure dose-response relationships in some way to assess the effect of odor on any given community.

Human responses to odor have often been expressed in terms of thresholds, and many procedures and devices have been described for such measurements.[15] The *absolute* or *detection threshold* is the minimum concentration of odorant that elicits a response, and the *recognition threshold* is the minimum concentration that can be identified by a subject. Since individuals differ in their responses, it has become customary to define the "50% thresholds" as the minimum concentrations at which half a population of subjects responds to or identifies an odor. This definition is parallel to the "median effective dose," ED_{50}, cited in the literature of toxicology.

In recent years it has become understood that this concept of thresholds ignores some fundamental aspects of human responses to stimuli. Such understanding has come about from the development of the theory of signal detection,[16] which recognizes human responses to be *decisions* as to whether a given sensation (which always includes some random background interference, or noise) consists of noise alone or contains some signal (odor) mixed with the noise. The responses, which take the form "Yes, I detected something" or "No, I don't detect anything," are not simply expressions of sensation but are also influenced by decisions based on other factors, such as expectations, the wish not to miss an odor that may be present, or the wish not to sound a "false alarm" by incorrectly detecting an absent odor. In this view, threshold values are not absolute properties of odorant substances but are dependent on the social and other human contexts in which they are experienced. Therefore reported threshold values must be regarded as functionally determined; that is, they depend on the procedure, the subjects, and the contexts of the investigation, and thus are not necessarily quantitatively transferable from one situation to another.

With these demurrers in mind, let us examine the first two sigmoid curves of Figure 4-1, the first of which represents the cumulative frequency of detection, and the second the cumulative frequency of identification or recognition. The absolute and recognition thresholds corresponding to

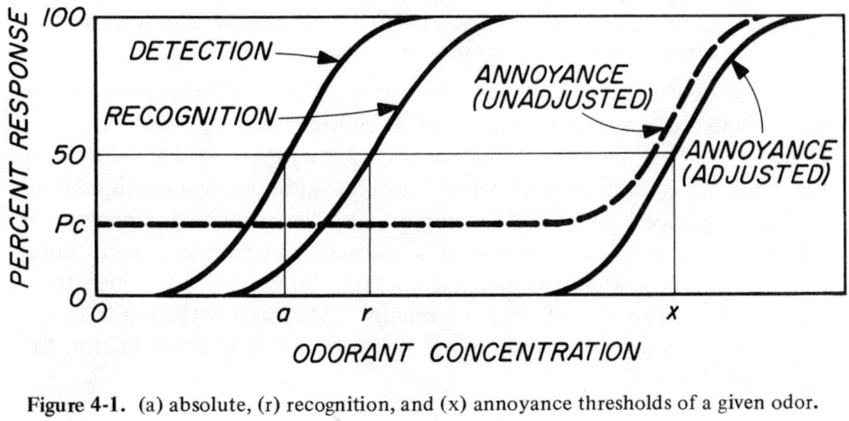

Figure 4-1. (a) absolute, (r) recognition, and (x) annoyance thresholds of a given odor.

ED_{50} are shown as intensities ***a*** and ***r***, respectively. Each of these intensities can be related to the concentrations of the odorant in an odor-free dose medium by means of the Stevens psychophysical power law.[17]

Ideally, the detection and recognition curves of Figure 4-1 should be established by exposing a large group of individuals simultaneously to odors of various concentrations, under such conditions that each subject is insulated from any influence from other subjects or other responses. For many reasons related to the difficulty of establishing uniform vapor concentrations over a large volume, or to the expense of providing an extensive network of piping and sniffing ports to service each individual, such an ideal procedure is impracticable.

Instead, the sigmoid detection and recognition curves of Figure 4-1 can be approximated by exposing many small groups of individuals to a different concentration of the odorous substance under investigation.

Detection and recognition thresholds can be used to monitor the effectiveness of various kinds of odor control equipment and procedures. Surveillance of ambient or emission samples by a few trained individuals is often adequate to determine whether the application of odor controls has successfully reduced emissions below some predetermined level. However, detection and recognition thresholds provide no information on the acceptability of odor to residents of a community. Yet, as was reported by a recent international symposium[18] on environmental health:

> The most important dimension of an odor is probably its acceptability, *e.g.*, what percentage of the population is annoyed by the smell.

The report goes on to state:

> The acceptability of an odor is probably in part dependent on the intensity, frequency, and duration of the experience, as well as the conditions under which exposure occurs. This probably also applies to pleasant odors. Practical experience will indicate that no odor is accepted in the long run if the intensity is too high or if the odor markedly differs from the "normal" odor sensations expected in a given context.

Thus it is argued that acceptance, not detection, is the proper basis for the establishment of an odor standard and its enforcement. Acceptance of a given odor is best measured by the degree of annoyance to that odor, with the minimum acceptable level defined in terms of an annoyance threshold.

An acceptance basis would be supported by the economist, among other behavioral scientists. The economist would seek an optimum level of odor which would exist when the marginal cost of odor control just equals the marginal benefit to the affected community. Whereas a detection basis may result in unnecessary costs of excessive control or business failure, an acceptance basis would not.

Let us now return to Figure 4-1, focusing our attention on the two sigmoid curves on the right. The dashed curve represents unadjusted cumulative frequency of expressed annoyance. The long horizontal portion that precedes the sigmoid section of this curve represents a constant level of expressed annoyance that is independent of odorant concentration, and that therefore must represent positive responses that are provoked by variables other than odor.

To correct for such irrelevant responses, we employ the following procedure:

P_t = % of test-area residents expressing annoyance to odors

P_c = % of control area residents expressing annoyance to odors

A = % of annoyance responses (adjusted)

The number of adjusted responses must be $P_t - P_c$, and the percentage must be based only on these individuals who are not responding inappropriately, that is, on $100 - P_c$. Therefore

$$A = \frac{P_t - P_c}{100 - P_c}$$

The adjusted annoyance curve of Figure 4-1 is based on the above equation, and the 50% annoyance threshold, x, is determined graphically from the curve. This correction is far from trivial. It is noted that P_c, the percentage of *control area* residents expressing annoyance to odors, was recently estimated by Copley International Corporation as 26% for the southern quarter of Los Angeles County. Even higher percentages have been encountered by researchers in control areas near a Swedish sulfate pulping plant.

REFERENCES

1. Copley International Corp. *National Survey of the Odor Problem.* Publication nos. PB-194 376 (text) and PB-194 377 (appendix), U.S. Dept. of Commerce, National Technical Information Service, Springfield, Va. (1970).
2. Leonardos, G. "A Critical Review of Regulations for the Control of Odors," *J. Air Pollution Control Assn.,* **24** (5) (May 1974).
3. Benforado, D. M., *et al.* "Odor Control Legislation," part V in "Odors: Evaluation, Utilization,aand Control," *Ann. N.Y. Academy of Science,* **237** (Sept. 27, 1974).
4. Friberg, L., *et al.* "Studies on Sanitary Nuisance of Smoke from a Sulphate Cellulose Factory (I). An Interview Study," *Nordisk Hygienisk Tidskrift,* **41**, 41-49 (1960).

5. Cederlof, R., *et al.* "Studies of Annoyance Connected with Offensive Smell from a Sulphate Cellulose Factory," *Nordisk Hygienisk Tidskrift,* **45**, 39-48 (1964).
6. Jonsson, E. "Annoyance Reactions to Environmental Odors," in *Human Responses to Environmental Odors*, A. Turk, J. W. Johnston, and D. Moulton, eds. (New York: Academic Press, 1974).
7. Jonsson, E., *et al.* "Community Reactions to Odors from Pulp Mills: A Pilot Study in Eureka, California," paper presented at the Conference on Methods for Measuring and Evaluating Odorous Air Pollutants at the Source and in the Ambient Air, Stockholm, Sweden, June 1-5, 1970.
8. Deane, M. and J. R. Goldsmith. "Health Effects and Annoyance Reactions to Pulp Mill Odor in a Rural Community," paper presented at the Conference on the Dose-Response Relationships Affecting Human Reactions to Odorous Compounds, Cambridge, Mass., April 26-27, 1971.
9. Copley International Corp. *A Study of the Social and Economic Impact of Odors, Phase II.* Publication no. PB-205-936, U.S. Dept. of Commerce, National Technical Information Service, Springfield, Va. (1971).
10. Ridker, R. G. and J. A. Henning. "The Determinants of Residential Property Values with Special Reference to Air Pollution," *Review of Economics and Statistics,* **49**, 246-257 (May 1967).
11. Anderson, R. J., Jr. and T. D. Crocker. "Air Pollution and Housing: Some Findings," paper prepared for the Institute for Research in the Behavioral, Economic and Management Sciences, Herman C. Krannart Graduate School of Industrial Administration, Purdue University, Jan. 1970.
12. Crocker, T. D. "Urban Air Pollution Damage Functions: Theory and Measurement," report prepared for the National Air Pollution Control Administration (1970).
13. Stradbeck, F. L., *et al.* "Professional Appraisers' Judgment of the Effect of Air Pollution on Property Values, " report prepared for the Public Health Service (Jan. 1969).
14. Copley International Corp. *A Study of the Social and Economic Impact of Odors, Phase III.* Publication no. EPA 650/5-73-001. U.S. Environmental Protection Agency, Air Pollution Technical Information Center, Research Triangle Park, N.C. (1973).
15. Moulton, D. G., A. Turk, and J. W. Johnston, eds. *Methods in Olfactory Research.* (New York: Academic Press, 1975).
16. Swets, J. A., W. B. Tanner, and T. G. Birdsall. "Decision Processes in Perception," in *Stimulus and Perception: Readings in Sensory Psychology*, W. S. Cain and L. E. Marks, eds. (Boston: Little, Brown, 1971).
17. Stevens, S. S. "The Surprising Simplicity of Sensory Metrics," in *Stimulus and Perception: Readings in Sensory Psychology,* W. S. Cain and L. E. Marks, eds. (Boston: Little, Brown, 1971).
18. The Third Karolinska Institute Symposium on Environmental Health. "Methods for Measuring and Evaluating Odorous Air Pollutants at the Source and in the Ambient Air," Stockholm, Sweden (June 1-5, 1970).

CHAPTER 5

ODOR PANELS

William P. Dolan
Rossnagel & Associates
Cherry Hill, N.J. 08003

There are no measuring devices or instruments available today which can provide an objective determination of odor level. Still, there are a myriad of codes and regulations stating what level of odor is not acceptable.

How does an agency or plant objectively test for odor level? An improvement over individual interpretations is the use of a group of individuals. A panel of individuals will result in a collective interpretation of the odor level. Accuracy and reproducibility of results are improved as the number of observers is increased. Panels for odor and taste evaluations have been in recorded use for hundreds of years. Regulatory agencies in their search for methods to measure the level of odor are resorting to the use of panels.

In 1956 the American Society for Testing and Materials published methods and techniques for the measurement of odor in atmospheres. This method is also referred to as the dilution method. This odor measurement technique is being used by the Los Angeles Air Pollution Control District as published in the *Air Pollution Engineering Manual*, available from the Public Health Service, U.S. Department of Health, Education and Welfare.

This method is based on the fact that the average subject can report the presence or absence of an odor with more certainty than he can determine objectionability. In this manner objectiveness is introduced into the test. It is essential that the panel members not attempt to give their subjective interpretation of the odor quality. A "yes" or a "no" is all that is required. A sample of the odorous gas is diluted with odor-free air until a dilution is achieved in which the odor is barely detectable by half of the odor panel.

The ratio of the total volume of the diluted sample to the volume of the original sample is a measure of the odor concentration.

Before discussing the actual performance of this type of testing, an understanding of certain definitions is essential (Table 5-1). For odor in air the ASTM defines an odor unit (o.u.) as one cubic foot of air at the odor threshold. More specifically, it is the quantity of any odor that, when dispersed in one cubic foot of odor-free air, produces a median threshold odor-detection response (positive response by 50% of the panel members).

Table 5-1. Definitions

Odor In Air:		
Odor unit (o.u.) (odor threshold)	–	Dimensionless unit representing the quantity of odor when dispersed in one cubic foot of odor-free air produces a positive response by 50% of the panel members
Odor concentration	–	The number of cubic feet that one cubic foot of odorous gas will occupy when diluted to the odor threshold
Odor In Water:		
Threshold odor number	–	Dilution factor
Odor intensity index (OII)	–	The number of times the sample had to be diluted in half to reach the threshold level

Odor concentration per the ASTM method is the number of cubic feet that one cubic foot of sample will occupy when diluted to the odor threshold. This can be expressed as the number of odor units in a cubic foot of sample. With this value the odor emission rate can be computed as follows:

$$N = ZQ$$

where:

N = odor emission rate, odor units per minute
Q = gas flow rate, scfm
Z = odor concentration, odor units per cubic foot

For odor in air the sampling procedure utilizes 250-cc gas glass sampling tubes for the collection of samples. Either a rubber bulb or a small vacuum pump is used to purge the sampling tube with the odorous gas. After sampling, the tubes are rushed to the laboratory for preparation and dilution before presentation to the odor panel. The need for expediting the samples to the odor panel cannot be stressed enough. Samples of the odorous gas can become diluted with odor-free air to the point where the odor can be detected by only 50% of the test panel.

The odor dilution apparatus is shown in Figure 5-1. A portion of the contaminated air sample (usually 50 ml or more) is transferred from the 250-ml gas collection tube to a 100-ml syringe equipped with an 18-gauge hypodermic needle. Mercury from a reservoir displaces the sample in the tube. The 18-gauge needle is held in place with a cork stopper. Transfer is made as rapidly as possible to minimize any chemical reaction with the mercury.

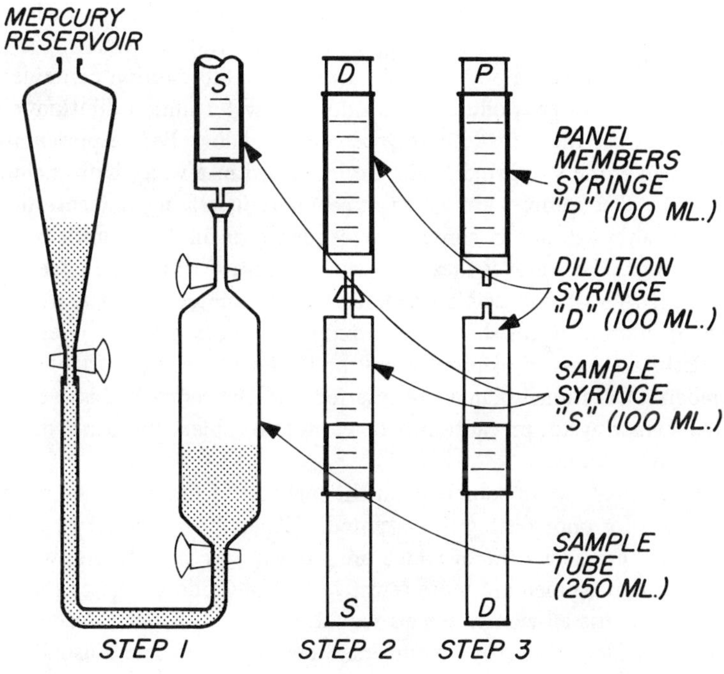

Figure 5-1. Equipment and techniques used by Rossnagel & Associates in sampling and conducting the odor test.

A 10-ml portion of the undiluted odor sample in the syringe is then transferred to a second 100-ml syringe using another 18-gauge needle. The seal between these two syringes is made by a cork stopper. The 10-ml sample in the second syringe is then diluted with fresh air to 100 ml, yielding an odor dilution factor of 10. Additional dilutions using 10 ml or more of the last diluted sample are then similarly made to desired concentrations.

The final dilution is made in the individual 100-ml syringe assigned to each panel member. When the panel member draws back the plunger, 10 ml of the diluted sample are injected into his syringe through an 18-gauge hypodermic needle. The panelist then draws back the plunger to the 100-ml mark and waits approximately 10 seconds for diffusion to be complete.

The sample is then expelled into the nostril, with breathing suspended, over a period of two to three seconds. A record (yes or no) is made as to whether an odor is perceived in the diluted sample. Successive dilutions of the original odor are presented to the panel in random order. The dilution at which only 50% of the panel can detect an odor corresponds to the odor threshold concentration. Precautions are taken not to destroy the odor sensitivity of the panelist by needlessly subjecting him to high odor concentrations or without allowing for adequate rest periods.

For odors in water, 500-cc flasks are used as the dilution containers. A sample of the water is diluted with odor-free water until a dilution is obtained that has the least definitely perceptible odor. Before presentation to the panel, the stoppered flasks are immersed in a water bath maintained at 104°F. The samples are tested in generally increasing concentration of odorant, although not in consecutive sequence of dilutions, until the odor is perceived. The odor is measured without regard to the presence of suspended matter or immiscible substances in the sample. In fact, if the dilutions of odorous water show any turbidity, color, immiscibility, etc., all the flasks are masked to prevent any prejudgments on the part of the panel members. Occasional blanks are inserted into the series of flasks to be tested. Once again, precaution is taken not to subject the panel members to high odor levels.

At this point some comment on the makeup of the panel and general conditions for odor testing is warranted. Generally, we prefer to use an eight-member panel, 50% of which are women. We have found, with experience, that women are more sensitive to many odorous gases than men. However, to use all-women panels would be unjustly strict. Certain stipulations are made to the panel members, especially the women, usually the day before the testing. For the men, we ask that no deodorant or after-shave lotion be used, and that at least one hour before the testing they refrain from smoking. For the women, we also request no deodorant, hair spray, perfume or flowers, and also no smoking for one hour before testing. Persons with head colds, allergies, etc., are eliminated from the panel. The individual preparing the syringes or flasks for presentation to the panel does not participate in the panel.

As mentioned before, the only response solicited from the panel members is a yes or no with regard to the presence of an odor. Typical test data for odor in air are displayed in Table 5-2. The dilution factor is noted and

Table 5-2. Odor-in-Air Dilution Test Responses

Location	Cyl. No.	Dilution Factor	No. of Panel Members Detecting Odor	Percent of Panel Members Detecting Odor
0-1	2-A	10	7	88
		100	4	50
		1,000	3	38
		10,000	1	13
0-2	9-A	10	7	88
		100	5	63
		1,000	1	13
0-3	10-A	10	5	63
		100	3	38
0-4	6-A	10	6	75
		100	2	25
0-5	8-A	10	4	50
		100	3	38
0-6	5-A	10	6	75
		100	4	50
		1,000	1	13
0-7	15-A	10	6	75
		100	5	63
		1,000	3	25

the percent of panel members detecting an odor computed. With this information an Odor Response Chart (Figure 5-2) is made. We use semilogarithmic paper since the data generally tend to follow a straight line. The line is drawn showing the average of the points. The intersection of the 50% panel response line is the threshold odor concentration (Z). With this information the odor emission rate can be computed.

For odors in water, the panel responses are displayed as shown in Table 5-3. With this method we can simply select the Odor Intensity Index (OII) with the corresponding 50% positive response by the panel. It is our opinion that the odor-in-water test is more reliable than the odor-in-air test. This may stem from the almost unlimited amount of sample to be sniffed from each dilution. The biggest problem from the odor-in-water testing is obtaining odor-free dilution water. We have on many occasions obtained positive responses from distilled water alone. To minimize this problem we pass distilled water (dilution water) slowly through a 3-ft 2-in. glass tube packed with charcoal.

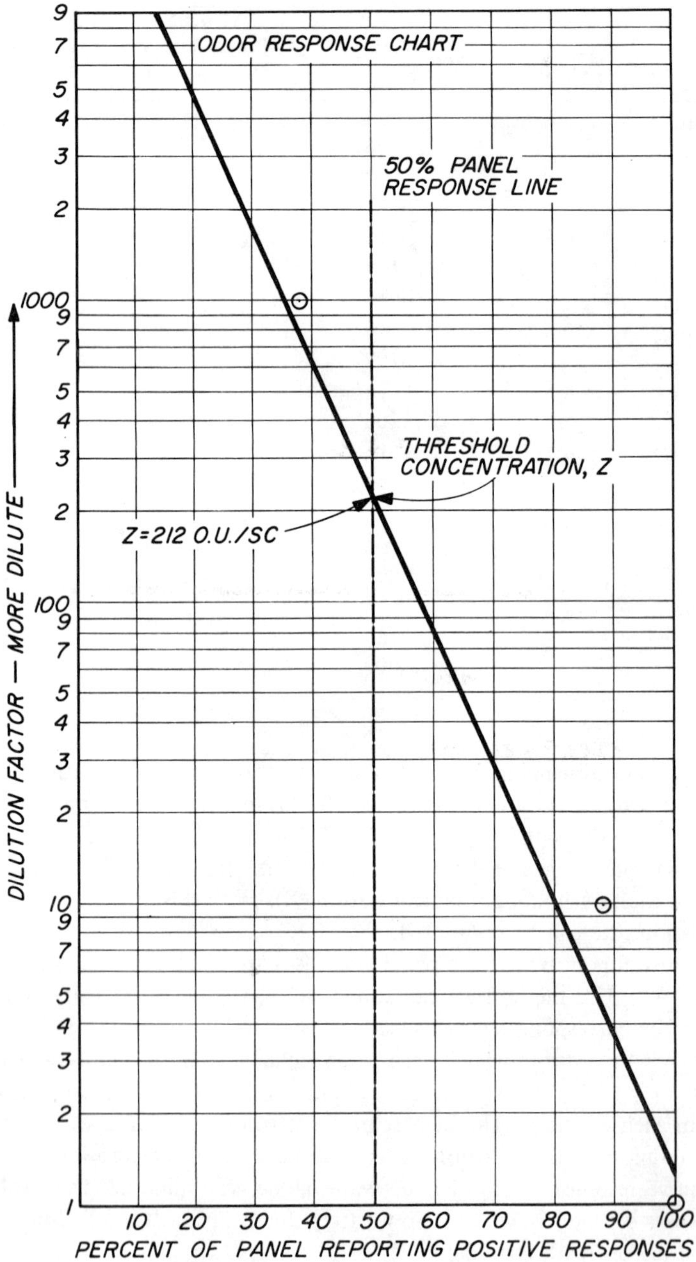

Figure 5-2. Odor response chart.

Table 5-3. Odor-in-Water Response Table

Sample II

Dilution	Volume Transferred to Odor Flask, ml	Threshold Odor Number (Dilution Factor)	OII	Test Panel Members							
				EJ	WR	FH	AS	SD	MH	BD	PD
Original Sample	200	1	0								
	100	2	1								
	50	4	2								
	25	8	3								
	12.5	16	4								
A	50	32	5	Yes	Yes	Yes	Yes	No	Yes	Yes	Yes
	25	64	6	Yes	No	Yes	Yes	Yes	Yes	Yes	Yes
	12.5	128	7	Yes	Yes	Yes	Yes	Yes	Yes	Yes	Yes
B	50	256	8	No	Yes	Yes	Yes	Yes	No	Yes	Yes
	25	512	9	No	No	Yes	No	Yes	No	Yes	Yes
	12.5	1,024	10	No	No	Yes	No	Yes	No	No	No
C	50	2,050	11	No	No	No	No	Yes	No	No	No
	25	4,100	12								
	12.5	8,200	13								
D	50	16,400	14								
	25	32,800	15								
	12.5	65,500	16								

Notes: Dilutions: A = 25 ml of original sample diluted to 200 ml; B = 25 ml of dilution A diluted to 200 ml; C = 25 ml of dilution B diluted to 200 ml; D = 25 ml of dilution C diluted to 200 ml.
Panel members were periodically given odor-free blanks to check their sensitivity.
Volume in odor flasks is made up to 200 ml with odor-free water.
OII = Odor Intensity Index.

One last point with regard to odor in air is the ability of this method to determine the odor removal efficiency of a control device. The reliability of the method is increased as gases from both the inlet and outlet are presented to the same odor panel.

REFERENCES

1. *Source Testing Manual.* Los Angeles Air Pollution Control District (1965), p. 108.
2. Summer, W. *Odour Pollution of Air.* (London: Leonard Hill, 1971).
3. *Air Pollution Engineering Manual.* U.S. Public Health Service (1967).
4. Jacobs, M. B. *The Chemical Analysis of Air Pollutants.* (New York: Interscience, 1960), p. 369.
5. ASTM Standard Method D1391-57, "Odor in Air."
6. ASTM Standard Method D1291-65, "Odor in Water."

CHAPTER 6

METHODS FOR THE CAPTURE AND IDENTIFICATION OF ORGANIC ODORS

W. A. Quebedeaux, M. F. Terraso, N. W. Tyler, Jr.
A. T. Hugh, J. Harris

Harris County Stream & APCD
Pasadena, Texas 77502

This chapter outlines two different methods that can be used for the capture and identification of organic odors. These methods have been successfully used in the identification of acrylate and amine odors. With slight modification, the same procedures can be used for other odoriferous compounds.

It must be realized that the intensity of compounds captured is not a requirement that must be met in order to win a legal case involving odor. However, the type of information which can be provided by these methods adds to the strength of the evidence and enables one to use a more scientific means of odor determination than is available if one had to depend upon the sense of smell alone.

CAPTURE OF ACRYLATE ODORS

Most odor nuisances that are dealt with from an enforcement viewpoint emanate from manufacturing or processing plants. However, to legally prove that an odor has created a nuisance, some conditions must be satisfied. The condition which takes priority is that the sample must be taken off the plant property from which the odor is emanating. The sample can be taken from adjacent plant property, public or private property.

Procedure

Sampling sites should be away from roadways or other sources of mobile interference. Upwind sample sites should be located directly upwind of the source being sampled according to the average wind direction. To provide the best evidence of an odor nuisance, the sample should be continuous for the period of sampling at the sample point selected and, if possible, identified by complaining citizens.

The downwind instrument is pointed in the direction of the source of the odor and the wind direction is noted. In the case of an upwind sample, the instrument is pointed into the wind and away from the source of odor. As long as the instrument is operating, it may be turned in any direction, including upside down, with little danger of losing any activated carbon.

The first step in taking an odor sample utilizing carbon adsorption is to mount an 8-in. by 10-in. fiberglass filter on a calibrated high-volume air sampler, such as those produced by Gelman, Mine Safety Appliance, Staplex, or Unico. The rough side of the filter should face out.

A high-volume air sampler is a requisite, due to the need of passing a relatively high volume of air through a relatively restricted sampling medium. Only glass fiber filters are suitable for the qualitative measurement of odors. Filters should be selected because of their low concentrations of metals and non-metals, such as sulfates, nitrates, fluorides, and hydrocarbon-soluble materials. Care must be taken not to support the instrument on its exhaust end. There is the possibility of damaging the air flow indicator normally located at the exhaust end of the instrument and/or stirring up dust which could be caught by the sample medium. If the filter is not placed on the hi-vol in the laboratory and carried into the field protected by a plastic bag, extreme care must be taken to protect the paper from contamination and the instrument from damage as the filter paper is inserted into the holder on the instrument.

To apply the activated carbon, the instrument is turned on and held in a position with the face of the instrument up and the filter paper horizontal. This will create a vacuum across the face of the filter paper. Next, a small amount of fine, activated carbon is sprinkled on the filter paper and evenly spread with a soft bristle brush until the entire face of the filter paper is covered with a thin, even layer of the material.

Care should be used to avoid using an excessive amount of the activated carbon or scrubbing the filter paper with the brush. If an excessive amount of activated carbon is used, the air flow through the instrument will be excessively and needlessly restricted. When using this sample medium, the volume of air normally drawn through a high-volume air sampler is only restricted 40 to 60%.

The instrument is usually operated for two hours if the odor is concentrated sufficiently to be adsorbed. The sample must be terminated if the wind direction changes appreciably or if weather conditions change. Dense fog or high humidity can cause the glass fiber filter and the activated carbon to become wet, thereby severely reducing the air flow through the instrument. The air which passes through a high-volume sampler also cools the electric motor in the instrument. If the flow is excessively restricted, it is possible to create temperatures high enough to burn out the motor.

The instrument should be monitored constantly throughout the period of the sample. If there is any major fluctuation in the volume of air passing through the instrument, it should be noted and the sample should be terminated immediately.

Upon completion of taking the sample, the air flow must be noted and recorded. The sampler, which is still running, is then faced upward, with the paper in a horizontal position. The instrument is then switched off and the air flow or vacuum is allowed to reach zero. The filter and carbon are then carefully removed from the instrument. The filter is folded lengthwise with the carbon to the inside. Care must be taken to avoid spilling any activated carbon.

The folded paper is then carefully rolled into a compact roll, again being careful not to lose any carbon. The roll is then placed in a paper extraction thimble, which in turn is placed in a small, amber glass jar and tightly capped. This is done to retain the sample in a small, enclosed atmosphere to reduce the amount of odorous materials which will escape from the activated carbon until equilibrium or saturation of the enclosed atmosphere occurs. The jar containing the sample is labeled with an indelible marker as to source, date, and time of the sample.

A situation which must be considered is the possibility of odors from other plants interfering with or shedding doubt upon the actual source of a particular odor. This problem can be corrected by taking an upwind sample between the interfering plant(s) and the source sampled. When this is not readily feasible, several samples with different wind directions can be taken and the interfering plant(s) recognized by plotting the wind sigmas to the results obtained. Naturally, there will be differences between the upwind and the downwind samples in direct relation to the different wind directions. The predominant odor identified with instrumentation from the composite samples will represent the source being sampled.

A clean technique must be used in taking a sample. The equipment must be clean and free of any minute residue from previous samplings. The technician must have clean hands and should avoid unnecessary contact with the actual sample medium. The power-supply generator must be downwind of the sample apparatus.

The continuity and completeness of sample data are of upmost importance from an enforcement as well as a scientific standpoint. A sample record is completed in the field or laboratory for each sample procedure. This record should contain a complete, permanent history of the sample. The following information should be logged on each sample record.

1. Name of the source sampled and the location of the source
2. Sample site location
3. Date and time the sampling was initiated
4. Initial air flow through the instrument
5. Wind velocity and wind direction
6. A map depicting the area of the source being sampled and the position of the sampler
7. Weather conditions that prevailed during the sampling period
8. Final air flow through the instrument
9. Date and time that sampling was completed.

The actual extraction of the sample for gas chromatographic analysis is straightforward. The extraction thimble containing the sample is put into a Soxhlet extraction apparatus and refluxed with a minimum volume (20.0 ml) of carbon disulfide (CS_2) for one hour. The resultant extract is compatible to gas chromatographic analysis using a flame ionization detector (FID). The extract is put into a clean vial, sealed, and labeled as to source, date, and time of the sample.

The mention of acrylate has not been made directly because this procedure can be used for the capture of many odors. However, to illustrate the technique a controlled experiment was conducted in the laboratory utilizing the aforementioned equipment. A mixture of alcohols, acetone, benzene, and various acrylates was placed in a beaker and stirred continuously at 50°C. Odors arising from the beaker were captured by a hi-vol instrument using paper treated with activated carbon. After one hour the sample taking was discontinued. The sample was extracted and analyzed.

Analysis of the charcoal extract was accomplished by gas chromatography. The column utilized was a 5-ft x 1/4-in. stainless steel coiled column containing 80 to 100 mesh Porapak Q. The mode of operation was isothermal with an oven temperature of 190°C. The injector temperature was 225°C and the H_2 flame ionization detector temperature was 280°C. The flow rate of N_2 was 50 cc/min. Column conditioning was done by flowing N_2 gas through the column for 24 hours with the oven temperature at 225°C.

In order to establish a "fingerprint" chromatogram, standard solutions of the compounds considered were made with CS_2 as the solvent. These compounds were methanol, ethanol, acetone, methyl acrylate, ethyl acrylate, methyl methacrylate, benzene, and n-butyl acrylate. They were run singularly and then compositely to establish their retention time(s). All reagents were either chromatographic or spectroscopic grade with the exception of the acrylates, which were reagent grade. No further purification was performed.

The results are displayed in Figure 6-1. The alcohols, acetone, and CS_2 elute within the first five minutes. Unfortunately, methyl methacrylate and ethyl acrylate overlap. Other columns or techniques must be used to differentiate between these two compounds. However, within the qualitative boundaries established for this technique the results are acceptable.

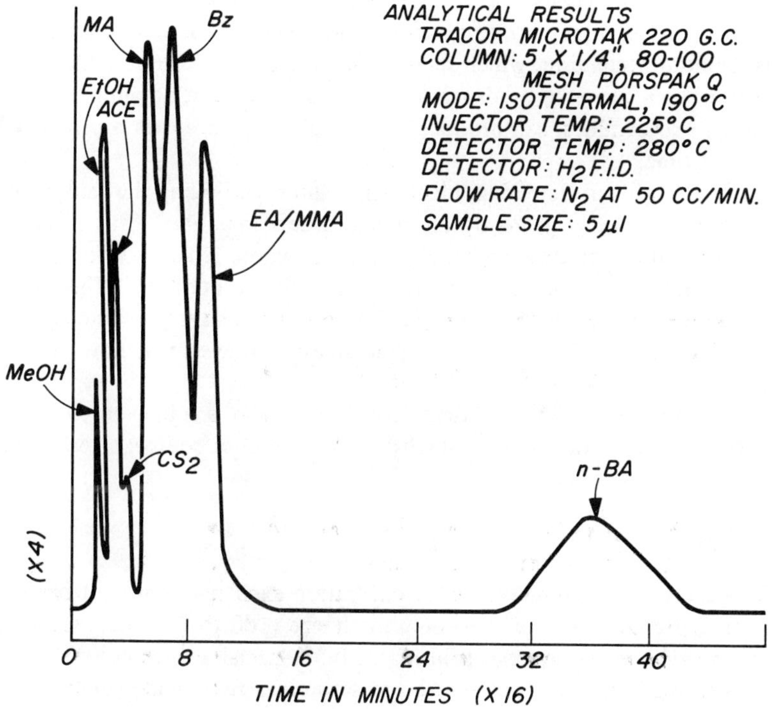

Figure 6-1. Analytical results.

What has been presented is a good, quick method compatible to the field and laboratory for the capture and detection of various acrylate odors. Quantification of the acrylate odor involves evaluation of the adsorption efficiency of the charcoal and solubility of the odors in carbon disulfide and other mechanisms, which have not been examined currently.

CAPTURE OF ORGANIC AMINES

The breathing of air permeated with amines, even in trace concentrations, is an experience not easily forgotten. This is especially true of air containing low quantities of substituted amines. People describe the odor as characteristic of dead and decaying fish. The odor is often detected in the air surrounding cattle feedlots, rendering companies, and chemical companies which produce or utilize amines.

The general odor threshold of amine compounds is in the part per million or microgram per liter of air range. In actual situations the concentration is generally higher than the threshold. Although the odor is initially offensive, most people become unaware of its presence after being subjected to it for a period of time, due to desensitization of the olfactory process. This defense mechanism, which reduces the unpleasant nature of the environment, can prove to be dangerous. Inhalation of certain amines, especially nitrosoamines, can produce physiological damage over a prolonged period of time.

Early attempts at detecting, identifying, and quantitizing the rank amine odors in the air relied primarily on olfaction determination. This determination consisted of subjecting a panel of people, whose sense of smell had been calibrated to a specific standard, to the odor in question. The odor was also gauged as to its Hedonic qualities, *i.e.,* pleasantness or unpleasantness. This test is still employed in some areas; however, it is now performed in conjunction with instrumental methods of analysis.

In an effort to support the field investigator, who is subjected to desensitization to amine odors, and establish a definite odor source, a resin air-capture system has been devised.

Procedure

The capture system employs a strong cation exchange resin (sulfonated polystyrene copolymer) of a medium mesh size (100 to 200, U.S. standard sieve) prepared in an aqueous solution of 0.5% glacial acetic acid. The resin is washed 10 times and allowed to settle in a five-to-one volume per weight ratio aliquot of the 0.5% acetic acid solution to remove any extraneous color and the fines. A final preparation, two to one, volume to weight ratio of 0.5% acetic acid to resin, is then slurried. A 50.0-ml volume of the suspension is poured into the capture apparatus. A piece of indicator (ninhydrin) impregnated filter paper is attached to the influent funnel. The system is transported to the field in this fashion (Figure 6-2).

When ready for use, the exhaust port is connected to a vacuum pump in series with a flow meter. The stopcock is turned to allow air at a rate of 20.0 l/min to be pulled through the resin suspension.

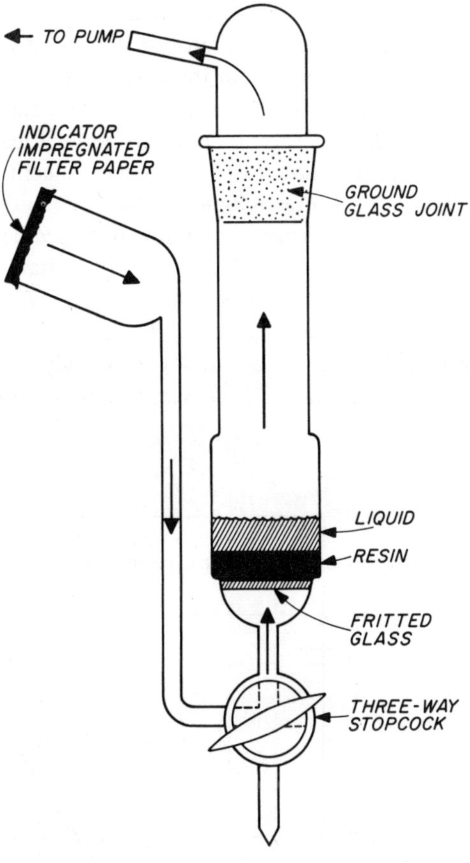

Figure 6-2. Capture system for organic amines.

Color production on the filter paper allows the investigator to determine if he is actually capturing the suspected compound and to keep suspended particles out of the system. The usual running time is one hour. When the sampling run is over, the stopcock is closed and the vacuum pump is disconnected. The system is brought intact to the laboratory.

A 20-ml volume of 0.5% acetic acid is allowed to flow through the capture apparatus, which serves as the chromatography column. Deionized distilled water is next flushed through until a neutral pH is reached, as indicated by pH paper. The amines are removed from the column with 0.1 N sodium hydroxide. A brown ring, which does not appear in controls, is observed to advance with the sodium hydroxide front. When the pH, as checked with indicator paper, is observed to rise from neutral, the eluant is collected and measured volumetrically. The usual volume of eluant is between 10 and 20 ml.

The initial detection of the presence of amines is observed on the filter paper indicator in the field. The subsequent detection and identification are performed in the laboratory and include the following: (1) the characteristic odor evolved from the basic eluant, (2) the color formation observed when the column eluant is spotted, dried, and sprayed with either ninhydrin or Dragendorff's reagent. Both are general tests for amines.[1]

A measured aliquot of the collected eluant is diluted with 0.01 N sodium hydroxide to a total volume of 100 ml. This allows a semiquantitization of the recovered samples, employing a specific ammonia electrode and a pH meter with a millivolt scale. A quantitative solution of a known amine is employed to produce a standard curve. A standard curve for trimethylamine is shown in Figure 6-3. The procedure is semiquantitative due to the fact that recovered samples are usually not pure and other ammonia compounds interfere. However, this allows for some estimation of the amount recovered.

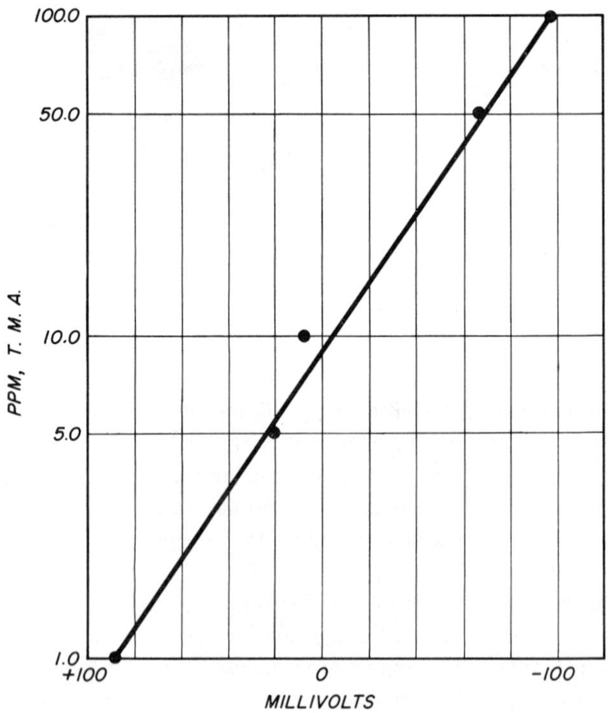

Figure 6-3. Trimethylamine using NH_4^+ electrode and Na^+ setting on orion meter.

The initial step in identification is thin-layer chromatography. The eluant is spotted on duplicate precoated and activated silica gel thin-layer plates, along with standards and reagents used in the capture and elution procedures. The plates are equilibrated in the chromatography chamber containing the solvent, a 4.5:4:1, v/v/v, mixture of chloroform:methanol:acetic acid for 10 minutes. The solvent front is allowed to ascend to the 15 cm mark. The plates are dried and developed, using ninhydrin on one and Dragendorff on the other. The unknown is then identified by color and R_f value (the distance from the starting point to the center of the substance spot divided by the distance from the starting point to the solvent front).[2] The R_f values obtained with this system remain fairly constant (Table 6-1). The

Table 6-1. R_f Values and Color for Amines Using Chloroform:Methanol:Acetic Acid in 4.5:4:1 Volume/Volume/Volume on Anasil-G

Amine*	R_f Value	Color	
		Ninhydrin	Dragendorff
1) methylamine	.55	purple	yellow
2) ethylamine	.57	purple	yellow
3) di-methylamine	.61	purple	yellow
4) tri-methylamine	.63	yellow-violet	gold
5) tri-ethylamine	.66	brown	orange
6) sec-butylamine	.78	pink	orange
7) di-iso-butylamine	.88	orange-pink	orange
8) iso-amylamine	.78	rose	orange

*Numbers correspond to spots on representative plates in Figures 6-3 and 6-4.

color production also allows identification of compounds demonstrating similar R_f values. Representations of the finished plates are illustrated in Figures 6-4 and 6-5.

There are several references in the chemical literature concerning the analysis of aqueous amines by gas chromatography using both gas-liquid (partition) chromatography and gas "solid" (porous polymer) chromatography. Some degree of success is achieved with either method.

However, due to the possible complexity of all air samples, a preliminary separation on the basis of the degree of substitution on the nitrogen is desirable, using the Hinsberg test for distinguishing between primary and secondary amines. The Hinsberg test involves addition of benzenesulfonyl chloride to an alkaline (NaOH) solution of amines to form amides. Formation of a precipitate or oil, which can be removed by filtration or centrifugation, indicates the presence of a secondary amine. The supernatant

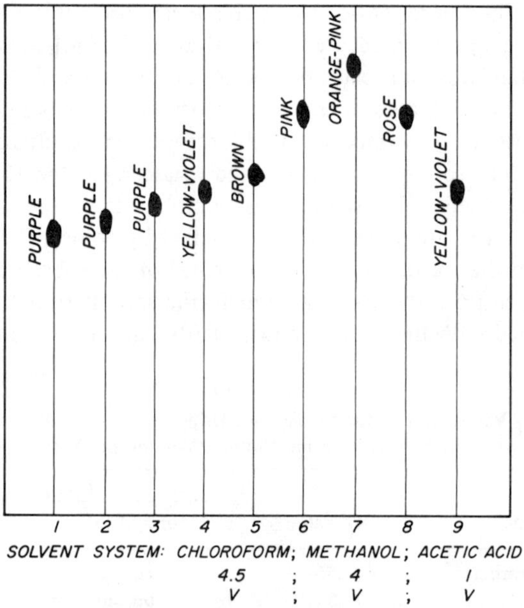

Figure 6-4. Silica gel T.L.C. with ninhydrin. Numbers correspond to amines listed in Table 6-1. Number 9 was an unknown, found to be trimethylamine.

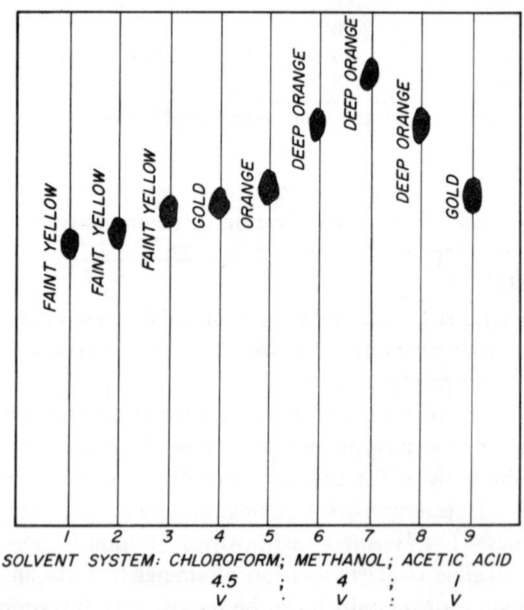

Figure 6-5. Silica gel T.L.C. with Dragendorff's reagent. Numbers correspond to amines listed in Table 6-1. Number 9 was an unknown, found to be trimethylamine.

solution is then acidified with hydrochloric acid. Formation of a precipitate or oil now indicates the presence of a primary amine. Again, removal of the precipitate or oil may be achieved by filtration or centrifugation. The supernatant now could contain any tertiary amines present as hydrochloride salts. Presence of tertiary amines can be verified by formation of methyl iodide, benzyl chloride, and other characteristic derivatives. Thin-layer chromatography of the resulting three groups of derivatives could provide both qualitative and quantitative information on the amines captured.

The results discussed in this chapter were obtained using simulated air samples and mixtures of known compounds. Field tests have been performed by a chemical company in an effort to combat an odor problem arising from the production of amines. Using information supplied on the fabrication and operation of this capture apparatus, it was able to capture, concentrate, detect, and identify concentrations of airborne amines below the threshold of odor at its plant site.

All methods of capturing amines in the field must meet the following requisites: (1) a concentration of the odor-causing agent must be possible and (2) the sample must be kept intact. Even systems employing the passage of air directly onto gas chromatography columns must concentrate the sample. The minimum level detectable by gas-liquid chromatography using a flame ionization detector is 10^{-10} g/l of air. This is a low concentration. However, even with the passage of air directly through a column, time must be allowed for an equilibrium to be set up between the agent to be identified and the liquid phase of the packing material. If present at a concentration of 1 mg/l of air, more than 60 mg of an amine can be captured in a normal running time of one hour.

The requisite of keeping the sample intact is met by this system. In addition, it provides enough sample for multiple runs and multiple methods of detection. This is not possible with on-column capture.

Besides meeting the two necessary requisites, this method has other enhancing features:

1. It is relatively inexpensive, since the resin can be regenerated.
2. It is possible to pull large volumes of air through the system.
3. Its use is not limited by temperature or humidity.
4. It is easy to use, self-contained, and can be operated with minimal training.
5. It can be applied to other odoriferous compounds by selecting the proper resin and solute conditions.

REFERENCES

1. *Thin-Layer Chromatography, a Laboratory Handbook*, E. Stahl, ed. (New York: Springer-Verlag, 1969).

2. *Progress in Thin-Layer Chromatography and Related Methods.* A. Niederwieser and G. Pataki, eds. (Ann Arbor, Mich.: Ann Arbor Science Publishers, 1970).

CHAPTER 7

GAS CHROMATOGRAPHIC ANALYSIS OF ORGANIC EMISSIONS

William P. Dolan

Rossnagel & Associates
Cherry Hill, N.J. 08003

Organic emissions may be categorized into two very broad classifications—condensible and noncondensible hydrocarbons. The basis for this is the vapor pressure of the compound at ambient temperatures, *i.e.*, high or low. For these reasons an understanding of the process which is the source of the pollutants is very desirable. A knowledge of the reactants, expected reactions, temperatures, concentrations, catalysts, and products will assist the pollution engineer in the proper selection or design of sampling methods.

Appropriate absorption reagents and sampling rates must be chosen. Also, consideration must be given to interaction of pollutants once they are absorbed or sealed in a container. When only noncondensible gases or small amounts of volatile solvents are present in the exhaust, it is only necessary to sample with a steel evacuated cylinder. This should, however, be an integrated sample (sample drawn slowly over a long period of time, 30 to 60 minutes). In some tests it is necessary to purge the sampling container with exhaust gas. However, if there are large amounts of hexane, for example, with a boiling point of 156°F, there may be some condensing of the vapors, particularly if the ambient temperatures are low. This condensation can cause low laboratory results since only the vapor phase will be measured.

The above example represents the borderline between sampling with evacuated cylinders alone and using a series of bubblers with a scrubbing liquid. Bubblers, immersed in an ice bath, are used to collect volatile condensible hydrocarbons. In Table 7-1, 212°F is arbitrarily set as the

Table 7-1. Breakdown of Organic Stack Gases

	Condensibles	
Noncondensibles	**Volatiles**	**Nonvolatiles**
Methane C_1	Acetone	Toluene C_7
Ethane C_2	Isopropanol	Xylenes C_8
Propane C_3	2-Butanone	Butylcellosolvent
Butane C_4	Hexane C_6	DOP
Pentane C_5	Benzene C_6	DIDP
Ethylene	2-Butanol	MIBK
Propylene	Petroleum Ether	Acetic Acid
Acetylene	Trichloroethylene	Mineral oil
Acetaldehyde		Kerosene

Note: The "C" numbers are basic categories. There can be more than one compound in a given "C" number.

dividing line between volatiles and nonvolatiles. This is based on the boiling point of the constituents. A volatile compound is one that boils below 212°F. Emissions in the nonvolatile column are generally sampled by using a series of impingers with proper sorption liquids.

Condensible hydrocarbons with boiling points 320°F and greater can be sampled with scrubbing liquids having a high collection efficiency. Vegetable oils, mineral oils, and plasticizers are examples of condensible hydrocarbon emissions in this category. When these high molecular weight materials are sampled from the exhaust of a gas oven or gas-fired fume incinerator, there will also be measurable amounts of unburned natural gas when sampled via an evacuated cylinder.

Other techniques used to sample condensible hydrocarbons are glass gas-sampling tubes or stainless steel U-shaped tubes immersed in dry ice. Glass gas-sampling tubes are used for rapid incremental sampling. They require minimum equipment, since the exhaust gas is purged through the tube by a rubber bulb. The condensible hydrocarbons will settle on the walls of the glass tube. A volatile solvent is then used to wash out the tube and a careful determination for nonvolatile residue is made. Samples of volatile hydrocarbons can be transferred to the gas chromatograph as follows:

1. Displacement of the gas, using mercury or brine
2. Diluting the gas with nitrogen to a positive pressure
3. Heating the glass tube to simulate stack temperatures and create positive pressure.

The stainless steel "U" tubes are immersed in a dry ice-acetone bath. Exhaust gas is pulled through the tube, which then acts as a trap condensing out all organics except methane and perhaps ethane. The "U" tube is maintained in the dry ice bath when transferred from the test site to the lab. It is then hooked up to the gas chromatograph. Nitrogen is passed through the tube while the tube is still immersed in the dry ice-acetone bath. In this manner any residual methane or ethane is swept into the chromatograph for total hydrocarbon determination. The "U" tube is then subjected to increasing temperatures as follows:

1. Dry ice-acetone bath
2. Ice water bath
3. Room temperature
4. Electrical heating of the tube.

At each temperature level, nitrogen gas is used to sweep out the corresponding level of condensed hydrocarbons which have revaporized.

Evacuated stainless steel cylinders readily lend themselves to integrated sampling when connected to an accurate low-level flowmeter. These steel cylinders are fairly expensive and require considerable care in cleaning to prepare them for reuse. A log of all steel cylinders should be maintained, recording the type of emission sampled, level of organic concentration, and the total hydrocarbon check after cleaning. For hydrocarbon constituents, gas chromatography is an invaluable technique. First of all, chromatography involves the separation of related compounds by allowing a mixture to slowly pass through an adsorbent (silica gel, clay, etc.) so that each compound will be adsorbed in a separate, often colored layer (hence the word chroma). Whether in the gas or liquid state, chromatography means separation by adsorption or partitioning.

Figure 7-1 depicts the various types of chromatography–gas (GC), liquid (LC), and thin layer (TLC). Liquid and gas chromatography involve a

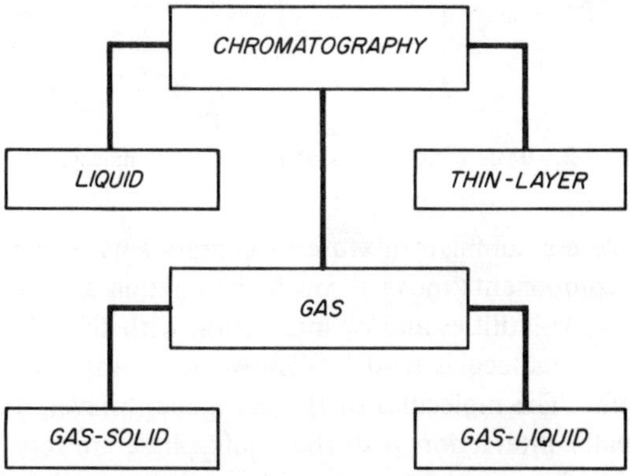

Figure 7-1. Chromatography types.

liquid-solid interface and a gas-liquid or gas-solid interface. Liquid chromatography in its simplest form is the passing of a liquid solution by gravity or pressure through a long tube containing an adsorbent. The compounds will separate out and exit at different times, therefore allowing themselves to be collected separately or be detected quantitatively. Thin-layer chromatography utilizes the same principle as liquid chromatography.

Gas chromatography can be divided into two major categories: gas-solid chromatography (GSC) and gas-liquid chromatography (GLC). In GSC the sorbent is a solid of large surface area. In GLC the sorbent is a nonvolatile liquid coated on an inert solid support. The moving gas phase, or carrier gas, is an inert gas such as nitrogen or helium, which is made to flow at a constant rate through a packed column. The packed column consists of a small-diameter tube (stainless steel, glass, etc.) containing the sorbent.

Figure 7-2 is a sketch of a gas chromatograph.

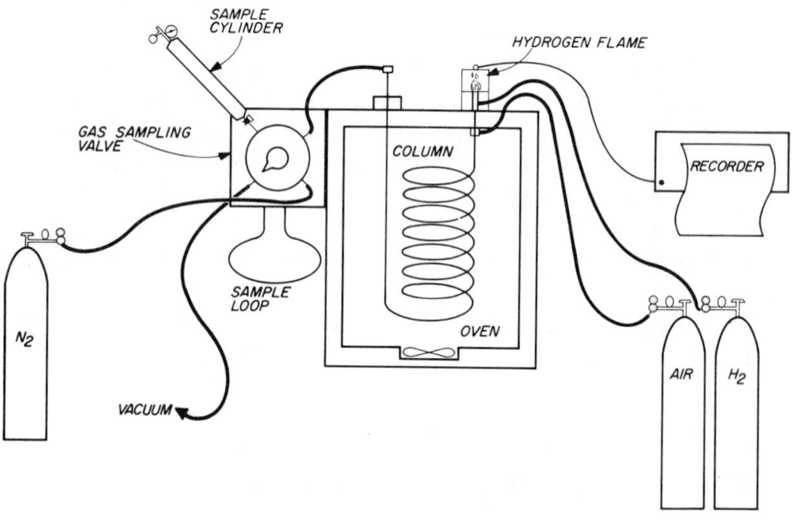

Figure 7-2. Basic components of FID gas chromatograph.

When a sample containing a mixture of components is injected into the gas stream, the components move through the column at rates dependent on their respective volatilities and/or interaction with the nonvolatile liquid phase. The various molecules tend to dissolve and revaporize as they pass down the column. The molecules of the component having the greater solubility or greater interaction with the liquid phase are retarded in their passage. With a column packing of suitable properties and sufficient length, the partition process between the gas and liquid phases results in a situation in which each component of the sample emerges in time from the end of the column as a binary mixture with the carrier gas.

To illustrate this phenomenon, a natural gas trace is depicted in Figure 7-3. Natural gas consists mostly of C_1-C_4 hydrocarbons, of which methane C_1 comprises about 85%, ethane 9%, propane 3%, nitrogen 2%, and butanes 1%. A gas chromatograph can easily, efficiently, and quickly separate these compounds and permit the quantitative detection of each down to the ppb range. By using standards under the same parameters of temperature,

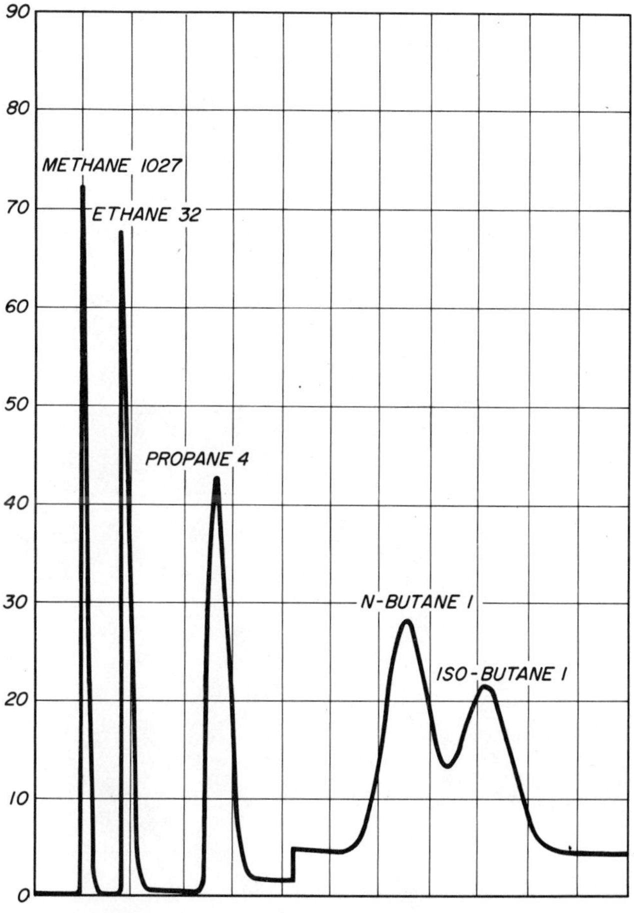

Figure 7-3. Natural gas trace.

column packing, length, etc., one can easily compute the concentrations of the unknowns. Once again it should be stressed that the identification of the constituents should be known prior to injection into the chromatograph. GC is a quantitative tool, and not necessarily a qualitative one.

If there is some question as to the presence of one or two compounds, proper interpretation of the retention times on chromatograms made under varying conditions could result in a 95% confirmation.

Gas chromatography instrumentation is almost a necessity for those involved in air pollution studies. Its use ranges from measuring constituents of flue gases to the chemical makeup of sewer gas to the myriad hydrocarbons from industrial sources, etc. Gas chromatography lends itself to the analysis of large numbers of samples with relative ease. Therefore it can be used for developing curves of the cyclical variations in concentrations of gases so common to many exhausts.

CHAPTER 8

RUBBER, PLASTICS AND GLASS INDUSTRIES ODORS

John Polhemus

Ball Corporation
Muncie, Ind. 47302

RUBBER AND PLASTICS

In the rubber industry there is a vast number of odors that may be present due to the chemicals used. A large number of the materials have complex molecular structures which will react, or decompose and react, in the production processes. One chemical could decompose and fractionate into many smaller molecular weight compounds, each of which could partially oxidize and react. Thus it is possible to produce small amounts of large numbers of chemicals, each of which could have a distinctive odor. The odor problem becomes more complex since most rubber odors are a mixture of odors rather than a pure odor.

Very simply, most odors from rubber industry sources fit into the following groups:

1. Nitro compounds
   a. Amines
   b. Oxides
2. Sulfur oxides
3. Plasticizers
4. Solvents
5. Partially oxidized hydrocarbons (aldehydes, ketones, phenol, alcohols, etc.)

To evaluate the problem in greater detail, the chemistry of rubber compounds must be studied along with their decomposition and reaction products. Table 8-1 contains a list of plastic homopolymers and a principal method

Table 8-1. Plastic Homopolymers

Family	Subgroups	Manufacturing Process
Acetal	Polyesters	Condensation of polyfunctional acids with polyfunctional alcohols
	Polyether	Polymerization of aldehydes
Acrylic		Monomer polymerization
Allyl		Allyl alcohol esters and dibastic acids
Amino	Urea-formaldehyde	Reaction of formaldehyde with amino compounds
	Melamine-formaldehyde	
	Aniline-formaldehyde	
Cellulosic		Chemical processing (mono acids and its anhydride) of cellulose
Epoxy		1. Epichlorohydrin and bisphenol A aliphatic polyols
		2. Polyolefins oxidized with peracetic acid
Fluoroplastic	Chlorotrifluoroethylene	Polymerization of $CF_2CF\ Cl$
	Fluorinated ethylene propylene (FEP)	Fluorination of ethylene propylene
	Perfluoroalkoxy resin (PFA)	Polymerization
	Polyvinylidene fluoride (PVF_2)	Polymerization of $CH_2\ CF_2$
	Polytetrafluoroethylene (TFE)	Polymerization of $CF_2CF_2C_2F_4$
Phenolic	Phenol formaldehyde	Condensation of phenol with aldehydes
	Phenol acetaldehyde	
	Phenol furfural	
Polyamide		Condensation of amino acids, or polycondensation of dibasic acids and diamine compounds
Polycarbonate		Action of phosgene on bisphenol A
Polyimide		Reaction products of dianhydrides and diamine
Polyolefin	Polyethylene	Polymerization of monomer
	Polypropylene	
Polyphenylene	Sulfide	
Polystyrene		Polymerization of styrene
Polysulfide		Ethylene dichloride and sodium polysulfide reaction
Polyurethane		Reaction of polyesters or polyethers with isocyanates
Polyvinyl	Polyvinyl acetate	Monomer polymerization
	Polyvinyl alcohol	
	Polyvinyl chloride	
	Polyvinylidene chloride	
Silicone		Polycondensation of polyfunctional silanols, formed by chlorosilane hydrolysis

of manufacture. The method of manufacture is important since residual monomer or reactant will often impart an odor on a neutral-smelling polymer. Table 8-2 contains a similar list of rubber homopolymers. A

Table 8-2. Rubber Homopolymers

Family	Manufacturing Process
Acrylonitrile	Acrylonitrile polymerization
Butyl rubber	Isobutylene polymerization
Polybutadiene	Butadiene polymerization
Polychloroprene (neoprene)	Chloroprene polymerization
cis-Isoprene	Isoprene polymerization

large variety of rubber and plastics today involves the copolymerization of homopolymers. A good example is acrylonitrile, which is copolymerized with chloroprene to form nitrile-chloroprene (a rubber) or butidiene and styrene to form acrylonitrilebutidienestyrene (ABS) a plastic. The odors of monomers and reactants used to produce these rubber and plastic polymers are found in Table 8-3.

A finisher rubber or plastic will contain dozens of processing compounds. Some of the major classes of compounding materials are listed in Table 8-4. Starting with organic accelerators in Table 8-5, the major chemical families of compounding material, their odors, and decomposition products shall be identified. These decomposition products may not necessarily be present during operating conditions; in fact, some are present only during emergency conditions.

All of the major organic accelerators are nitrogen groups, and the odors in these groups can vary. The thiazole group can vary from odorless to that of a mercaptan (skunk-like odor). Because all of the groups contain nitrogen, nitrous oxides (sharp pungent odor) are characteristic decomposition products. The decomposition products of any group are greatly dependent on physical conditions, namely temperature and available oxygen. Decomposition can vary so greatly that no attempt has been made to identify these complex chemicals other than as "intermediate products." Certainly the greatest family contained here would be the amines. In general, all amines possess an odor that can only be described as "amine." The lower molecular weight amines have an ammonial odor that may resemble a fish smell. The odor decreases as the molecular weight increases.

Most of the organic accelerators contain sulfur, which will decompose to sulfur gases. Sulfur dioxide (SO_2), the major component, has a marked pungent odor. Other sulfur gases include CS_2, H_2S, and SO_3.

Table 8-3. Monomers and Reactants Used in Plastic and Rubber Production

Family	Characteristic Smell
Acetaldehyde	Pungent, fruity[7]
Acrylonitrile	Practically odorless,[6] slight peach
Analine	Strongly amine-like[6]
Bisphenol A	Mild phenolic[7]
Butadiene	Slight aromatic[6]
Chlorotrifluorethylene	Faint ethereal[8]
Dichloroethylene	Pleasant[8]
Difluorethylene	Faint ethereal[8]
Epichlorhydrin	Irritating chloroform[6]
Ethylene	Sweet odor[8]
Ethylene urea	Odorless[8]
Formaldehyde	Irritating, strong[6]
Furfural	Bitter almonds[6]
Isocynates	Irritating[6]
Methyl methacrylate (an acrylic monomer)	Sweetish characteristic[6]
Phenol	Characteristic[6]
Phosgene	New-mowed hay[7]
Silicone intermediates	Irritating[6]
Styrene	Unpleasant, characteristic[6]
Urea	Almost odorless[8]
Vinyl acetate	Sweetish, rather pleasant[6]
Vinyl chloride	Sweetish, not very marked[6]
Vinylidene chloride	Sweetish, chloroform-like[9]

Table 8-4. Compounding Materials

Accelerators	Dispersing agents
Anti-cracking agents	Extending oils
Anti-foaming agents	Fillers, insulating and reinforcing agents
Anti-oxidants	Flame retardants
Antiseptics, germicides, bacteriostatic agents	Foaming agents
Anti-static agents	Inhibitors
Bloom inhibitors	Lubricants and stabilizers
Catalysts	Odorants
Coloring agents	Plasticizers
Curing agents	Retarders
Detergents and emulsifiers	Solvents
	Vulcanizing agents

Table 8-5. Organic Accelerators

Common Types Used	Characteristic Odors	Characteristic Decomposition Products
Aldehydeamines		
Reaction products of aldehydes and aniline (*i.e.*, butyraldehyde-aniline)	Mild, aromatic, amine	NH_3 NO_x Intermediate products
Dithiocarbamates		
Zinc, lead, sodium, potassium cupric, bismuth, cadmium selenium, salts of dialkyl dithiocarbamates (*i.e.*, zinc dibutyl dithiocarbamate)	Mild, Pleasant	NH_3 NO_x Intermediate products (amines) Sulfur gases Metallic oxides
Guanidines		
Di, tri aromatic guanidines (*i.e.*, diphenyl guanidines)	Partially odorless, aromatic	NH_3 NO_x Intermediate products
Sulfenamide		
Saturated -2- benzotmiazole-sulfenamides (*i.e.*, n-cyclo-hexly benzothiazole-sulfenamide)	Amine	Sulfur gases NO_x Intermediate products
Thiazole		
Usually mercapto or benzo thiazoles, sometimes modified (*i.e.*, 2-mercapto benzothiazole)	None, mild, mercaptan	Sulfur gas NO_x Intermediate products
Thiurams		
Tetra alkyl thiuram mono/di sulfide (*i.e.*, tetramethyl thiuram di-sulfide)	Practically odorless	Sulfur gases NO_x NH_3 Intermediate products

The majority of the anti-cracking agents are waxes. Most of these are petroleum-based waxes, which would have a paraffin wax odor. Major groups of anti-foam agents and their odors are found in Table 8-6.

Table 8-7 contains the major groups of anti-oxidants. Like the accelerators, these groups are complex chemicals. It has already been established that the decomposition of amines will produce other amines and nitrous gases. Phenolic decomposition may result in a variety of partially oxidized products, including alcohols and aldehydes, each with its characteristic odor.

Anti-static agents fall into two main groups: amines and quaternary ammonium compounds. The latter group ranges from odorless to ammonia-like. Major groups of catalysts include peroxides, azo types (nitrile), and

Table 8-6. Anti-Foaming Agents

Group	Characteristic Odor
Oils, Glycerides, Fatty Acids	
Land oil	Peculiar[8]
Pine oil	Piney[8]
Vegetable oils	Varied
Fatty acids	Slight, varied
Alcohols	
Capryl alcohol	Aromatic[8]
Cetyl alcohol	Faint[8]
Cyclohexanol	Camphor-like[8]
Lauryl alcohol	Floral[8]
Phosphates	
Tricresyl	Mild to odorless[8]
Tributyl	Odorless[8]
Silicones	Mild, varied but irritating

Table 8-7. Anti-Oxidants

Group	Characteristic Odor
Alkylated phenols and bisphenols	Phenolic
Alkylidene bis, tris, and polyphenols	Phenolic
(di) Thio, bis tris, and poly alkylated phenols	Phenolic
Amines	Amines
Esters	Odorless to mildly fruity
Organic phosphites and phosphates	Mild

halides (metallic and nonmetallic). The peroxides will be discussed in the vulcanization section. The metallic halides are capable of releasing hydrochloric acid, which has a strong irritating odor.

Inorganic coloring agents or pigments include chromates, ferrycyanides, sulfides, selenides, oxides, silicates, aluminates, and titanates. Organic coloring agents are mostly nitrogen-containing compounds and would include azos, phthalocyanines, and quinacridones.

The two largest groups of curing agents are amines and anhydrides (Table 8-8). Detergents and emulsifiers are largely composed of soaps and fatty acids and have corresponding odors. Most extenders are petroleum-based oils. Extender oil is a mixture that may contain paraffin, naphthene, and aromatic-type petroleum. The components in each group are too numerous to list, but one should be familiar with the odor associated with each type.

Table 8-8. Amine Curing Agents for Epoxy Resins

Ethylene diamine	Dicyandiamide
Diethylenetriamine	Propylene oxide-amine adducts (typical)
Triethylenetetramine	Acrylonitrile-amine adducts (typical)
Tetraethylenepentamine	Diethanolamine
Dimethylamino propylamine	Piperidine
Diethylaminopropylamine	Phthalic
Ethanolamine	Maleic
p,p-Methylenedianiline	Tetrahydrophthalic
Aminoethylpiperazine	Methyl nadic
Diaminodiphenylsulfone	Dodecenyl succinic
Menthanediamine	Nonyl succinic
meta-Xylene diamine	

Fillers and insulating and reinforcing agents are largely composed of inorganics, mainly minerals, and have minimal odor problems. Halides (bromine and chlorine) and phosphorous are the active components in flame retardants. The odor of these flame retardants is varied because it depends upon the chemical group to which these active elements are attached. In general, alkyl dianyl or trianyl (*i.e.*, cresyl diphenyl/triphenyl) phosphates have little if any odor. The halide compounds can range from odorless to chloroform-like, or that of its basic chemical carrier.

Foaming or blowing agents are mostly nitrogen compounds, as seen in Table 8-9,[5] and release a variety of nitrogen gases.

Lubricants and stabilizers include medium molecular weight (200-2,000) salts, esters, amides, esters, and ketone. The main chain molecule is usually a fatty acid derivative and may be steric, lauric, ricinolic, oleic, napthenic, etc. Metallic salts include calcium, magnesium, sodium, lead, cadmium, barium, zinc, etc. The metallic salts of fatty acids have a slight to strong fatty acid smell. A stabilizer, like cadmium naphthanate, may even impart an odor on the finished product. Butyl stearate, a typical fatty acid ester, has no odor,[6] while methoxyethyl oleate has an oily odor.[6] Some blends will contain a perfume to mask the odors in a finished product.

Table 8-9. Commercial Organic Foaming Agents

Chemical Description
Azo-Compounds
1. Azobisformamide (azodicarbonamide)
2. Azobisisobutyronitrile
3. Diazoaminobenzene
N-Nitroso-Compounds
1. N,N-Dimethyl-N,N-dinitroso-terephthalamide
2. N,N-Dinitrosopentamethylentetramine
Sulfonyl Hydrazides
1. Benzenesulfonylhydrazide
2. Benzene-1,3-disulfonylhydrazide
3. Diphenylsulfon-3,3-disulfonyl hydrazide
4. 4,4 Oxybis (benzenesulfonyl hydrazide)

An extensive number of plasticizers or processing oils is used in the rubber and plastics industry. Some of the common plasticizers are found in Table 8-10. The odor descriptions are from the manufacturer's trade literature. Cases of more than one odor description are due to differences in the way manufacturers describe their products.

The odor of a plastisol, in fact, can be imparted by other trace materials in the product. Some materials, like phthalate plastisols, can be described as having a characteristic odor, but such a description serves little purpose.

Table 8-11 contains a list of solvents used in the rubber and plastics industry. Some of these solvents are used in batches while others are used in material processing or auxiliary operations. The final column contains available odor thresholds.

Vulcanizing agents are contained in three main categories:

1. Sulfur and sulfur chloride
2. Oxo compounds
   a. Peroxides
   b. Phenols
   c. Dioximes and quinones
3. Metal
   a. Lead oxide
   b. Selenium
   c. Tellurium

Sulfur chloride has a strong, penetrating odor while sulfur has a distinct odor. Dicumyl peroxide has an acrid, fruity odor and will yield acetophenone

Table 8-10. Plastisols

Type	Odor
	Adipates
Di(butoxyethyl) (BCA)	Mild[14]
Diisobutyl (DIBA)	Mild[14]
Diisodecyl (DIDA)	Mild[14,15,16]
Diisooctyl (DIOA)	Mild,[14,15] slightly aromatic[6]
Dimethyl (DMA)	Mild, pleasant[14]
Di(n-octyl, n-decyl) (DNODA)	Mild[14,15]
Dinonyl (DNA)	Mild[14]
Dioctyl (DOA)	Mild,[14,15,16] slightly aromatic[6]
	Phosphates
Octyl diphenyl	Slight[6]
Tri-(butoxyethyl)	Mild[16]
Tricresyl (TCP)	Mild[15]
Triphenyl (TPP)	Very faint aromatic[16]
	Phthalates
Butyl benzyl (BBP)	Slight[14]
Butyl cyclohexyl (BCHP)	Mild[14]
Butyl decyl (BDP)	Mild[14,15]
Butyl octyl (BOP)	Mild[14]
Dibutyl (DBP)	Mild,[14,15] practically odorless[6]
Di-(butoxyethyl) (BCP)	Mild,[14] slightly aromatic[6]
Dicapyl (DCP)	Mild,[14] slight[16]
Dicyclohexyl (DCHP)	Essentially odorless,[14] moderately aromatic[16]
Diethyl (DEP)	Slightly aromatic[14,16]
Dihexyl (DHP)	Mild[15]
Diisobutyl (DIBP)	Mild,[14] essentially odorless[18]
Diisodecyl (DIDP)	Mild,[14] slight[16]
Diisohexyl (DIHP)	Mild[14]
Diisooctyl (DIOP)	Mild,[14,15] slight,[16] slightly aromatic[6]
Di(methoxyethyl)	Mild, slightly aromatic[18]
Dimethyl (DMP)	Slightly aromatic[6,14,16]
Dioctyl (DOP)	Mild, slight
Ditridecyl (DTDP)	Mild,[14,15,17] slight[16]
n-octyl, n-decyl (NODP)	Mild,[14,15,17] slight[16]
	Sebacates
Dibutyl (DBS)	Mild[14,15]
Diisooctyl (DIOS)	Mild[14]
Dimethyl (DMS)	Mild[14]
Dioctyl (DOS)	Mild,[14,15] practically nil[6]
	Other Name
Dioctyl azelate (DOZ)	Practically odorless[18]
Glycerol triacetate	Essentially odorless[18]
Ethyl phthalyl ethyl glycolate	Slight[16]
Butyl phthalyl butyl glycolate	Slight[16]

Table 8-11. Solvents Used in the Rubber Industry

Name	Characteristic Odor	Odor Threshold (ppm)
Acetone	Chemical sweet, pungent[28]	100,[19] 320[20]
Amyl acetate	Pear-like[19]	
Benzene	Sweet-solvent[28]	3,[21] 5,[19] 60[20]
Butyl acetate	Fruity[8]	7,[20] 0.6[21]
Butyl alcohol	Vinous[8]	1[24]
Butyl ether	Ethereal[9]	<300[9]
Capruyl alcohol	Pungent aromatic,[7] sweet pungent[28]	
Carbon tetrachloride	Heavy ethereal[7]	100,[19] 200[20]
Chloroform	Sweet, ethereal[14]	<300[9]
Chlorotoulene	Pungent, irritating	0.047
Cyclohexane	Bland, fatty, benzene,[19] pungent[7]	<300[9]
Cyclohexanol	Camphor-like[7]	160[22]
Cyclohexanone	Acetone-like[7]	730
Diacetone alcohol	Faint, pleasant[7]	<100[9]
Dichloroethyl ether	Chloroform-like[8]	15[9]
Di-iso-butyl ketone	Mild[8]	<25[9]
Dipentene	Lemon-like[19]	
Ethyl acetate	Fragrant[19]	50[20]
Ethyl alcohol	Sweet floral[28]	10,[19] 50[20]
Ethyl amyl ketone	Mild fruity[9]	5[9]
Ethylene dichloride	Chloroform-like[8]	100[23]
Heptane	Faint	220[20]
Hexanes	Faint,[8] sweet	
Hexyleneglycol	Mild	
Isobutyl alcohol	Sweet	40[20]
Isoheptanes	Faint	
Isohexanes	Faint, sweet	
Isooctanes	Faint, pleasant, sweet	
Isopentanes	Pleasant	
Isoporone	Mild	
Isopropyl acetate	Pear-like[19]	30[20]
Isopropyl alcohol	Pleasant[8]	40[20]
Methyl acetate	Fragrant[19]	200[20]
Methyl alcohol	Sweet, fruity[28]	100,[19] 5,900[20]
Methyl amyl acetate	Pleasant[7]	
Methyl chloride	Ethereal[19]	>10[19]
Methyl ethyl ketone	Sweet,[28] acetone-like[7]	10,[19] 25[20]
Methyl isoamyl ketone	Pleasant[8]	
Methyl isobutyl ketone	Sweet, floral, fruity[28]	0.47,[19] 8[20]
Methyl isobutyl carbinol	Faint, ethereal	
Methylene chloride	Ethereal[19]	214,[19] 150[20]
Monochlorobenzene	Chlorinated, benzene-like, moth balls,[28] mild[19]	0.21[19]
Naphtha	Camphor-like	
Nitro-ethane	Pleasant[8]	
Orthodichlorobenzene	Pleasant[8]	
Pentanes	Pleasant[8]	<5000[9]

Table 8-11, continued.

Name	Characteristic Odor	Odor Threshold (ppm)
Perchloroethylene	Sweet, chlorinated,[28] chloroform-like[7]	4.7[19]
Propylene dichloride	Chloroform-like[8]	
Pyridine	Burnt, pungent, rank, diamine[28]	0.82,[24] 0.021,[19] 0.012,[26] 0.23[25]
Toluene	Sweet, solventy[28]	0.25[27]
Trichlorobenzene		
Trichloroethane	Pleasant,[7] sweetish[9]	<500[9]
Trichloroethylene	Sweet, solventy,[28] chloroform-like[7]	21.4,[19] 80,[20] 25[26]
Triglycol dichloride	Mild	
Xylene	Sweet, oily, moth balls[28]	1[24]

(sweet pungent odor), methane (odorless), alcohols (varied odors), and methylstyrene. a,a bis (t-butylperoxy) diisopropylbenzene, another organic peroxide, has an acrid, medicinal odor. It decomposes to phenols (distinctive odors), acetone (sweetish odor), methane and ethylene (sweet odor). Decomposition of the oxo vulcanization agents will, in general, leave fragmented, partially oxidized allyl and alkyl compounds (phenols, aldehydes, ketones, alcohols, etc.). Acrolein (CH_2CHCHO), a toxic material with an irritating odor, resembling a snuffed candle, is often found in rubber operations.[19]

Thermal Decomposition

As already mentioned, thermal decomposition may be an important odor source. Heat and available oxygen conditions are often excessive and are not necessary plant conditions. The pyrolysis products of a polyester (acetal group) will contain modified maleic acid anhydride, unsaturated alkyl and monomeric triallyl/cyanurate, hydoquinone and tertbutyl perbenzoate.[29] Amino-type plastics, urea melamine-, and aniline-formaldehyde will produce hydrogen cyanide (HCN) and ammonia (NH_3) when they burn incompletely.[30]

Teflon (TFE resin) has been found to produce small quantities of silicon tetrafluoride, hexafluoroethane, octafluorocyclobutane, octafluoroisobutylene, and hexafluoropropylene at temperatures of 572 to 1,022°F.[31] However, the quantity of these materials may be too small to produce a detectable odor. Pyrolyzing polyvinylidene fluoride (PVF_2) has been found to yield hydrogen fluoride.[32]

Phenol-formaldehyde resins, when heated, were also found to yield phenol and formaldehyde.[30] Unsaturated compounds, such as aldehydes, may be produced when polyethylene, polyolefin burns.[34]

Polyphenylene sulfide pyrolyzes at high temperatures (700°F) to produce diphenyl phenylene sulfide, sulfur dioxide, and carbonyl sulfide.[35]

In summary, odors can often be detected in rubber and plastics plants from the following operations:

1. Compounding and formulation
2. Material handling
3. Banbury
4. Mills
5. Curing, vulcanizing operations
6. Forming operations
   a. Extruders
   b. Presses

GLASS INDUSTRY ODORS

The glass industry is a relatively minor source of industrial odors. The compounds used in glass formulation are obtained almost exclusively from the mineral kingdom and have little if any odor. Odors that exist in glass container plants can be put in the following categories:

1. Furnace products
2. Mold lubrication maintenance
3. Coating
4. Miscellaneous operations

Table 8-12 contains compositions of flint and amber glass furnace emissions.

Table 8-12. Emissions from the Manufacture of Glass[41]

Emissions	Flint Glass Furnace	Amber Glass Furnace
Solids, gr/scf	0.029	0.041
Solids, lb/hr	2.1	5.4
Sulfur trioxide, ppm,vol	17	15
Sulfur dioxide, ppm	250	315
Fluorine, ppm	2.2	1.9
Chlorine, ppm	4.9	4.1
Nitrogen oxides, ppm	340	640
Carbon monoxide, ppm	375	40

Odorous emissions include sulfur oxides (sharp, pungent odors), nitrogen oxides (sharp, irritating odor), halide gases (pungent, irritating odor), and minor concentrations of gases with trace metals.

Petroleum-base mixtures are the most widely used type of mold lubricants. Thermal decomposition produces an array of fractionated organic material, dependent upon temperatures and oxygen availability. The odor can be described as that of oil or oil smoke.

Elemental sulfur has been used as a reducing agent on oxidized sections of molds. This process can liberate large quantities of sulfur oxides. Metallic halide gases such as stannic chloride ($SnCl_4$) or titanium tetrachloride ($TiCl_4$) are used to impart a scratch-resistant coating on hot glass.

These "hot end coatings" hydrolyze rapidly to produce hydrochloric acid, which has a marked irritating odor.

"Cold end coatings" are used to provide lubricity for the glass container conveyance. Most of these coatings are odorless, while some may have a very mild odor.

A variety of odors may exist about a glass plant, not from the glass processing but from support operations. Monochlorobenzene (chloronated benzene-type odor)[28] and other solvents are used in quality control labs for their refractive index, as well as a variety of oils used for their density. Glues are used for labels and box assembly. Hydrofluoric acid (HF), which has an odor similar to hydrochloric acid, is used for glass etching. The machine shop is a source of oil and lubricant-type odors. Aromatic odors are sometimes present because aromatics are used as carriers in Teflon coating. Table 8-13 is a list of solvents used in vapor degreasing and their odors.

Table 8-13. Degreasing Solvents

Agent	Odor
1,1,1 trichloroethane	Sweet, pleasant
Trichloroethylene	Sweet, chloroform-like
Perchloroethylene	Sweet, chloroform-like
Methylene chloride	Ethereal

REFERENCES

1. *Modern Plastics Encyclopedia*, 1974-75. (New York: McGraw-Hill, 1975).
2. *Rubber Red Book*, 24th ed. (New York: Palmerton Publishing Co.)

3. Kaufman, M. *Giant Molecules, the Technology of Plastics, Fibers, and Rubber* (Garden City, N.Y.: Doubleday, 1968).
4. Stern, H. J. *Rubber: Natural and Synthetic,* 2nd ed. (New York: Palmerton Publishing Co., 1967).
5. Simonds, H. R., J. M. Church. *The Encyclopedia of Basic Materials for Plastics.* (New York: Reinhold, 1967).
6. Lefaux, R. *Practical Toxicology of Plastics.* (London: Iliffe Books, 1968).
7. Sax, N. I. *Dangerous Properties of Industrial Material.* (New York: Van Nostrand Reinhold, 1968).
8. *Condensed Chemical Dictionary,* 8th ed. (New York: Van Nostrand Reinhold, 1971).
9. Patty, F. A. *Industrial Hygiene and Toxicology,* 2nd ed. (New York: John Wiley, 1963).
10. Pennwalt Company: technical literature.
11. R. T. Vanderbilt Co.: technical literature.
12. Dupont, Elastomer Chemical Department: technical literature.
13. American Cyanamid Co.: technical literature.
14. W. R. Grace, Hatco Chemical Division: technical literature.
15. United States Steel, Chemical Division: technical literature.
16. Monsanto: technical literature.
17. Allied Chemical Co.: technical literature.
18. Eastman: technical literature.
19. Sullivan, R. J., *et al. Air Pollution Aspects of Odorous Compounds,* Litton Systems, Inc., Report to National Air Pollution Control Administration, Department HEW (Sept. 1969).
20. May, J. "Odor Thresholds of Solvents for Assessment of Solvent Odors in Air," *Staub* **26** (9), 34 (1966).
21. Stern, A. C., ed. *Air Pollution,* vol. II, 2nd ed. (New York: Academic Press, 1968).
22. Matsushita, H., *et al.* "Determination of Threshold Values for Olfactory Perception of Primary Odour Substances," *Industrial Health,* **5**, 221 (1967).
23. *Air Pollution Abatement Manual.* (Manufacturing Chemists Assoc., Inc., 1955).
24. Rosen, A. A., J. B. Peter, and F. M. Middleton. "Odor Thresholds of Mixed Organic Chemicals," *J. Water Poll. Control Federation,* **34** (1), 7 (1962).
25. Summer, W. *Methods of Air Deodorization.* (Amsterdam: Elsevier, 1963).
26. *International Critical Tables of Numerical Data, Physics, Chemistry and Technology*, vol I, National Research Council of U.S.A. (New York: McGraw-Hill, 1926).
27. Stalker, W. W. "Defining the Odor Problem in a Community," *J. Am. Ind. Hyg. Assoc.,* **24**, 600 (1963).
28. *Research on Chemical Odors, Part I: Odor Thresholds for 53 Commercial Chemicals.* (Arthur D. Little for Manufacturing Chemists Assoc., 1968).
29. Madorsky, S. L. and S. Straus. "Stability of Thermoset Plastics at High Temperature," *Modern Plastics,* **38**, 138-140 (1961).

30. Dufour, R. E. "Survey of Available Information on the Toxicity of the Combustion and Thermal Decomposition Products of Certain Building Materials under Fire Conditions," Bulletin of Research No. 53, Underwriters' Laboratories.
31. Zapp, J. A., Jr., G. Limperos, and K. C. Brinker. "Toxicity of Pyrolysis Products of Teflon Tetrafluoroethylene Resin," paper presented at meeting of AIHA (April 26, 1955).
32. Madorshy, S. L. and S. Straus. "Thermal Degradation of Polymers at High Temperatures," *Journal of Research of the National Bureau of Standards,* **63A**, 261-8 (1959).
33. Dupont, unpublished data, private communication.
34. Zapp, J. A., Jr. "Toxic and Health Effects of Plastics and Resins," *Archives of Environmental Health,* **4**, 125-36 (1962).
35. Phillips Petroleum Co., unpublished data, private communication.
36. Fulks, B. D., R. J. Morris, Jr. and W. J. Lamdein. "Polystyrene Resins, Identification of Combustion Products from Self-Extinguishing Formulations," Union Carbide Plastics Co., unpublished report (1960).
37. Yarborough, V. A. and J. F. Haskin. "Spectrometric Study of the Gaseous Products from the Combustion of 'Vinylite' Sheeting," Union Carbide Plastics Co., unpublished report (Sept. 28 and Oct. 26, 1950).
38. Crider, L. B. "Thermal Decomposition Products from Poly (Vinyl Chloride)," B. F. Goodrich Chemical Co., Development Center, Avon Lake, Ohio, unpublished report (1962).
39. Himes, J. B. *The Determination of Organic Materials in Stack Emissions*, presentation before the Amer. Chem. Soc., Rubber Group, Toronto, Canada (May 11, 1974).
40. Smith, Dr. R. University of Michigan, private correspondence.
41. Stockham, J. D. "The Composition of Glass Furnace Emissions," *J. Air Pollut. Control Assoc.,* **21** (11) (Nov. 1971).

CHAPTER 9

DOMESTIC SEWAGE AND REFUSE ODOR CONTROL

Thomas J. Olenik

New Jersey Institute of Technology
Newark, N.J. 07102

The problem of annoying odors in the vicinity of sewage treatment plants (STP) and refuse disposal sites (RDS) has intensified sharply in the United States. Land adjoining these sites becomes subdivided and subsequently developed for industrial and residential purposes. This observation is especially true in metropolitan areas, where land is markedly scarce. It is not unusual to find STPs and RDSs adjacent to new housing developments.

Many states have codes for minimum distances between housing developments and sewage and refuse treatment areas. However, even though a new STP may meet the minimum distance requirement, along with some additional landscaping to avoid visual objections from the surrounding residents, it is entirely possible that the odors produced can become the largest nuisance in the neighborhood. This thought also applies to an RDS, and in many cases may be intensified. It is certain that most of the general public has been exposed to the smell of decaying refuse either on their own premises or by passing a poorly operated landfill or open dump. Therefore, it is the intent of this chapter to discuss in detail the odor problems associated with STPs and RDSs.

ODORS ORIGINATING FROM DOMESTIC SEWAGE

Chemical Origin of Odors

Odors from domestic wastewater usually result from some sort of biological activity in the sewer collection system and wastewater treatment plants. The gases that arise from domestic sewage are inorganic. Sewage containing industrial wastes may have their odor problem compounded by organic gases from any waste chemicals added to the sewer system.

Of the inorganic gases, hydrogen sulfide (H_2S) and ammonia (NH_3) are odorous while dissolved oxygen (DO), carbon dioxide (CO_2), methane (CH_4), nitrogen (N_2) and hydrogen (H_2) are odorless. Gases originating from organic sources by anaerobic decomposition usually contain nitrogen and sulfur. It must be remembered that the many different combinations of gases that can occur at one time can compound or change the description of the odor that would exist individually. Tables 9-1 and 9-2 summarize the substances that are the odor-causing agents.

Table 9-1. Odor Characteristics and Threshold Concentrations[21]

Substance	Formula	Threshold Odor (mg/l)	Remarks
Allyl mercaptan	$CH_2{:}CH{\cdot}CH_2{\cdot}SH$	0.00005	Very disagreeable, garliclike
Ammonia	NH_3	0.037	Sharp, pungent odor
Benzyl mercaptan	$C_6H_5CH_2{\cdot}SH$	0.00019	Unpleasant odor
Chlorine	Cl_2	0.010	Pungent, irritating odor
Chlorophenol	$Cl{\cdot}C_6H_4{\cdot}OH$	0.00018	Medicinal odor
Crotyl mercaptan	$CH_3{\cdot}CH{:}CH{\cdot}CH_2SH$	0.000029	Skunk odor
Diphenyl sulfide	$(C_6H_5)_2S$	0.000048	Unpleasant odor
Ethyl mercaptan	$CH_3CH_2{\cdot}SH$	0.00019	Odor of decayed cabbage
Ethyl sulfide	$(C_2H_5)_2S$	0.000025	Nauseating odor
Hydrogen sulfide	H_2S	0.0011	Rotten egg odor
Methyl mercaptan	CH_3SH	0.0011	Odor of decayed cabbage
Methyl sulfide	$(CH_3)_2S$	0.0011	Odor of decayed vegetables
Pyridine	C_6H_5N	0.0037	Disagreeable, irritating odor
Skatole	C_9H_9N	0.0012	Fecal odor, nauseating
Sulfur dioxide	SO_2	0.009	Pungent, irritating odor
Thiocresol	$CH_3C_6H_4{\cdot}SH$	0.0001	Rancid, skunklike odor
Thiophenol	C_6H_5SH	0.000062	Putrid, nauseating odor

Table 9-2. Selected Odor Threshold Concentrations[36]

Chemical	Symbol	Concentration for Threshold Odor In Parts Per Million By Volume
Carbon disulfide	CS_2	0.21
Acetaldehyde	$C_2H\ O$	0.21
Hydrogen sulfide	H_2S	0.00047
Nitrogen compounds		0.00021-100.0
Skatole	C_6H_6N	0.019
Mercaptans		
Ethyl	C_2H_6S	0.001-0.00026
Methyl	CH_4S	0.041-0.0021
Chlorine	Cl_2	0.314
Ammonia	NH_3	46.8
Perchloroethylene	C_2Cl_6	4.68
Phenol	C_6H_6O	0.6

Odor-Producing Mechanisms

Fresh domestic sewage (containing dissolved oxygen) has a slight odor, usually described as musty. Natural aerobic breakdown of the sewage, whether it is occurring in a sewer system or wastewater treatment plant, depletes the DO. Of course, in an STP the DO is being replaced by mechanical means. Only in the digestion tanks should the odorous process of anaerobic decay be allowed. Once sewage is depleted of DO, the odors of H_2S and other malodorous compounds are given off and a general septic condition exists. Any added chemicals from industrial processes will cause even greater problems. Since most of the odorous gases that arise from sewage are caused by bacteria, it is appropriate to discuss in detail the process that allows these odors to escape into the atmosphere.

Bacteria attack the organic matter to gain energy for their normal life sources and new cell production. The first step taken by the bacteria is to remove hydrogen atoms from the organic matter and thus gain energy (dehydrogenation). An inorganic or organic substance then becomes a hydrogen acceptor, causing the following reactions to occur in the order shown:[9]

H Acceptor				Reduced Product	
O_2^+	+	$4H^+$	=	$2\ H_2O$	(1)
$2NO_3^-$	+	$12H^+$	=	$N_2 + 6H_2O$	(2)
SO_4	+	$10H^+$	=	$H_2S + 4H_2O$	(3)
Oxidized Organics	+	xH^+	=	Reduced organics	(4)
CO_2	+	$8H^+$	=	$CH_4 + 2H_2O$	(5)

The sources of odors are created in equations 3 and 4. The odor coming from equation 4 depends on what product produced it (organic acids, aldehydes, ketones, amines, sulfides, mercaptans, indoles, skatoles, etc.). Also, there are many different types of bacteria involved in each reaction. Equation 1 is the only aerobic reaction. The result of the reactions can be caused by either anaerobic or facultative organisms.

Hydrogen sulfide gas, the most prevalent odor associated with domestic sewage, has been thoroughly studied in the last 30 years. Besides odor production, crown corrosion of concrete sewers is another effect. Sulfact ions normally exist in domestic wastewater in concentrations ranging from 30 to 60 mg/l. Higher quantities may exist, depending on how many times the water has been reused.

The anaerobic process of creating H_2S starts out with the sulfur, contained in proteins, being broken down to form SO_4 ions. The reaction continues as shown in equations 6 and 7 and Figure 9-1.

$$SO_4^{-2} + \text{organic matter} \xrightarrow[\text{bacteria*}]{\text{anaerobic}} S^{-2} + H_2O + CO_2 \qquad (6)$$

$$S^{-2} + 2H^+ \rightleftharpoons H_2S \qquad (7)$$

As expected in such anaerobic reactions, the amount of bacteria, the pH, and the temperature of the wastes are crucial factors. Heukelekian has found that the number of sulfate-reducing bacteria ranges from 60 to 600 per milliliter (ml) of sewage and up to 25,000 per ml for sludges.[16] The effects of pH should be apparent from observation of equation 7. Since the gaseous state causes the odor, an increase in H^+ ions will cause more H_2S production. This effect can easily be summarized in Figure 9-2, where the optimum pH range is 7.5 to 8.0. Studies by Baumgartner[3] have shown that the optimum temperature for H_2S generation is around 30°C.

*Typical sulfate-reducing bacteria are *Vibrio desulfuricans*, Spirillum desulphurican, and *Microspira aestauri.*

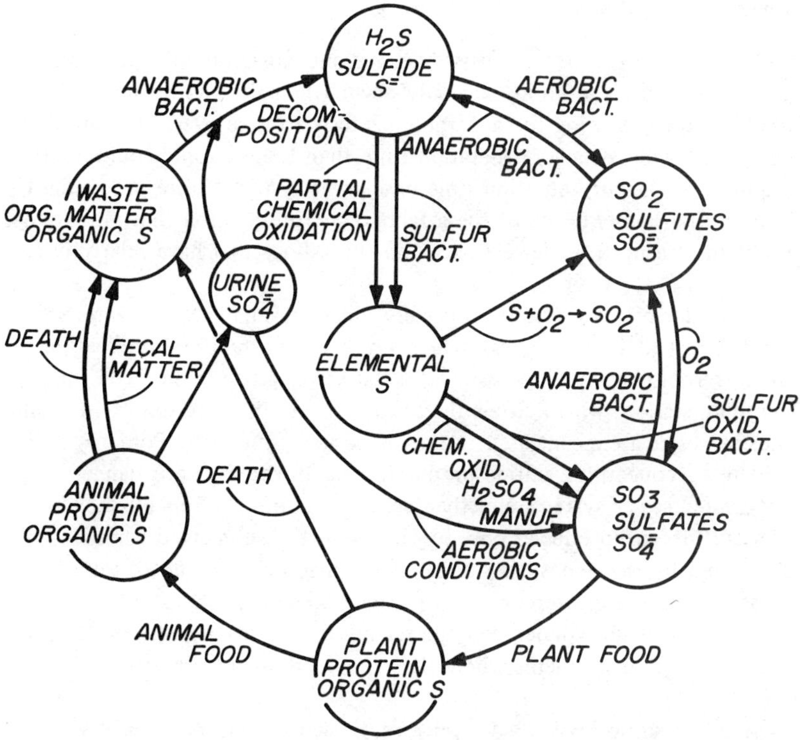

Figure 9-1. The sulfur cycle.[32]

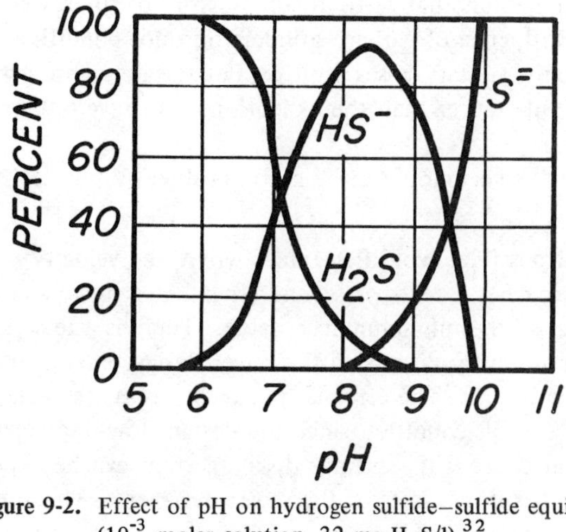

Figure 9-2. Effect of pH on hydrogen sulfide–sulfide equilibrium (10^{-3} molar solution, 32 mg H_2S/l).[32]

Sewer System Odors and Their Control

Raw sewage can become very odorous once anaerobic conditions occur. These anaerobic conditions can only occur when the DO of the sewage is used up, thus allowing the anaerobic bacteria to take over. The anaerobic bacteria, having a longer generation time than their aerobic counterparts, require a sufficient detention time in a sewer system before producing H_2S. The natural consequence of these facts is the observation that most odor problems arising from sewers occur on those lines that have relatively flat grades.

The odor effect in sewers is usually most prevalent during warm weather conditions (20°C and higher), which favor a more rapid multiplication of the anaerobic bacteria. A result of a flat sewer grade is the need for a pumping station, which stores the sewage for a longer period of time and thus serves as a concentrated source of odors. Studies by Pomeroy and Bowlus,[27] concerning sulfide production, reinforce the odor danger of flat-graded sewer systems and their resulting pump stations by showing that sulfides are produced only by slimes on the submerged surface of the sewer and by the deposited sludge. Since a sewer constructed on a flat slope will have a greater flow depth versus a steeper-sloped sewer of the same size, it is no wonder that the technical journals are filled with articles concerning odor problems where the phrase "flat-graded sewer" always appears.[7,8,15,19,30]

A pump station will then receive septic sewage and provide it with additional surface area in the wet well for further decomposition of the slimes. Not to be forgotten is the increased popularity of garbage grinders, which provide even more solids that settle to the bottom of the sewer. A prime example of the influence of garbage grinders on odor conditions can be found on the many military bases built on flat terrain where garbage grinders are the rule rather than the exception, with severe odor problems being the result.

The control of sewer odors can be accomplished by physical and/or chemicals means.

1. If at all possible, avoid flat grades where the velocity of a sewer flowing full approaches 2 feet per second (fps). Most sewers are designed with this number as the minimum acceptable. There is one serious error that occurs by constantly discussing the sewer design parameters at full or even half-full conditions. The error is that most sewers, especially laterals, never approach the full condition since the design flow-rate capacity of the sewer at 2 fps far exceeds the average discharge that can be expected.

Therefore it is important for the sewer designer to check the velocity of the sewage at the average flow rate (usually two to three times less than the design flow rate) through the use of a partial flow diagram that is available in most references dealing with sewer design. Even the average flow rate may be meaningless on a time basis for many sewers. It is quite possible that some approximation can be made concerning the velocity at the minimum flow rate.

On the other side of odor control through good sewer design should be the avoidance of excessive turbulent conditions such as deep manholes. At these locations the previously tranquil sewer flow is energized by a sudden change in elevation, thus allowing the sewage to release any stored-up gases. Vortex manholes will avoid this condition by allowing for a nonturbulent change of elevation.

2. Sewer cleaning programs may need to be increased to reduce odor problems by removing the sludge and slime layers where the H_2S and other gases originate. Pump stations especially need to clean the slime that builds up on wall surfaces. Small, packaged pump stations are rarely cleaned, but they should be monitored for odors more frequently as they receive a low sewage flow rate.

3. Trapping of gases has been successful in Dallas, where "U" sections are provided for house connections and turned-up bends for mains and laterals.

4. Venting of sewage by natural and forced draft is practical. The only problem is the need to obtain a disposal point for the ejected gases that will not itself become a point source of odor.

5. The addition of air to sewage force mains will prevent large columns of odor-laden gas at the free discharge end. This problem can be especially acute if the force main is long, followed by a discharge into a flat-graded gravity sewer. Air injection at the wet well of a pumping station will tend to keep the slime formation from producing gases.

Chemical control of odors can be accomplished by the following methods. Economic considerations and individual problems are the controlling elements of choice.

1. Chlorination. As in most sewage treatment applications, chlorine has been found to be the most popular chemical for inhibiting or disinfecting the anaerobic bacteria causing H_2S generation in sewers. Since chlorine is a strong oxidizing agent, there are many possible reactions. The main ones occur with H_2S, NH_3, and C_6H_5OH (phenol).

$$H_2S + 4Cl_2 + 4H_2O \longrightarrow H_2SO_4 + 8HCl \quad (8)$$

$$NH_3 + Cl_2 \longrightarrow NH_2Cl + HCl \quad (9)$$

$$HN_2Cl + Cl_2 \longrightarrow HNCl_2 + HCl \qquad (10)$$

$$HNCl_2 + Cl_2 \longrightarrow NCl_3 + HCl \qquad (11)$$

Equations 9, 10, and 11 are the typical textbook equations associated with chlorination in the presence of free ammonia. Ammonia is an odor-causing agent, but it must be noted that the reaction should not be allowed to proceed to the trichloramine form (NCl_3), which is malodorous.

Also, it is quite common for chlorine to react with other organic compounds contributed from industrial sources, such as phenol, and create an intolerable product such as chlorophenol, with its very low threshold odor number.

The effectiveness of chlorine and its economic desirability lies in the fact that inhibition of the odor-causing bacteria is all that is needed, and not complete disinfection. The usual range of chlorine dosage for inhibition purposes is 10 to 50 mg/l, with 8.87 mg of chlorine required to completely oxidize one mg of sulfide as 9.22. In many cases a much higher dose of chlorine is needed to effectively treat denser slimes and sludges on the sewer invert. Mechanical sewer cleaning, combined with chlorine or one of the other oxidizing agents discussed below, is the usual method of attack. Care must be taken to decide where on the sewer system is the best starting point. By applying the chlorine at points farthest away from the treatment plant, the most effective use of the chemical is achieved.

2. One of the more widely used commercial products is a chlorinated hydrocarbon (orthodichlorobenzene), known as Cloroben.* Cloroben contains an emulsifying ingredient that allows the insoluble orthodichlorobenzene to make intimate contact with the odor-producing agents and thus inhibit their growth.

Data concerning particular odor problems and the use of Cloroben to solve them can be found in the many articles published concerning Cloroben.[7,8,15,19,30] The most general method of a thorough chemical cleaning using Cloroben is the Cloroben Up-Sewer Chemical Rotation Method. With this method, accumulated grease deposits and slime formations are dissolved; high sulfide production is eliminated.

The Cloroben Up-Sewer Rotation Method consists of two major steps. The first is a shock treatment of isolated local areas where excessive accumulation of grease and slime may occur. Such areas are cleaned one by one, using shock-dose Cloroben applications. In most cases the amount of Cloroben used in these local areas during such initial cleaning is equal to, or even stronger than, that normally applied to the entire flow. This strong Cloroben dilution is kept in contact with sewer walls, wet wells, and force

*Cloroben Chemical Corp., South Kearny, N.J.

mains for one to two hours, then released—either all at once, if final treatment is primary, or gradually if the disposal method consists of both primary and secondary treatment.

When such shock treatment of local up-sewer trouble spots takes place in connection with activated sludge or trickling filter plants, care should be taken to ensure that the amount of Cloroben released over any extended time does not exceed 20 ppm of plant capacity during that time.

After the first step (gradual shock dosing and cleaning of trouble spots) has been accomplished, the rotation method changes to a maintenance program using the principle of periodic shock dosing of local up-sewer lines; *i.e.*, the entire amount of Cloroben intended for stabilizing the flow and relieving final treatment is applied from one or two points up-sewer. The Cloroben passing through those local lines will be highly concentrated, but by dilution of down-sewer flows it may only amount to 1 or 2 ppm when reaching the plant.

By shifting and rotating the up-sewer dosing points, laterals, branch sewers, lift stations, wet wells, and force mains will periodically receive an application sufficiently strong to prevent any further excessive formation of grease and slime deposits (Figure 9-3).

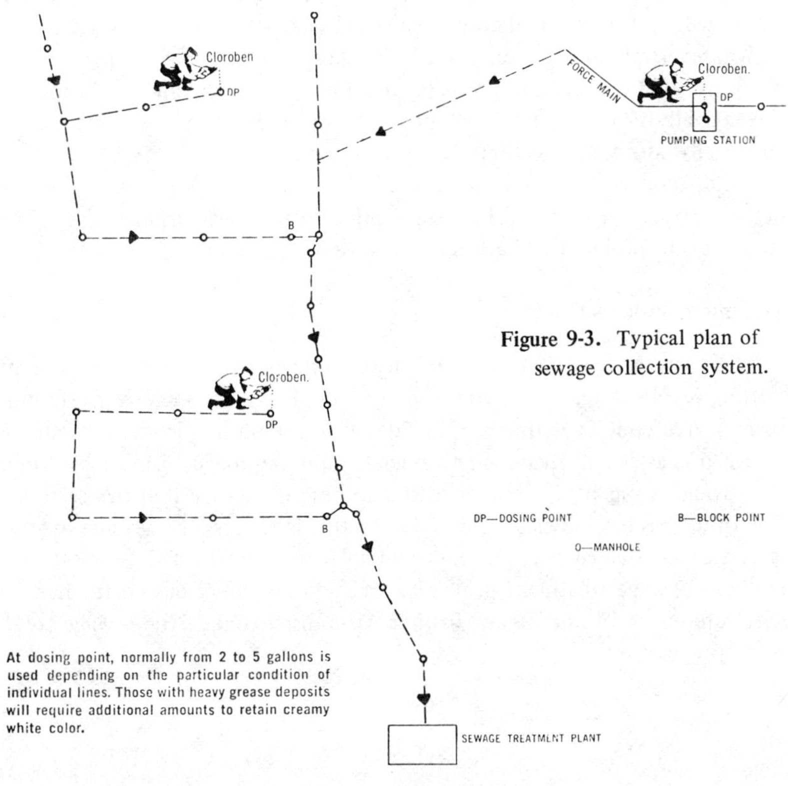

Figure 9-3. Typical plan of sewage collection system.

3. Sodium hydroxide can be applied to control H_2S odor by causing a shift in the equilibrium of equation 3. This method is suggested for use at a temporary storage area such as a pump station wet well. A lime slurry can also be just as effective when applied to sewer lines. This method has been used in Los Angeles at a rate of 1,000 to 4,000 gph of lime slurry for a wastewater flow of 3,500 gpm.[31]

4. A commercial product called Oxinite, containing a proprietary mixture of air, oxides of nitrogen, and other gases, has also been applied to pump stations and mildly sloped sewers. The mixture of gases is produced in a self-contained unit by passing air through a series of baffles in which an electrical current is maintained to produce the Oxinite. The manufacturer claims that the BOD of the raw sewage can be reduced substantially, with one application showing a reduction from 120 to 32 mg/l.

CONTROL OF ODORS AT SEWAGE TREATMENT PLANTS

Anyone familiar with treatment plant operations realizes that there are many sources within the plant where foul odors can be generated. Many times the odor problem can be solved by simple cleaning operations. Other plants will require more elaborate physical and/or chemical methods.

Whatever methods are employed, the design and operating engineer must be aware of the increasing problem of odors. The problem should increase substantially in the future as many plants are hydraulically overloaded, leading to odorous conditions.

Other modern treatment devices, such as ammonia stripping, will be an obvious source of odors. The design engineer must be careful to avoid an air pollution problem while treating a water-related one.

Treatment Process Odors

Basic sewage treatment mechanisms beyond primary treatment or simple settling revolve around allowing a volume of sewage to achieve maximum surface area contact with air. This idea is accomplished in the activated sludge process by diffusing air through a bubbling mechanism or by forcing air through a mechanical mixer into a volume of sewage that has settled. The other basic, biological alternative to this is to pass the sewage over a large surface area such as gravel (trickling filter). Both of these methods are basic sewage treatment principles and will not be discussed further. (See references 14 and 39 for further basic discussion of the sewage treatment process.)

By forcing air into the sewage, small bubbles in the order of 1,000 to 1,500 microns (one micron = 10^{-4} centimeters) are produced (see Figures 9-4 and 9-5). These bubbles burst and dispense small droplets into the air. Most of these droplets will have their water part evaporated, leaving a small (10 to 15 micron) solid particle suspended in air.[18] Depending on weather conditions, these solid particles can be spread over the surrounding area, causing respiratory irritation and possibly infection.

Gases and vapors are the odor-causing agents in an STP. For many years, design engineers never considered the sewage treatment process as giving off odorous gases and vapors. This may be a theory-and-practice problem, but is nevertheless a reality. A survey of gases and vapors, described in reference 29, shows the main odor-causing agents in order of significance to be methyl mercaptans, methyl sulfides, amines, indoles, skatoles, and H_2S. Certain hydrocarbons, such as the methane produced by sludge digestion, can be a large point source of air pollution if they are not collected and disposed of properly. Any addition of industrial waste will, of course, increase the production of gases. Figures 9-6 and 9-7 show the sources of odors in typical treatment plants.

Bacterial Emissions from an STP

A complementary subject to odors emanating from an STP is bacterial emissions. It is common knowledge that human fecal waste contains etiologic agents that cause intestinal disease. What happens to these bacteria when they are dispersed in the air has not received much attention. It must be remembered that a milliliter of sewage can contain from 10^6 to 10^7 organisms which can become airborne, as described in the previous section. Ledbetter conducted tests for airborne bacteria emissions at an activated sludge plant.[18] His results, exhibited as Figures 9-8 and 9-9, show the effects of downwind distance and wind velocity. He also found that half the bacteria collected were enteric, with some being a genera that is considered pathogenic.

Trickling filters were tested for coliform counts in Utah by Adams and Spendlove.[1] *E. coli* were used as an indicating organism, in much the same way they are used to check the contamination of water and sewage. Two municipal plants of 6 and 25 mgd capacity were studied using special samplers. Samples were taken during daytime and nighttime hours in periods ranging from five minutes to one hour at points near the trickling filters to a distance of 0.8 miles downwind. The results showed a large number of coliforms originating from the trickling filters, with the larger plant naturally giving off more coliforms. It was shown that, in general, high wind velocities, high relative humidity, reduced sunlight and darkness,

Figure 9-4. Mechanical aeration unit.

Figure 9-5. Diffused aeration tank.

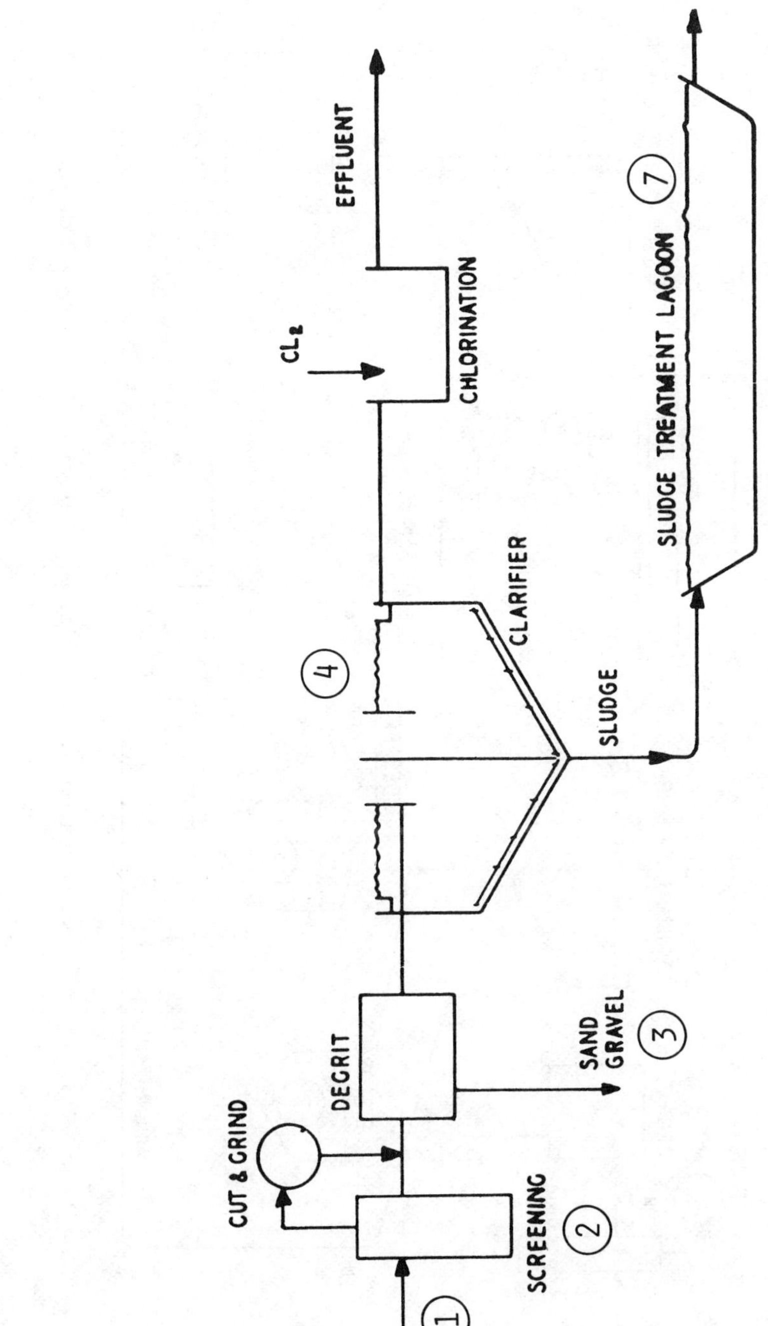

Figure 9-6. Simple wastewater treatment plant.[38]

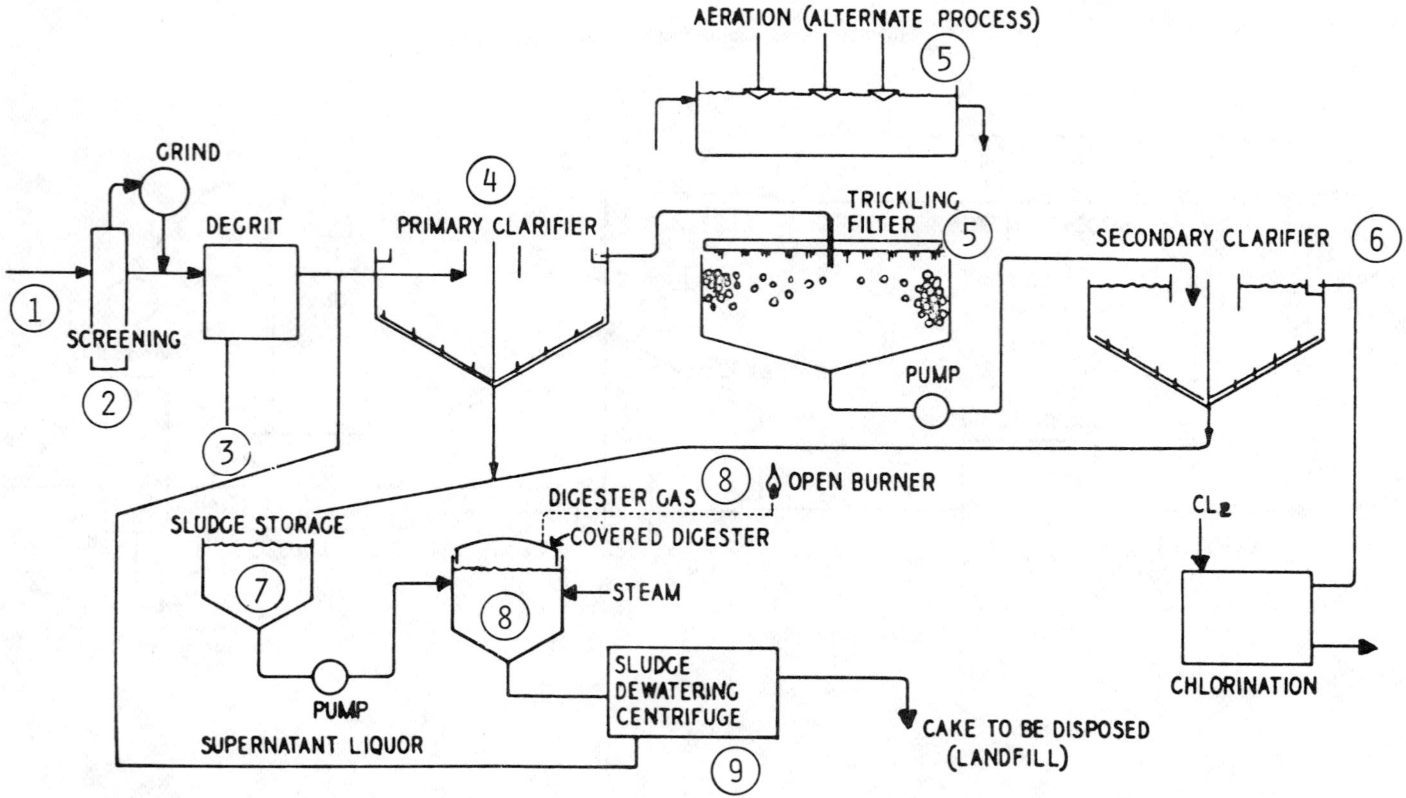

Figure 9-7. Primary and secondary treatment with sludge digestion.[38] (Diagram courtesy of Envirotech Corp.)

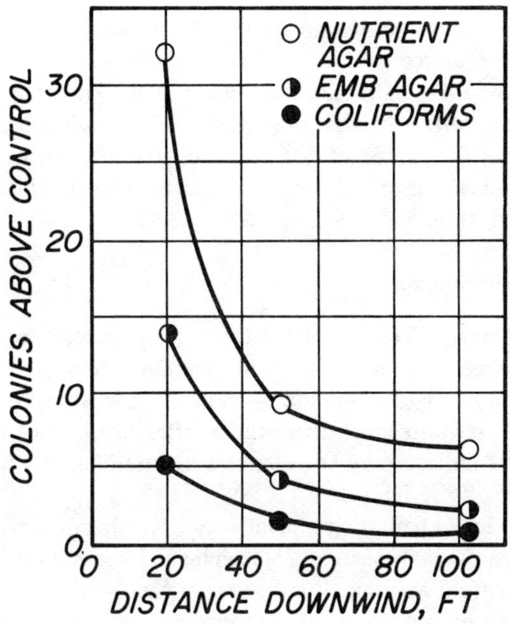

Figure 9-8. Effect of distance on airborne bacteria emissions.[18]

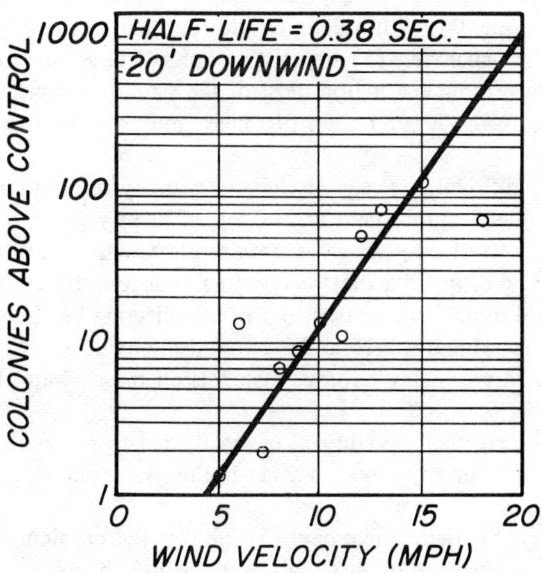

Figure 9-9. Effect of wind velocity on airborne bacteria emissions.[18]

and low temperatures gave the greatest coliform recoveries at all points downwind of the source.

These studies, along with Albrecht's work at the University of Florida,[2] along with a few earlier articles, show the lack of consideration allotted to the possibility of the spread of disease by an STP. While the above reference articles are by no means conclusive, it would be wise for this subject to be given greater scrutiny.

Odor Generation Sources

The odors from an STP arise from three main sources: odorous raw sewage, overloaded secondary treatment facilities, and sludge treatment devices. Referring to Figures 9-6 and 9-7 and the discussion below, a drop of sewage may be traced through these three different types of plants, with sources of odors indicated by the numbers in parentheses, which correspond to the numbers on the figures.

The raw sewage (1) entering a plant may have already developed odors, as discussed in a previous section. Screening and comminuting devices (2) can create odor if an accumulation of grease is allowed to be built and if the objects removed by screens are not disposed of quickly. Smaller plants that use bar screens may not have a self-cleaning attachment. The accumulated debris may be coated with or be made of organic material that may quickly give off an odor. Usually these objects are land-filled once a day or even less frequently, with a definite odor problem having a good chance to develop. Grit chambers (3) are another immediate point source of odors because of the organic coating on most of the sand, metal pieces, etc., that are collected. Again, it is a matter of timely housekeeping to avoid an odor problem.

When the bulk of the sewage reaches the primary settling tank (4), there are usually influent and effluent weirs. The effluent weir especially can cause the emission of odorous gases from septic sewage. The weir allows the sewage to be dropped a foot or two to a collecting trough, and then possibly an additional foot or two to a main collecting bin between two parallel tanks (see Figures 9-10 and 9-11). Also, any scum collection device will concentrate highly odorous materials on the air-water face of the tank.

The sewage may then be pumped or gravity fed to the next unit process with little chance for any more odor generation, since the sewage is usually confined in an enclosed vessel.

If secondary, biological treatment (5) follows, the problem of odor-causing aerosols has been discussed already. It must be added, however, that in the activated process, water spraying of dry foam formations is a

Figure 9-10. Settling tank overflow.

Figure 9-11. Settling tank collection trough.

common practice. By quickly bursting the bubbles that form, a more rapid formation of gases is achieved and the overall quantity of odorous gases may be increased.

Secondary settling tanks (6) may cause odor problems if the sludge is not removed on schedule. There should be little odor production at this point in the process, since aerobic treatment preceded the settling process.

The sludge treatment process that can include thickening, digestion, and sludge dewatering (7) is a viable and many times vigorous source of odors. The sludge is constantly in an anaerobic state, except in certain sludge dewatering processes such as vacuum filtration. Sludge thickeners usually receive both septic primary and secondary sludges.

The digestion process gives off gases that contain methane, volatile organics, and other odorous gases such as H_2S. In most plants these gases are confined and flared to destruction (8). Since the sludge must be heated to about 90 to 95°F for efficient anaerobic digestion, a transfer of the odor-laden sludge by plunger-type pumps gives additional opportunities for point sources of odorous compounds.

Digested sludge that is either air dried, mechanically dewatered, or burned (9) contains ammonia as its predominant odor, with small amounts of volatile organic acids adding to the problem. Odor problems increase in sludges if they are improperly treated and then exposed to the atmosphere. Lime treatment of raw sludge can push the equilibrium equation for ammonia toward the gaseous state, with an odorous condition resulting.

Sludge processing can also include incineration, with multiple hearth furnaces being most popular. This operation consists of first having the moisture removed in the upper furnace and the next several hearths operating at higher temperatures to oxidize the remaining organic materials. If the combustion furnace is operated at temperatures ranging from 1,200 to 1,400°F, no odors will be given off. Any particulate matter is normally removed by a wet scrubber.

Oxidation ponds, serving as independent and complete treatment units or as tertiary treatment devices, can be a source of a multitude of odors. The organisms employed–algae, bacteria, and actinomycetes–oxidize the organic matter discharged into the pond for food. The aerobic and anaerobic processes going on in the pond can transfer to the atmosphere algae, bacteria, oxides of sulfur and nitrogen, ammonia, methane, H_2S, and hydrocarbons.[23] Hydrocarbons can reach the atmosphere by simple evaporation before the biological breakdown process takes place. Methane can be prevalent, and may be produced at a rate of 10 cu ft/lb of BOD destroyed.[23]

Odors from nidole, sulfides, skatole, mercaptans, cadavarine, amines, metabolic products of algae, and actinomycetes have been associated with oxidation ponds by other authors.[6,12,29] A search of the literature concerning oxidation ponds shows that algae are probably the main problem organisms with regard to odor production. The algae involved–diatons, blue-green algae, and pigmented flagellates–can in themselves cause the odor problem or provide the food necessary for the odor-causing actinomycetes. The actinomycetes produce odor-causing by-products such as aromatic amines, aldehydes, saturated fatty acids, ketones, and unsaturated aromatics.

Methods of Odor Prevention or Removal

General

Treatment plant prevention or removal of odors can involve a multitude of methods. There are many proprietary methods described in the literature, most of which will be discussed below. In most cases it is very difficult to identify the cause and nature of an odor, and trial-and-error methods have to be employed. The many factors in selecting the best method involve consideration of local weather conditions, treatment method, previous experience with odors, ability of the plant operators, duration and frequency of each type of odor, and the available funds for correcting the problem.

As suggested by the U.S. EPA, there are three general areas of odor control in wastewater treatment plants:[45]

1. Changes in present plant operation, with consideration given to new treatment techniques
2. Chemical treatment
3. Collection and treatment of odorous gases.

Changes in Present Plant Operation

Plant operation is probably the cheapest and, in many cases, the most effective means of odor control. Timely maintenance, that prevents offensive odors from reaching the surrounding areas, may be all that is needed for most small plants. The following items should be part of any normal operation considerations.

Grit Chamber. Most of the odors in this unit process are from H_2S. When odors become frequent, the operator should check flow-through velocities in the system, determine the total volatile solids of the grit, analyze the raw sewage for total and dissolved sulfides, hand-lead acetate /H_2S/ tiles in the chamber, and check for debris accumulation, including grease. Corrective methods can include a daily washing of the grit chamber to remove any overabundance of organic solids, occasional dosing of the chamber with hyperchloride or other chemical preparation, and sealing the chamber so the odors can be removed by a gas scrubbing apparatus.

Primary Settling Tanks. The presence of H_2S gas will require tests for sulfides, pH, and DO. The most important factor will be the amount of total solids in the tank. Overloaded primary tanks are common, as they often receive solids from other plant processes.

Prechlorination and pH control are suggested for minor odor problems. More intense odors are probably due to excess solids. These odors can be controlled by making sure that excess total solids from other plant processes do not exceed 5% of the daily inflow.[44] Regular removal of sludge and scum layers may be all that is needed for most settling tanks for odor control.

Aeration Tanks. Any odor problems in this process would be due to excessive turbulence in the bulk of the water. Under windy conditions, an odorous spray can be spread throughout the area. Also, water spraying of foam buildup will make more gas pockets available, but any odor problem should be in the immediate vicinity of the tank.

Trickling Filters. Odor problems caused by H_2S can be significant because as the slime layer builds up around the filter rocks, it is natural that the inner layer of organic material becomes anaerobic. This anaerobic condition eventually causes a sloughing off of the built-up layer with a release of gases. By increasing the recirculation rate in a high-rate filter, the DO will be increased and quicker sloughing will occur. The vents and underdrain system should be checked to avoid concentrated odors after one of these items becomes clogged. More expensive measures include covering the entire bed, with the odors fed to a wet scrubber.

Sludge. There are very few changes that can be performed on most modern anaerobic sludge digestors and associated unit processes. Most well-designed digestors destroy the odors by burning off the noxious gases. Most odors come from sludge thickening tanks and from discharging incompletely digested sludge to sand drying beds or dewatering units. With incineration of raw sludge, odors from sludge handling will be increased.

Oxidation Ponds. Odors from oxidation ponds are caused by H_2S from an accumulation of solids on the bottom of the pond. If the pond is used as a tertiary unit, odors should be negligible, unless there is a substantial accumulation of blue-green algae. Ponds used as secondary treatment devices will contain substantially more solids and can give off odors, especially during warm weather conditions. In either case, when odor problems arise the pond should be checked for scum accumulation, pH of influent and effluent, DO of the pond at several points and depths, and total and dissolved sulfides in the pond. Control measures can involve the use of mechanical aerators, prechlorination, addition of sodium nitrate, and other odor-masking chemicals.

Chemical Treatment

Chemical treatment can provide a convenient method of odor control in an STP, especially when the odors originate from organic sources. This odor removal mechanism usually involves oxidation coupled with a pH adjustment. The chemicals used for odor control are chlorine and its related compounds, potassium permanganate, nitrates, metallic ions, spray counteractants, and proprietary chemicals under various trade names. The adjustment of pH to prevent H_2S formation is usually done by lime, and in some cases, by caustics.

Chlorination is the most popular chemical odor treatment method because it is already available at most plants for treating sewage. As in sewers, the chlorine retards biological activity and oxidizes odorous sulfur compounds to the free colloidal sulfur. A dose of 10 mg/l at the maximum flow is recommended for prechlorination.[45] Table 9-3 gives typical values for the chlorine dose to be used in raw wastewater.

Table 9-3. Chlorine Odor Reduction[45]

Chlorine Dosage	Detention Time in Minutes	
	1	120
mg/l	Sulfide, mg/l	0 Sulfide, mg/l
0	1.7	1.7
5	0.5	0.7
10	0.2	0.2
25	0.2	0.2
50	0.2	0.2

Chlorine dioxide has been used on an emergency basis at various plants when the aeration tank did not supply sufficient oxygen due to overloading or maintenance.

Nitrates have been applied to tidal basins and lagoons for suppression of odors. The mechanism of odor control chemically combined oxygen which prevents the reduction of sulfates. As long as some oxygen and nitrates are present, other more odorous decomposition products are not allowed to form. The cost effectiveness of nitrate addition is low, with a plant at Houston reporting a cost of \$40,000/year for the addition of 10 lb of sodium nitrate per pound of sulfide present.[5]

Metallic ions such as copper, mercury, zinc, and iron will combine with sulfide to form insoluble precipitates. Use of these metallic ions has been researched since the 1930s. Pomeroy and Bowlus[27] found that ferric iron (Fe^{+3}) is more effective for sulfide precipitation than ferrous (Fe^{+2}). When some oxygen was present in the wastewater, a combination of Fe^{+3} and Fe^{+2} was most effective. It was also found that Fe^{+2} works best when the water is devoid of oxygen. However, unless there is a cheap source of iron, this method is not economical.

Counteraction of odors involves overwhelming a foul odor with one that is supposedly more pleasant. Complete counteraction of an odor is unlikely because of the many variables involved. Attempts at counteraction often lead to masking, which may present a new problem, depending on the chemical used. Odor counteractants are vaporized and usually introduced at ground level, on the roof of the plant structure, and in the vicinity of the odor source to provide complete vertical coverage of the odor. The counteractant is vaporized by drawing the chemical from a drum and spraying it through a compressor. In one installation at a 12-mgd activated sludge plant, two nozzles were mounted in the area of the aeration tank, along with another nozzle at the grit chamber.[28]

At Chatham, New Jersey, spray devices were placed along the fence line at a 3-mgd trickling filter plant. The nearest residential area was approximately 700 ft away. Odor problems were noticeable only during the summer months. The spraying masked the odor with a sweet, perfume-like fragrance. Unfortunately, the masking agent soon became as offensive as the pungent odors, so the spraying was discontinued. Effective plant operation has taken its place.

Proprietary chemicals are added to the sewage flow. Chemicals are usually dripped unto the flow at the head of the plant, as shown in Figure 9-12. These chemicals can also be sprayed on any odor-causing source with effectiveness. They have, for example, been applied to digestion tanks to prevent foul odors on sludge drying beds and to break up scum accumulation in the tanks.

Collection and Treatment of Odorous Gases

Confinement of odors through the use of domes, enclosures, etc., has become a more popular practice for odor control at large treatment plants. Physical confinement involves covering most unit processes, since odor creation will vary with location and time.

The interior atmospheric pressure must be kept slightly below that of the exterior to ensure positive collection of the odors. Ventilation ducts,

Figure 9-12. Deodorizing chemicals being added to the grit chamber.

needed in areas where workmen perform constant maintenance, have to be connected to the central collection system. Examples of physical confinement of odors can be found at the North San Mateo County Sanitation District water pollution control plant at Hayward, California, at Los Angeles, San Diego, New York City, and many other locations. The major cost of this type of odor treatment is the confinement structure. Recently, formed-in-place, low-cost styrofoam domes have been constructed at Cedar Rapids, Iowa, and Elmira, New York.[10]

Once the gases are confined, there are three general methods of treatment: (1) simple or catalytic combustion, (2) ozonation, and (3) chemical oxidation or absorption.

Combustion is one of the more desirable techniques of odor control. Complete combustion produces carbon dioxide, oxides of nitrogen, water, and sulfur dioxide. If complete combustion is not achieved, the end products can be more odorous than the original gas. Simple combustion usually involves a separate system when large volumes of gas need to be destroyed.

The key to complete combustion is maintaining the proper temperature and contact time in the combustion chamber. Temperatures between 1,350 and 1,500°F have been used at many plants, with contact times ranging from 0.3 to 3 seconds. At San Diego, a fume incinerator is supplied with odors taken from sedimentation basins. Temperatures are maintained between 1,350 and 1,500°F in a furnace that uses two-thirds fuel and one-third digester gas.[38] If a multiple hearth furnace is employed at the plant for sludge incineration, the odorous gases can be used as combustion air for the furnaces.

Catalytic combustion allows the use of a substantially lower incineration temperature (600°F) through the use of a platinum-metalic screen which accelerates the combustion process. Operating costs can be high since natural gas may be required for complete combustion if there is not sufficient digestor gas available. Austin, Texas, uses this process on a unit mounted over a 3-million-gallon sewage holding tank (Figures 9-13 and 9-14). The catalyst at this installation will last for 5,000 hr of operation before the inorganic dust has to be removed.[41]

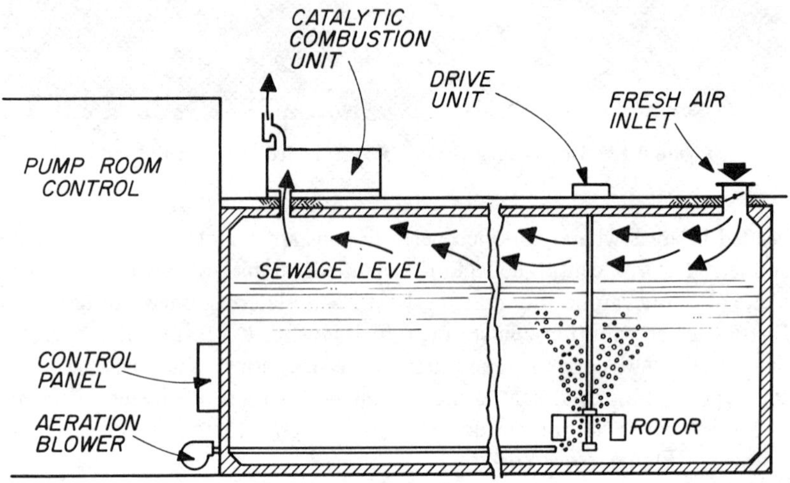

Figure 9-13. Elevation view of the installation at Austin showing the location of the various elements.[40]

Ozone has been used with success in domestic wastewater. Due to its high generation costs and its high reactivity, the addition of ozone to a large flow is not practical. However, ozone can be applied economically to odors that are collected by covered facilities. Basically, ozone will oxidize sulfides and amines to nonodorous sulfoxides and amino oxides. A dose of 1 mg/l of ozone by volume is needed to handle up to 10 mg/l of sulfide.[42] Detention time in the blower assembly should be at least 15 sec, with longer times preferable.

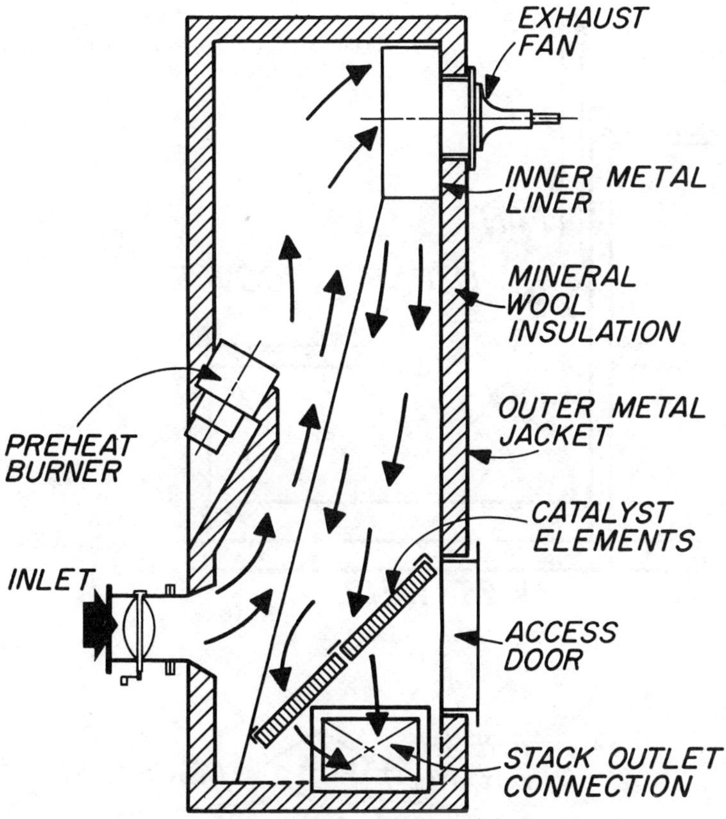

Figure 9-14. Plan view of the Austin oxidation unit.[40]

Ozone has been applied to grit and screening chambers, aeration tanks, vacuum filters, and sludge storage tanks. Figure 9-15 shows a typical installation over a wet well, while Figure 9-16 shows the use of ozone over a sludge storage tank. Odor problems in these unit processes were controlled by the use of 1 mg/l of ozone by volume.[42] Any efficient control of odors by ozone usually demands that the plant be under one roof with exhausts going through a single stack.

Single units are easily sized by knowing the frequency of air change and the dose of ozone needed. Units are available capable of producing 1,000 lb/day. It is difficult in designing an ozonation system to determine the conditions that have led to the maximum odor intensity. Excess exhausting of ozone, which has a pungent odor, must be avoided. Installation of the

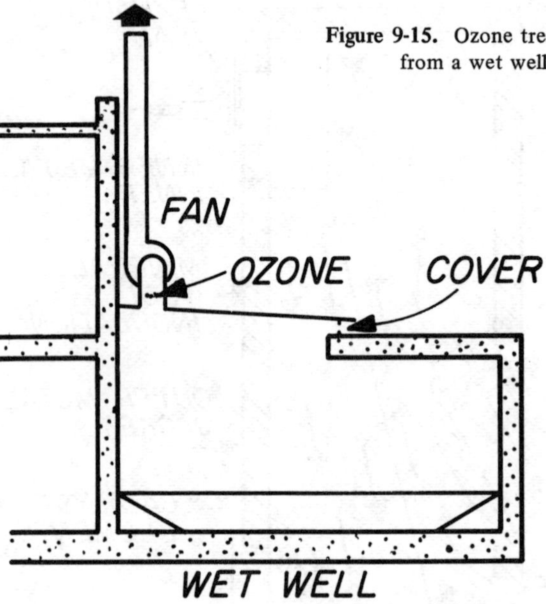

Figure 9-15. Ozone treatment of exhaust gases from a wet well.[41]

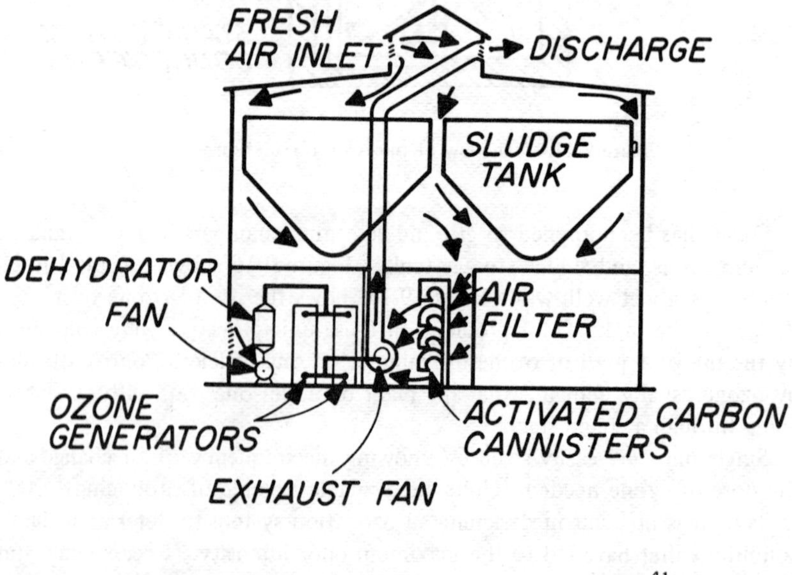

Figure 9-16. Ozone treatment of odors from sludge tank.[41]

unit involves a source of cooling water, electricity, a drain for cooling water, and piping to distribute the ozone. Electricity requirements amount to approximately 10 kwh/lb of ozone produced.[42] The Owl's Head plant in Brooklyn has been adding ozone to all of the gases taken from the top of the settling tank, along with the gases from all other covered units.

Absorption of odorous gases can be applied to various covered unit processes. Wet scrubbing of gases with spray devices or scrubbing equipment, packed columns, adding ozone to the gas, or passing the gas through activated carbon have all been used. The liquid used in the wet scrubbing process will depend upon the gas involved, with alkaline solutions preferred for H_2S removal. Hypochlorite, lime, permanganate, and chelated iron have been used in scrubbing, but their use tends to raise costs considerably.

ODORS FROM REFUSE DISPOSAL

Refuse disposal in the United States consists almost entirely of dumping on land. Recycling and incineration are used, but at higher costs. The only feasible method of odor control for an open dump or sanitary landfill is chemical spraying.

Sanitary Landfill Odor Control

Most metropolitan refuse consists of cardboard and paper (80%). The remaining refuse consists of tin cans (10-15%) and food and vegetable wastes (5-10%). After three or four days this buried material produces temperatures of 130 to 150°F, due to normal bacterial action. These temperatures destroy all the insects and larvae which infest the covered refuse. Since the protein content is so low, it is not considered necessary to add deodorants to the refuse before covering in the northern communities.

Although the sanitary landfill method can be an effective process, there is still a need for a chemical which will control all odor and pest nuisances. By proper application of deodorizing chemicals such as Ozene,* proper odor control can be achieved. During the summer, daily applications of a 1% dilution at the rate of 300 to 1,000 gallons per acre will control odors and any insects breeding in the cover. Only occasional spraying of the covered fill is needed during the cooler months. In southern communities, where higher temperatures promote rapid decomposition and the refuse may contain larger amounts of food and vegetable wastes, control of odors and pests may require modified methods, such as spraying, before covering the refuse.

---

*A product of Allied Chemical Co.

When choosing a landfill site, the community should consider various factors. Soil that consists of a sandy loam base is considered excellent for landfill purposes. Proximity to residential areas is a prime consideration. Also, haul distance, drainage, prevailing winds, roads, and the underground water table must be considered.

REFERENCES

1. Adams, A. P. and J. C. Spendlove. "Coliform Aerosols Emitted by Sewage Treatment Plants," *Science,* 1218-1220 (September 18, 1970).
2. Albrecht, C. R. "Bacterial Air Pollution Associated with the Sewage Treatment Process," University of Florida (1958).
3. Baumgartner, W. H. "Effect of Temperature and Seeding on Hydrogen Sulfide Formation in Sewage," *Sewage Works Journal,* **6**, 399 (1934).
4. Beardsley, C. W. "Suppression of Sewer Slimes," *Sewage Works Journal,* **21**, 1 (1949).
5. Bryan, A. C. "Experiences with Odor Control at Houstin, Texas," *Sewage and Industrial Wastes,* **28** (12), 1512 (1956).
6. Clare, H. C., J. K. Neel, and C. A. Monday. "Studies of Raw Sewage Lagoons at Fayette, Missouri, 1958-59 with a Resume of 1957-58 Operations," *Proceedings* of Symposium on Waste Stabilization Lagoons, USPHS (1960).
7. Cloroben Chemical Corporation. *Landfill Manual.* (South Kearny, N.J.: Cloroben Chemical Corp., 1959).
8. Cloroben Chemical Corporation. *The Cloroben Up-Sewer Chemical Rotation Method.* (South Kearny, N.J.: Cloroben Chemical Corp., 1959).
9. Dague, Richard R. "Fundamentals of Odor Control," *J. Water Poll. Control Fed.* (April 1972).
10. *Dow Domes–Environmental Enclosures.* (Midland, Mich.: Dow Chemical Co., 1968).
11. Eliassen, Rolf. "The Effects of Chlorinated Hydrocarbons on Hydrogen Sulfide Production," *Sewage Works Journal* (May 1944).
12. Eliassen, Rolf and C. A. Vath. "Air Pollution Control in Sewage Treatment Plants," *J. Water Poll. Control Fed.,* **32** (4), 424 (1960).
13. "Environmental Wastes Control Manual," *Public Works* (1973).
14. Fair, G. M., J. C. Geyer, and D. A. Okun. *Water and Wastewater Engineering.* (New York: Wiley, 1968).
15. Gary, L. "Controlling Sewer Insects and Sewer Odors," *Public Works* (June 1950).
16. Heukelekian, H. "Some Bacteriological Aspects of Hydrogen Sulfide Production from Sewage," *Sewage Works Journal,* **20**, 490 (1948).
17. Kellum, J. "Sewer Odor Control," *Water and Waste Engineering* (July 1970).
18. Ledbetter, J. O. "Air Pollution from Wastewater Treatment," *Water and Sewage Works* (February 1966).
19. Lusk, G. "Chemical Control of Grease and Odors," *American City* (July 1958).

20. Maass, A. F. "Successful Odor Control at Michigan Waste Water Treatment Plant," *Water and Sewage Works* (September 1967).
21. McCord, C. P. and W. N. Whiteridge. *Odors, Physiology and Control.* (New York: McGraw-Hill, 1949).
22. Murphy, R. S. and I. W. Santry. "Activated Carbon Removal of Hydrogen Sulfide," *Journal Sanitary Engineering Division.* Proceedings American Society of Civil Engineers, **86** (SA6), 1 (1960).
23. Nelson, R. Y. and J. O. Ledbetter. "Atmospheric Emissions from Oxidation Ponds," *J. Air Poll. Control Assoc.,* 14 (February 1964).
24. "Odors and Their Control," *J. Air Poll. Control Assoc.,* **36**, 1424-1426 (1964).
25. Pateman, S. G. "Oxnard Sewage Plant Solves Odor Control Problem," *Water and Sewage Works* (September 1969).
26. Paino, B. and C. E. Olsen. "Odor and Insect Control at an Open Landfill Operation," *Public Works Magazine* (April 1959).
27. Pomeroy, R. and F. D. Bowlus. "Progress Report on Sulfide Control Research," *Sewage Works Journal,* **28**, 597 (1946).
28. Post, N. "Counteraction of Sewage Odors," *Sewage and Industrial Wastes,* **28** (2), 221 (1956).
29. Post, N. "The Odor Phase of Waste Disposal," *Water and Sewage Works,* R377-379 (1960).
30. Price, L. and H. Curran. "Experiments with Odor Control in Sewers," *Public Works Magazine* (1958).
31. Santry, I. W. "Hydrogen Sulfide Odor Control Measures," *J. Water Poll. Control Fed.* (March 1966).
32. Sawyer, C. N. *Chemistry for Sanitary Engineers.* (New York: McGraw-Hill, 1966).
33. *Sewage Treatment Plant Design.* (American Society of Civil Engineers, 1959).
34. "Stabilized Chlorine Dioxide for Odor Control," *Public Works,* 166 (April 1968).
35. *Standard Methods for the Examination of Water and Wastewater,* 13th ed. (APHA, AWWA, WPCF, 1971).
36. Stone, R. "Sewage Treatment System Odors and Air Pollutants," *J. Sanitary Engineering Division,* ASCE, SA4 (August 1970).
37. "Suggested Basic Operator Training Course Outline–Odors and Their Control," *J. Water Pollut. Fed.* (November 1964).
38. Sutton, G. P. "Odors and Air Pollution from the Treatment of Municipal Waste Water," (Salt Lake City, Utah: Envirotech Corp., 1971).
39. *The 1974 Environmental Wastes Control Manual.* (Ridgewood, N.J.: 1974).
40. Ullrich, A. H. and R. J. Ruff. "Oxidation of Sewage Odors," *Water and Sewage Works.* (September 1959).
41. Unangst, P. C. and C. A. Nebel. "Ozone Treatment of Sewage Plant Odors," *Water and Sewage Works,* **118**, R42 (1971).
42. U. S. Department of the Interior. *A Study of Sludge Handling and Disposal* (May 1968).
43. U. S. Environmental Protection Agency. *Odors Emitted from Raw and Digested Sewage Sludge.* (Washington, D.C.: December 1973).

44. U. S. Environmental Protection Agency. *Procedures for Evaluating Performance of Wastewater Treatment.* (Washington, D.C.)
45. U. S. Environmental Protection Agency. *Process Design Manual for Upgrading Wastewater Treatment Plants.* (Washington, D.C.: October 1971).
46. Voorhees, J. "Economical Waste-Water Disposal and Odor Control," *Food Processing* (December 1959).
47. *Wastewater Engineering.* (New York: McGraw-Hill, Metcalf and Eddy, 1972).
48. *Water and Wastes Engineering,* 28 (September 1969).
49. Wechter, W. H. "Sewage Odors at Military Posts Corrected by Chemical Treatment," *Wastes Engineering* (August 1958).

CHAPTER 10

INVESTIGATION OF ODOR CONTROL IN THE RENDERING INDUSTRY

Richard H. Snow

IIT Research Institute
Chicago, Ill. 60616

The objective of this program was to develop an effective and economic means of odor control. This project involved several phases, which are summarized in this chapter.

ODOR IDENTIFICATION

Odor samples were taken at several rendering plants and their strengths were measured by odor panel tests. The odorous components were measured by a combination of gas-liquid chromatography (GLC) and mass spectroscopy. This information made it possible to more scientifically investigate the effectiveness of various odor control means, since for the first time the chemical and physical properties of the odor were known. Previously, only subjective evaluations could be made of the effectiveness of various odor control methods, although the dilution method allowed performance to be measured in odor units or reduction ratios.

Identification was done by sampling odors from six plants before and after scrubber and incinerator operations. Computer regression of the odor

---

The work on which this paper is based was performed pursuant to Contract No. 68-02-0260 with the U.S. Environmental Protection Agency. The investigations were made under the sponsorship of the Fats and Proteins Research Foundation, Inc., which received partial financial support for this study. Those who contributed to the project were H. G. Reilich, David Jones, Dr. A. Dravnieks, and Dr. Del Doty.

intensity with the GLC peaks present determined which peaks could be responsible for the odors. An effort was then made to identify these peaks by mass spectroscopy. The important peaks included various sulfides, disulfides (and probably mercaptans, C4 to C7 aldehydes, trimethyl amine and various C4 amines, quinoline, dimethyl pyrazine and other pyrazines, C3 to C6 acids). Other compounds that are present, but are probably not so important as far as odor is concerned, are C4 to C7 alcohols, ketones, and hydrocarbons, as well as many common aromatic compounds. Some of these compounds contain double bonds.

ENGINEERING AND ECONOMIC ANALYSIS OF CONTROL METHODS

Before undertaking laboratory experiments, known and potential odor control methods were compared, based on limited data from the literature and reports of performance and cost from member companies of the Fats and Proteins Research Foundation. This comparison indicated that scrubbing was probably the most economic method, especially if one or more effective scrubbing reagents could be found. Incineration was also feasible, but was more costly for most applications. The study also indicated that commercially available control equipment was not designed on a reliable basis, partly because of the lack of quantitative data on the odorous compounds in rendering odors. Rendering companies were not in a position to rationally select the control equipment offered by manufacturers.

Although the experimental work concentrated on scrubbers, the design analysis was extended to several types of incinerators, scrubbers, and carbon beds. More novel approaches were also considered but were not studied extensively. A design computer program was written for each control method (except carbon beds), and these programs can be used for preliminary estimates of the cost of each method for the particular conditions of any rendering plant.

Liquid scrubbing was selected as the method to be further investigated experimentally.

Screening Experiments

Screening experiments were carried out to determine the most promising scrubbing liquids for removing the pure odorants identified in rendering plant air. A bubbler technique was developed for these experiments, which eliminated mass transfer and kinetics as variables, and focused attention on the capacity of the reagents to react with odorants. The basic assumption was made that most odorants must be chemically neutralized, and not merely dissolved in the scrubbing liquid.

In these experiments one or more odorant compounds were mixed with water in a gas bubbler bottle. An excess of a scrubbing agent was added, air was bubbled through the mixture, and the odorant concentration in the air was compared before and after adding the reagent. This method eliminated both mass transfer and any reaction rate effects, since the contact time was long. Concentrations were measured by chromatograph, by calibrating the area under the chromatograph peak with known amounts of the pure compounds.

Packed Column Scrubbing Experiments

In these experiments pure odorants were scrubbed in a 6-in.-diameter laboratory packed column, using the most promising scrubbing liquids found in the screening tests. This scrubber was of standard configuration, with known mass transfer properties. Scrubbing reagents tested included types that have already been used for odor control, plus some not previously tried. These experiments measured reaction rate effects, in contrast with the bubbler experiments.

Water was found to be effective for many highly soluble odorants. Sodium carbonate was no more effective than water, and sodium hydroxide was little better. Strong oxidizing agents were the most effective. Sodium hypochlorite was effective against a spectrum of odorants. Potassium permanganate reacted more slowly but attacked some odorants not handled by hypochlorite. Hydrogen peroxide was almost as effective, as was sodium persulfate. Sodium bisulfite was effective only against aldehydes.

It was suggested that a two-stage scrubber might remove a wider range of odorants than a one-stage scrubber. Two oxidizing agents might be the most effective combination, although other combinations could be tried. Tests on combinations should be done with the normal odorous emissions in a rendering plant.

Engineering Studies of Scrubbers

Further preliminary engineering design calculations were done for scrubbers, using time-sharing computer programs developed for the purpose.

As a result of the information presented, we can conclude that scrubbers offer the most economical control method, provided that effective reagents can be found. An incinerator may be more economic for a small, highly concentrated process air stream, since removal of the higher concentration only requires incineration at a higher temperature and/or longer contact time. The contact times and temperatures conventionally used for design of such incinerators are inadequate for odor control, however. Catalytic incineration does not appear to be economically competitive.

A novel commercially available spray scrubbing system is calculated to control odors at a cost one-fourth that of other available equipment. A typical unit cost is 16¢ per 1,000 cfm hr of operation. Previously available spray scrubbers were not effective and the reasons for the improvement are suggested by the calculations. There have been several installations of this type of scrubber in rendering plants; an 85% removal of odor was reported in one case. Much higher removals would be necessary in practice, and this might be achieved through a more rational choice of operating conditions and selection of the best scrubbing liquids.

IDENTIFICATION OF ODORANTS

Samples were taken from a variety of rendering plants using specially developed chromatographic adsorber samplers. Some of these were measured for odor level by a trained panel while others were analyzed by gas chromatograph. Then a computer regression analysis was done to determine which peaks could be correlated with the odor intensity from different plants. The GLC peaks were sniffed by the analyst and described, and this information is summarized in the first few columns of Table 10-1. An asterisk indicates the most significant odorous peaks while a plus sign indicates other significant odorous peaks.

Vapor samples were passed through a 10-ft SP-1000 column and major peaks were introduced into the mass spectrometer. The process was repeated using a Pennwalt 223 column, which is suited to separations of amines. A number of compounds were tentatively identified.

These mass spectrograph identifications were far more successful than any previously done and most of the odorous peaks appear to be identified. The following sulfides were found to be important: methyl sulfide, methyl disulfide, and propylene sulfide. The following nitrogen compounds were found to be important: trimethylamine, butylamine, quinoline, dimethyl, methylethyl, and trimethyl pyrazines. Other peaks with amine-like odors were not identified. A number of C4 to C7 aldehydes, ketones, and alcohols were also found, but specific compounds could not be definitely identified due to interference of hydrocarbons. Acids were not identified, although several odorous peaks occur at Kovats indices where acids are expected. The acids may have been present in concentrations too low for the mass spectrometer. Several acids were tentatively identified earlier in the project. Butyric acid was included as one of the odorants being studied.

Table 10-1. Identification of Major Odor Components in Rendering Odors

Important Odorous Peaks					Mass Spectrometer Identification		Other Possibility	
Odor Note	Time, Sec.	Odor Panel Regression	Peak No.	Kovats Index	Kovats Index	Compound	Kovats	Compound
Sulfide	30		52		830	Dimethyl sulfide		
Grassy	237		16	1062	1010	Pentanal		
Sulfur	236	*	18	1073	1035-1105	Dimethyl disulfide		
Sulfur	173	+	20	1028	970	(A sulfur compound)		
Cabbage	880	*	14	1325	1370	(Dipropylene sulfide)		
Sour	52		50	804	800	Trimethylamine		
					880	C_4 aldehyde		
Yeast	53		44	804	800	Vinyl butyl ether		
Amine	107		26	933				
Amine	107		28	1028	980	(An amine)	950	(Butylamine)
Amine	248	+	36	1070	1015	(Butanedione)		
Amine	303		30	1115			1112	(Isobutylamine)
Amine	374		32	1161				
Amine	447		34	1187				
Oil	131		38	984	945-1000	Benzene		
Refinery	538		24	1227	1210	C_3 benzene		
Burnt	1233	+	40	1440	1390	(Dimethyl pyrazine)		
					1470	(Trimethyl pyrazine)		
					1510	(Dimethyl-ethyl-pyrazine)		
Rendering	1335	+	22	1488	1410	(Nonadienal)	1570	(Butyric acid)
Burnt	1687	*	42	1618	1700	Quinoline		
					1660	(A sulfur compound)		
Bitter	655	+	46	1251			1205	(Heptanol)
Unpleasant	474		56	1201	1150	(A sulfur compound		
Unpleasant	1860	*	54	1634	1620	Me-indole or -indene	(1650)	(Isovaleric acid)
Cheesy	2002	+	48	1686			1690	(Valeric acid)
Unpleasant	2582	+	58	1812			(1810)	(Hexanoic acid)
Popcorn	810		60	1306	1260	Difurfuryl ether		

Note: Tentative identifications are in parentheses; estimated Kovats indices are in parentheses.
*Indicates most significant odorous peaks
+Indicates other significant odorous peaks

DESIGN AND COST OF DIRECT-FIRED INCINERATORS

A design procedure for incineration by direct flame was developed and programmed on a time-sharing computer. In the case of incinerators the rate is assumed to be governed by an overall combustion reaction, the rate constants for which are determined as a function of temperature from available plant data. Based on these concepts, the programs are able to explore or optimize the operating conditions to find the design having minimum annual cost. Cost data are supplied to the program as input parameters for cost equations in the program. Both investment amortization and operating costs are included in simple form.

The reaction can be considered first-order with respect to the pollutant in the plant air stream and first-order with respect to oxygen. If excess oxygen is present, the oxygen variation will be small enough to consider it constant.

For first-order kinetics and a constant volume reactor,[2]

$$\ln Y'_o/Y_e = k\, t_R$$

or (1)

$$k = \ln (Y'_o/Y_e)/t_R \ \text{sec}^{-1}$$

where

Y'_o is the inlet concentration of the pollutant
Y_e is the exit concentration of the pollutant
t_R is the residence time of the pollutant.

The source of combustion air is plant air.

The products of flame combustion are mixed with the remaining plant air to result in the calculated temperature necessary to incinerate the odor.

The constants A and ΔE in the Arrhenius equation were not previously known for the reaction of rendering odorants in combustion products.

Data were provided by one rendering company on the lowering of odor units by an incinerator. The resulting coefficients are

$$\Delta E = 9560$$
$$A = 1900$$

These data should be checked by additional measurements on plant incinerators. They are the best values available and they were used as the basis of design calculations.

There are two independent variables in the incinerator design that determine the extent of oxidation of pollutants: the reactor volume and the mean gas temperature. A separate design calculation was done for each of several reactor volumes and the design having minimum cost was picked. For each chosen reactor volume there is a temperature which will give the desired reduction of odorant concentration. This temperature is determined by substituting the desired reduction ratio and the contact time corresponding to the reactor volume into the Arrhenius equation which gives the effect of temperature on the reaction rate.

The calculation of heat requirement can be separated into two parts: a flame calculation and a heat balance on the mixing process. The heat balance requires us to integrate the heat capacities of two mixed streams: the plant air stream and the products of combustion stream.

These calculations were programmed and the total fuel requirements were calculated.

The reactor volume is based on the actual flow rate and the assumed residence time. The design calculations are repeated for several residence times by a program loop.

The installed cost of an incinerator is expressed in terms of its volume. It can be estimated from a correlation of data found in the literature.

The operating cost includes fuel, fan power, and maintenance costs.

The summary of costs is given in Table 10-2. These data are summarized from computer results for costs at various reactor temperatures and represent the optimum reactor temperature with regard to total annual cost. The fuel costs are comparable to those estimated by others, except for a 10% allowance for heat loss for the reactor. Probably this allowance should be neglected for a large system.

CATALYTIC BED AFTERBURNER DESIGN EQUATIONS

Since a direct flame incinerator has a relatively high cost due to the high operating temperatures, use of a catalyst to reduce the incineration energy requirement offers a potential advantage.

The catalyst allows incineration at a lower temperature. Catalysts are normally porous particles and are frequently used in a fixed bed contactor. The design of a catalyst bed reactor depends in part on diffusion through these pores. For certain types of catalyst in which the pore passages are short, pore diffusion can be neglected. In this case the controlling factor is mass transfer to the gross surface of the catalyst. The present results are limited to this type of catalyst.

It is necessary to preheat the raw plant effluent gases to a temperature of approximately 800°F before they are treated by the catalytic unit.

The design method and computer program for the preheater are the same as those used for the direct-fired incinerator.

Mass transfer characteristics are known only for a reactor using a honeycomb type of catalyst support with flow in the longitudinal direction. The

Table 10-2. Incinerator Cost Summary

Flow, CFM	Reduction Ratio Odor Concentration	Reactor			Costs			
		Temp., °F	Residence Time, Sec	Volume, Cu Ft	Investment, $	Fuel $/Year*	Total $/Year	$/1,000 CFM-Hr
1,000	100	1069	0.7	33	12,500	3,150	6,400	1.28
	1,000	1186	0.7	36	13,000	3,570	6,950	1.39
	100,000	1236	1.0	52	15,000	3,753	7,703	1.55
5,000	100	1069	0.7	164	24,500	15,700	22,540	0.90
	1,000	1082	1.0	237	28,500	16,000	23,800	0.95
	100,000	1111	1.5	362	34,000	16,500	25,700	1.03
25,000	100	1069	0.7	885	49,500	87,000	103,000	0.83
	1,000	1082	1.0	1276	57,600	88,500	106,500	0.85
	100,000	1111	1.5	1952	68,800	91,300	112,300	0.90

*Based on 5,000 hours operation per year. Gas cost $0.50/1,000 cu ft.

honeycomb catalyst support can be obtained in 1-ft square units of short lengths. They may be stacked to any desired depth.

The flow through the honeycomb passages can be treated like flow in a pipe. A standard j-factor correlation was used for design.

Capital costs include the cost of housing for the catalyst bed, insulation, a fan and motor. A typical catalyst price is \$450/cu ft of bed required. Total annual costs were estimated, and they include amortization, fuel, fan power, and maintenance.

Design and cost calculations were carried out for a catalytic incinerator system. Results are given for three different flow rates and four different reduction ratios in Table 10-3. By use of a catalyst bed, the temperature necessary for incineration of odor vapors can be reduced from 1,200 to about 800°F, thus reducing the fuel cost. At the same time, there is an added investment cost for the catalyst bed and additional fan cost due to pressure drop through the bed. The calculations showed that these costs are about equal to the savings in fuel costs, so that catalytic incineration offers little advantage over direct-fired incineration. Slight improvement could be obtained by finding catalysts with less pressure drop, or which operate reliably at lower temperature. However, it is clear that these improvements will be only marginal.

Only if the plant effluent gas going to the incinerator is already partly heated will a catalytic incinerator show an economic advantage, because the increment of heat necessary to reach the catalyst temperature is much less than that required to reach a direct-fired incinerator temperature. The gaseous effluent from rendering plants is not expected to exceed 120 to 140°F, and this temperature is not sufficient to affect the heat economy. If a condenser is used, as may be necessary to scrub out fatty particles that may interfere with the catalyst, the temperature will be even lower. We conclude that catalytic incineration is not promising for this application.

CARBON BED ADSORPTION

The economics of carbon bed adsorption is different from the other methods considered and may have a place in certain applications. The cost of the carbon and its regeneration depends strongly on the odorant concentration to be removed, so that carbon can be considered for treatment of very dilute odors, such as ventilation air. Design computer programs have not been developed but a rough example calculation was done. The design is based on equilibrium capacity.

The example is to treat 100,000 cfm of ventilation air containing only 10 ppm of odorant. It is assumed that a 100-fold reduction in concentration is sufficient for ventilation air, although much greater reductions are

Table 10-3. Catalytic Incineration Design and Cost Estimates

TEMPERATURE DEGREES F

Inlet	Outlet
32	800

UNIT COSTS OF CAPITAL EQUIPMENT

Catalyst	Ducting	Insulation	Blower	
450	15.00	10.	Coefficient 11.00	Exponent .680

UNIT COSTS OF OPERATION

Fuel Gas $/mcf	Power $/kwh	Maintenance $/cfm-yr	Amortization Factor	Annual Hours in Operation
.500	.015	.080	.250	5000

CALCULATED OPERATING CONDITIONS

Space Vel. 1/hp	Reynolds No.	Schmidt No.	Pressure Drop In. of Water	Horsepower	Number of Units
72013	589	2.529	28.12	105.43	5.

Flow Rate	Odorant Concentration		Length	Capital Cost		Annual Operating Cost			Total Operating Cost	
cfm	Inlet	Outlet	Feet	Installed	Annual	Fuel	Power	Maint.	Annual	$/1000 cfm-hr
1000.	10	1	.636	10053	2513.	2712.	203	210	5637	1.13
1000.	100	1	1.402	10721	2680	2712	447	210	6048	1.21
1000	1000	1	2.169	11388	2847	2712	691	210	6459	1.29
1000.	100000	1	3.702	12723	3181	2712	1180	210	7282	1.46

CALCULATED OPERATING CONDITIONS

Space Vel. 1/hp	Reynolds No.	Schmidt No.	Pressure Drop In. of Water	Horsepower	Number of Units
72013	589	2.529	28.12	527.15	25

Flow Rate	Odorant Concentration		Length	Capital Cost		Annual Operating Cost			Total Operating Cost	
cfm	Inlet	Outlet	Feet	Installed	Annual	Fuel	Power	Maint.	Annual	$/1000 cfm-hr
5000	10	1	.636	23770	5943	13558	1013	1049	21562	.86
5000	100	1	1.402	27108	6777	13558	2234	1049	23618	.94
5000	1000	1	2.169	30446	7612	13558	3456	1049	25674	1.03
5000	100000	1	3.702	37123	9281	13558	5899	1049	29786	1.19

CALCULATED OPERATING CONDITIONS

Space Vel. 1/hp	Reynolds No.	Schmidt No.	Pressure Drop In. of Water	Horsepower	Number of Units
72013	589	2.529	28.12	527.15	25

Flow Rate	Odorant Concentration		Length	Capital Cost		Annual Operating Cost			Total Operating Cost	
cfm	Inlet	Outlet	Feet	Installed	Annual	Fuel	Power	Maint.	Annual	$/1000 cfm-hr
25000	10	1	.636	62128	15532	67789	5064	5243	93628	.75
25000	100	1	1.402	78818	19705	67789	11172	5243	103909	.83
25000	1000	1	2.169	95509	23877	67789	17279	5243	114189	.91
25000	100000	1	3.702	128889	32222	67789	29494	5243	134749	1.08

needed for process air. The amount of carbon needed is calculated from equilibrium based on the capacity of the carbon. Data are available from Calgon Corp. on the capacity of carbon for isopropyl alcohol. Odorous materials of higher molecular weight and less polarity will be more readily adsorbed, and therefore this is a safe method of design.

The total quantity of odorant removed depends on the flow rate and the concentration. Ten ppm corresponds to

$$\frac{10 \times 10^{-6}\ \text{ft}^3 \quad 60.1\ \text{lb} \quad 1000}{\text{ft}^3 \quad 392\ \text{ft}^3 \quad 1000} = \frac{0.0015\ \text{lb isopropanol}}{1000\ \text{ft}^3\ \text{air}}$$

This uses up

$$\frac{0.0015 \times 100}{0.7} = \frac{0.216\ \text{ft}^3\ \text{bed}}{\text{min}}$$

For 100,000 cfm and a 16-hr day, this requires a 208 ft^3 bed.

The second major cost of carbon bed treatment is the cost of forcing the air through the bed. This depends on the thickness of the bed. If the bed is made too thin, mass transfer rate might limit the adsorption process. However, mass transfer data on packed beds indicate that a bed 10 particles thick will remove several orders of magnitude of odorant concentration. Therefore a bed design is chosen spreading the 208 ft^3 over an area to give a bed thickness of one in. and the bed area is 2,496 sq ft. The air velocity is 40 ft/min. The pressure drop for a 6-16 mesh carbon bed is four inches water pressure.

The final major cost is the cost of regeneration. The most logical way to regenerate a carbon bed is by hot air, since the effluent from the bed can be economically purified by a direct-fired incinerator. Part of the incinerator effluent can be mixed with fresh air and recycled to perform the regeneration.

The regeneration would take place at the end of the day and would require two hrs. If an incinerator is also used for process air treatment, the same incinerator could be used for the regeneration air. The ability to regenerate depends on the change in adsorption isotherms for the higher gas temperature. Assuming a 100 to 1 concentration effect due to regenerating between 400 and 500°F, the volume of regeneration air is 1/100 the volume of air deodorized. This means that in two hr a flow of 8,000 cfm will regenerate the bed which treated 100,000 cfm for 16 hr. From the incinerator cost in Table 10-2, the investment cost is about $40,000. The fuel cost is less than that in the table because the time of operation is only two hr per day. This calculation requires that the flow during regeneration be countercurrent to the normal flow during use, since otherwise additional time would be required to thoroughly purge the final portion of the bed near the outlet.

Total and Annual Costs

The total and annual costs are listed in Table 10-4. The total cost per 1,000 cfm-hr is only 11¢ lower than the other methods considered. However, this is only for ventilation air. A similar calculation was done for process air, assuming 1,000-fold reduction in concentration. The bed capacity and regeneration costs rise sharply.

Table 10-4. Costs for Carbon Bed Treatment of Ventilation Air With Low Odor Concentration

	Total	Annual*
Carbon Bed Investment		
Carbon, 208 ft^3	$ 3,000	
Housing	40,000	
Fan	10,000	
Ducting, installation	30,000	
Total	$83,000	$21,000
Incinerator Investment		
8000 cfm	40,000	10,000
Operating Costs		
Fan power		5,000
Incinerator (fuel)		4,000
Maintenance		15,000
Total		$55,000
Cost per 1000 cfm-hr		$ 0.11

*Based on 5,000 hours operation per year.

It should be emphasized that these costs are only rough ones and are based on data for compounds other than actual odorants. They suggest that carbon beds can be considered for ventilation air but not for process air.

DESIGN AND COST OF COUNTERCURRENT PACKED TOWER GAS SCRUBBING SYSTEM

A design procedure has been completed for countercurrent packed scrubbing towers. The design of scrubbers is based on the idea that reagents have been found that react rapidly with the odorants, once they reach the liquid phase, and that the rate of removal is controlled by mass transfer from the gas to the liquid. This design method has been programmed on a time-sharing computer system and was used to design the experimental scrubber for use on the project.

The cost data have been entered in the form of equations. To use the program, the user specifies the air flow capacity, the fraction odor removal, and the unit cost of items, such as electricity. The program performs a series of design calculations to find a liquid flow rate and column diameter for the lowest cost. It then prints out dimensions of the tower, flow rates, and costs.

A series of packed scrubber designs was calculated using the method described above. These calculations are based on the assumption that two scrubbing liquids can be found that will, between them, remove all important odors. The design is also based on a conventional type of packed tower using 1½-inch saddles. Other scrubber configurations may prove to be more economical.

Preliminary calculations showed that smaller packing is less economical. Plastic Tellerette packing material may be more economic, and calculations could be done with this packing.

It was found economical to use the maximum gas rate that the column can handle without flooding and the maximum liquid recirculation rate for which experimental correlations are available. Under these conditions the cost of scrubbing is less than that of incineration, except when the inlet odor concentration is as high as 1,000 ppm. In this case the chemicals' cost raises the cost of scrubbing above that of incineration, because the incinerator can reduce a higher odor concentration simply by operating at a higher temperature. Results are summarized in Table 10-5.

DESIGN AND COST OF HORIZONTAL SPRAY SCRUBBER

A new type of spray scrubber was designed by Air Conditioning Corporation, Greensboro, North Carolina. This scrubber has been installed in a number of rendering plants and is found to give satisfactory odor reduction. This scrubber consists of a long duct into which the plant air flows. A bank of spray nozzles is mounted near the entrance of the duct, and the scrubbing liquid is atomized by pumping the liquid under pressure through the nozzles. At the end of the scrubber is an effective mist eliminator. Two or more units can be connected in series to make use of several scrubbing liquids. It is possible to calculate the performance by using penetration theory to estimate the mass transfer to the drops in the scrubber.

Table 10-5. Scrubber Cost Summary
(Pentanal in Air) (1½-in. Intalox Saddles) Two Scrubbers in Series

Flow cfm	Reduction Ratio Odor Concentration	Flow Rates, lb/hr sq ft Gas	Flow Rates, lb/hr sq ft Liquid	Packing Depth, ft	Tower Diameter, ft	Pressure Drop in. Water	Costs: Investment, $	Costs: Chemicals* $/Year	Costs: Total** $/Year	Costs: $/1000 cfm hr
1,000	100	569	7500	5.2	3.2	0.46	8,200	7	2.370	0.49
	1,000	570	7500	7.8	3.2	0.69	12,180	72	3,450	0.69
	100,000	570	7500	12.9	3.2	1.15	20,120	7,200	12,600	2.50
5,000	100	570	7500	5.1	7.1	0.49	22,600	35	7,200	0.29
	1,000	570	7500	7.7	7.1	0.74	33,600	350	10,400	0.41
	100,000	571	7500	12.8	7.1	1.23	55,600	35,000	50,000	2.00
25,000	100	571	7500	5.1	15.8	0.49	67,400	175	24,800	0.19
	1,000	571	7500	7.7	15.8	0.74	99,800	1,750	35,200	0.28
	100,000	571	7500	12.8	15.8	1.23	164,680	175,000	225,000	1.80

*Chemical cost based on effluent concentration 0.01 ppm by volume. Typical chemical cost based on reaction with NaOH at 20 cents/lb.

**Based on 5,000 hours operation per year.

The success of this scrubber depends on the use of low air velocities such that the droplets remain in the scrubber for a second. Previously designed spray scrubbers either have insufficient droplet area/unit volume or insufficient contact time. Although the nozzles point upstream, the stopping distance of droplets is such that they soon acquire the air velocity and drift down the scrubber to the exit. Therefore the droplet contact time is only slightly longer than the air contact time in the scrubber.

Nozzles are used which give a relatively coarse spray, with a mean size of at least 400 μm. The size depends slightly on water pressure and significantly on size of nozzle.

According to the methods of calculation used, each droplet acts independently of the others. Actually the droplets can collide and coalesce. This effect was computed to see if it is significant, but it was not compensated for. In addition the droplets can settle and collect on the floor of the scrubber. The terminal settling velocity of 400 μm droplets in air is 3 ft/sec. This velocity cannot be directly applied to the turbulent flow conditions of the scrubber, but it suggests that the performance will be reduced if the contact time is more than a few seconds.

The equation for mass transfer is the same as that for a packed scrubbing tower. Only the method of calculating the mass transfer coefficient is different.

The mass transfer coefficient is predicted from penetration theory based on the transient diffusion of odorant through the gas layer adjacent to each drop. The result is

$$k_g = \frac{2}{RT} \sqrt{\frac{\text{Diff}}{\pi \theta_c}} \quad \frac{\text{lb-mol}}{\text{hr-ft}^2\text{-atm}}$$

The contact time, θ_c, of droplets in the air depends on the length of scrubber X_L and the gas (and droplet) velocity.

The resulting design equation is

$$\frac{X_L\, k_g\, a\, P_T}{G_m} = \frac{2\, T\, a}{460} \sqrt{\frac{\text{Diff}\; X_L\; W\; Z}{60\pi\; \text{cfm}}}$$

We now design the scrubber by fixing the length X_L = 29 ft, as is done by Air Conditioning Corporation, and vary the height Z. In doing so, we allow the number of nozzles per unit of cross-sectional area N_{oza} to remain constant; the total number increases with height Z. The liquid flow per nozzle L_{noz} is also fixed; so changing the height changes the liquid flow and increases the droplet area. It also decreases the air velocity and increases the contact time. As a result, the capacity depends on $Z^{3/2}$. These substitutions are included in the design computer program.

The flow rate per nozzle L_{noz} is another variable that affects the design, and hence the scrubbing cost. The program provides for reading in four values of L_{noz}; it computes the scrubber design and cost for each, and chooses the most economic one. Other variables, such as spray nozzle pressure, affect droplet size and power. These can be explored by trial calculations with the program.

Costs of each component of equipment and operating costs are similar to those used for the packed tower scrubber, except the coefficients are different. These equations are included in the computer program.

The drop collision rate is estimated by applying equations from the kinetic gas theory.

In most cases the decrease in number of drops due to collision is small, but it could be large if a high droplet concentration is produced.

Tables 10-6, 10-7, and 10-8 present results of calculations with the design computer program. All of these examples are for two-stage units. The cost of this method of scrubbing appears to be about one-fourth that of other methods when expressed in $/1,000 cfm-hr. The effect of flow capacity on unit cost is very small, as shown in the tables. These tables correspond to cost data kindly provided by Air Conditioning Corporation, and they check these values within 15%. The cost decreases somewhat with lower reduction ratios, and the method even competes with carbon adsorption for treating ventilation air.

EXPERIMENTAL SCRUBBING INVESTIGATIONS

A preliminary economic analysis of various odor removal methods, based on literature data and reports of member companies, indicated that scrubbing with reactive liquids was the most promising method. This method was therefore chosen for experimental study.

The two types of scrubbing reaction studies were an exploratory bubbler experiment method and a packed column scrubber method. One focuses attention on the capacity of the reagent to consume odorant, and the other measures the rate of reaction with a relatively fresh reagent. Both criteria are important for selecting reagents.

In the bubbler exploratory method, a mixture of several odorants is dissolved in 200 ml of water in a glass bubbler equipped with a fritted glass disperser. Air is drawn through the bubbler at a rate of about 50 ml/min for one minute. This air is drawn through the sampling loop by means of an aspirator. After one minute, the air sample in the loop is transferred to the chromatograph. The process is repeated several times to establish the effluent odorant concentrations with no reagent present. Then the bubbler is opened and 50 ml of scrubbing liquid is added.

Table 10-6. Design and Cost of Spray Scrubbing

Nozzles/Sq Ft	Length	Units/Series	DDrop μM	Nozzle Psi
4.0	29.0	2.0	400.0	15.0

GPM/Nozzle			
1.000	2.200	5.000	10.000

COEFFICIENTS AND EXPONENTS FOR COST CORRELATIONS

Scrubber			Fan	Pump		Nozzles
176.000	0.800	0.290	0.800	11.500	0.600	5.000

UNIT COSTS FOR SCRUBBER SYSTEM

Power $/kwh	Maintenance $/cfm	Chemicals $/cfm-yr	Nozzles $ ea	Amortization Factor
0.015	0.005	0.014	5.000	0.250

Flow Rate cfm	Concentration Inlet	Concentration Outlet	Henry	Schmidt No.	Temperature Degrees F	Hrs/Yr
25000	1000.000	1.000	0	0	70	5500

INTERMEDIATE VALUES DURING OPTIMIZATION

Annual Cost	Capital	Time-Sec	Height	Gpm/Noz	DROPLETS Ft2/Ft3	No	Ni
21858.9	50438.0	6.2	11.1	1.0	7.8	1367018	1150861
20486.2	37340.9	3.7	6.6	2.2	10.1	1777898	1349237
20918.9	28629.5	2.1	3.8	5.0	13.3	2337459	1508854
22569.7	23846.0	1.3	2.4	10.0	16.7	2944946	1573407
20918.9	28629.5	2.1	3.8	5.0	13.3	2337459	1508854

OPERATING CONDITIONS OF TOWER-CALCULATED
THE MATERIAL OF CONSTRUCTION IS STAINLESS STEEL

Scrubber Length Feet	Scrubber Height Feet	Total Flow Rates Gas(cfm)	Liquid (gpm)
29.0	3.8	25000.0	607.3

Nozzles	Banks	DDrop μM	Units/Series
121	1	400	2

Pressure Drop	Fan Horsepower	Pump Horsepower
3.000	16.821	20

CAPITAL COSTS FOR SCRUBBER SYSTEM

Shell	Nozzles	Fan	Pump	Externals	Total
14541	1215	1025	1076	10773	28630

ANNUAL COST FOR SCRUBBER SYSTEM

Amortization	Power	Maintenance	Chemicals	Total	$/1000 cfm-hr
7157	2000	1000	700	10857	0.17

Table 10-7. Effect of Air Flow

UNIT COSTS FOR SCRUBBER SYSTEM

Power $/kwh	Maintenance $/cfm	Chemicals $/cfm-yr	Nozzles $ ea	Amortization Factor
0.015	0.005	0.014	5.000	0.250

Flow Rate cfm	Concentration Inlet	Outlet	Henry	Schmidt No.	Temperature Degrees F
38000	1000.000	1.000	0	0	70

INTERMEDIATE VALUES DURING OPTIMIZATION

Annual Cost	Capital	Time-Sec	Height	Gpm/Noz	DROPLETS Ft2/Ft3	No	Ni
32227.4	72169.5	6.2	16.9	1.0	7.8	1367056	1150877
30463.5	53668.5	3.7	10.0	2.2	10.1	1777935	1349251
31332.4	41364.4	2.1	5.8	5.0	13.3	2337508	1508863
33954.4	34604.5	1.3	3.6	10.0	16.7	2945008	1573410
31332.4	41364.4	2.1	5.8	5.0	13.3	2337508	1508863

OPERATING CONDITIONS OF TOWER-CALCULATED
THE MATERIAL OF CONSTRUCTION IS STAINLESS STEEL

Scrubber Length Feet	Scrubber Height Feet	Total Flow Rates Gas (cfm)	Liquid (gpm)
29.0	5.8	38000.0	923.1

Nozzles	Banks	DDrop μM	Units/Series
185	1	400	2

Pressure Drop	Fan Horsepower	Pump Horsepower
3.000	25.569	30

CAPITAL COSTS FOR SCRUBBER SYSTEM

Shell	Nozzles	Fan	Pump	Externals	Total
20328	1846	1433	1383	16374	41364

ANNUAL COST FOR SCRUBBER SYSTEM

Amortization	Power	Maintenance	Chemicals	Total	$/1000 cfm-hr
10341	4000	1520	1064	15557	0.16

Table 10-8 Effect of Air Flow

UNIT COSTS FOR SCRUBBER SYSTEM

Power $/kwh	Maintenance $/cfm	Chemicals $/cfm-yr	Nozzles $ ea	Amortization Factor
0.015	0.005	0.014	5.000	0.250

Flow Rate cfm	Concentration Inlet	Outlet	Henry	Schmidt No.	Temperature Degrees F
66000	1000.000	1.000	0	0	70

INTERMEDIATE VALUES DURING OPTIMIZATION

Annual Cost	Capital	Time-Sec	Height	Gpm/Noz	DROPLETS Ft2/Ft3	No	Ni
54116.9	116035.8	6.2	29.3	1.0	7.8	1367094	1150899
77324							
51677.2	86832.3	3.7	17.3	2.2	10.1	1777984	1349271
69044							
53588.8	67417.2	2.1	10.0	5.0	13.3	2337573	1508875
67072							
58353.5	56746.0	1.3	6.3	10.0	16.7	2945089	1573413
53588.8	67417.2	2.1	10.0	5.0	13.3	2337573	1508875
67072							

OPERATING CONDITIONS OF TOWER-CALCULATED

THE MATERIAL OF CONSTRUCTION IS STAINLESS STEEL

Scrubber Length Feet	Scrubber Height Feet	Total Flow Rates Gas(cfm)	Liquid (gpm)
29.0	10.0	66000.0	1603.3

Nozzles	Banks	DDrop μM	Units/Series
321	1	400	2

Pressure Drop	Fan Horsepower	Pump Horsepower
3.000	44.409	55

CAPITAL COSTS FOR SCRUBBER SYSTEM

Shell	Nozzles	Fan	Pump	Externals	Total
31616	3207	2229	1926	28439	67417

ANNUAL COST FOR SCRUBBER SYSTEM

Amortization	Power	Maintenance	Chemicals	Total	$/1000 cfm-hr
16854	6000	2640	1848	27342	0.16

Sampling is repeated several times. The ratio of effluent concentration before and after adding scrubbing liquid is a measure of the reactivity of the liquid for each odorant. The mass transfer behavior of the bubbler is effectively cancelled out by this procedure, and only the reactivity is measured.

A list of odorants used in bubbler or scrubber experiments is given in Table 10-9.

Table 10-9. List of Odorants for Scrubbing Tests

	MW	Density, g/cc	MP, °C	BP, °C
SULFIDES				
n-Propyl sulfide (dipropyl sulfide)	118.2	0.839		142-3
Dimethyl disulfide (methyl sulfide)	94.19	1.057		116-8
Propylene sulfide				
AMINES				
Trimethylamine	59.11	gas	-117	2.9
Tert-butylamine	73.14	0.698	-67	45.2
n-Butylamine	73.14	0.739	-50	77.8
Dimethyl pyrazine	108.14	0.990	15	155
Quinoline	129.15	1.095	-15.6	237
ACIDS				
Butyric	88.1	0.949	-47	155
ALDEHYDES				
n-Pentanal (n-valeric aldehyde)	86.13	0.819	-92	103.4
ALCOHOL				
n-Pentanol (n-amyl alcohol)	88.15	0.818	-78.9	138.1
KETONE				
Butandione (diacetyl)	86.19	0.981	- 4	87-8
UNSATURATED ALKANE				
Heptadiene 2,5 (simulant for dimethyl pentadiene)	96.17	0.733		107

A list of scrubbing liquids was also selected and is shown in Table 10-10. This list includes a number of liquids previously suggested by others, some that are in actual use and a number that we have thought of. We have included several that are used as oxidizing or reducing agents in paper pulp bleaching or other industrial purification operations.

Table 10-10. List of Scrubbing Solutions Screened

Sodium hydroxide	Hydrochloric acid
Sodium carbonate	Nitric acid
Sodium bisulfite	Ozone solution
Sodium hypochlorite (representing free chlorine)	Copper sulfate
Hydrogen peroxide	Ferric chloride
Potassium permanganate	Barium chloride
Potassium persulfate	Aluminum chloride

The quantity of each odorant added to the bubbler depends on the solubility of the odorant in water. A quantity is added which gives a sufficient vapor pressure to obtain a reasonable peak size with the chromatograph. Thus amyl alcohol is relatively soluble and requires 20 mg, while propyl sulfide is relatively insoluble and requires only a few μg. In fact, the quantity of propyl sulfide used is so small that an appreciable fraction of it is vaporized by the air stream and its effluent concentration decreases with time even when there is no reaction. In the presence of this gradual decrease, only large reductions by reaction can be definitely identified.

At the other extreme is butyric acid, which is miscible with water in all proportions. One gram of butyric acid in 250 ml water has so low a vapor pressure that it cannot be measured by chromatograph. Although the bubbler experiment cannot be done with butyric acid, this result shows that water alone is an effective scrubbing liquid and no other is needed.

Results of all the bubbler experiments are summarized in Table 10-11. Water is the first and simplest reagent. Butyric acid and amyl alcohol are so soluble that water is an effective scrubbing liquid for them. Various amines appear to react extensively in water, so it may effectively scrub them too. Dimethyl pyrazine reacts in water and could not be detected after the bubbler. Quinoline appeared to react in the bubbler, or the sample loop of the chromatograph, giving a very broad peak, but the result is uncertain.

Acids such as hydrochloric, nitric, and probably sulfuric react with amines and have a slight effect on dipropyl sulfide and amyl alcohol. Acids cannot be considered effective scrubbing liquids. Likewise alkalis are poor scrubbing liquids. Although sodium carbonate is used industrially for odor scrubbing, its action may be attributed almost entirely to the water rather than to the alkali. Sodium hydroxide will be converted to carbonate by absorbing CO_2 from the scrubbed air.

The most effective reagents are the strong oxidizers: sodium hypochlorite, hydrogen peroxide, potassium permanganate, and sodium persulfate. All

Table 10-11. Summary of Bubbler Experiment Results–Percent Removal by Reagent

		Reagent								
		Hydrochloric Acid	Nitric Acid	Sodium Hydroxide	Sodium Carbonate	Sodium Hypochlorite			Potassium Permanganate	
Odorant	Water	5%	5%	5%	5%	2.5%	1%	0.5%	0.6% Neutral	+1% NaOH
Di-n-propyl sulfide	Vaporizes*	50-90	90	90	Uncertain		″99+	>′99	>″′99	>99
Dimethyl disulfide	Vaporizes*	None		None		>99.9			>99.9	
Propylene sulfide	Vaporizes*	None		None		>99.9			>99.5	
n-Pentanal (n-valeric aldehyde)	*	None	Slight	Slight	None		90	75	99+	98
n-Pentanol (n-amyl alcohol)	Soluble*	30	40	Increased	None		None	Uncertain	None	Increased
Butanedione										
Heptadiene 2,5	Vaporizes*	None		None		None			80	
Butyric acid	>99.9									
Dimethyl pyrazine	>90									
n-Butylamine	*									
tert-butylamine	**	>99					>95			
Di-n-butylamine	*	>95					95			
Quinoline	Uncertain									
Isobutylamine	Reacts									
sec-Butylamine	Reacts									

Table 10-11, continued

	Reagent										
	Hydrogen Peroxide	Ozone Water			Sodium Persulfate	Sodium Bisulfite	Ferric Chloride	Copper Sulfate		Barium Chloride	Aluminum Chloride
Odorant	5%	~1%	pH 1	pH 12	Satd.	5%	5%	5%	5% pH 3	5%	5%
Di-n-propyl sulfide	> 99	None	None		~99	None	None	80	Uncertain	None	None
Dimethyl disulfide	60	None	None	None							
Propylene sulfide	> 99.9	None	None	None							
n-Pentanal (n-valeric aldehyde)	90	None			20	99.5	None	70	20	None	None
n-Pentanol (n-amyl alcohol)	None	None	None	None	None	None	None	70	Uncertain	None	None
Butanedione											
Heptadiene 2,5	None	None	None	None							
Butyric acid											
Dimethyl pyrazine											
n-Butylamine											
tert-butylamine	> 99										
Di-n-butylamine	90										
Quinoline											
Isobutylamine											
sec-Butylamine											

*Used as standard for comparison.
' = Product peak appeared.

of these remove sulfides, although hydrogen peroxide is less effective on dimethyl disulfide. Permanganate is more effective on pentanal than the other oxidants, and is the only oxidant to react with heptadiene 2,5. Permanganate is more effective in neutral than in alkaline solution.

Sodium hypochlorite is the cheapest of these oxidants and it should therefore be the first choice. Chlorine water might also be considered; hypochlorite is a solution of chlorine in alkali. Chlorine was tried, but it oxidized the chromatograph column and prevented analysis of odorants. With special precautions, it could be tried in a scrubber test. Hypochlorite has the advantage that the chlorine is not lost in the effluent gas. Various metal salts were also tried; copper sulfate was the only one that offered any promise.

A hypochlorite solution can be made by releasing chlorine gas into the scrubber while a sodium hydroxide solution is being circulated. The oxidizing power of the solution can be maintained by bleeding chlorine gas into the scrubber continuously. Chlorine in bulk is very inexpensive. If the chlorine is not added continuously, the hypochlorite will soon decompose, and it will not be effective. Only occasionally will it be necessary to make up a fresh solution, if this method of replenishment is followed. The chlorine can also be produced on the site by electrolysis of salt solution; equipment for this process is commercially available.

Packed Laboratory Scrubber Experiments

We chose a conventional packing for the scrubber column (½-in. Intalox saddles) because there are extensive data on the mass transfer coefficients for this packing and we wished to compare our results to other published data for which chemical reaction rate is not a limitation.

The data obtained from this experimental system will be useful for designing other types of scrubbers, even though ½-in. Intalox saddles may not be the most economical packing.

The column was designed using these data and the gas scrubbing computer program described. With the height of packing used and the chosen flow rates, the column should be capable of reducing any given odorant concentration by 90% if the reagent reacts very fast with the odorant, so that only mass transfer from the gas to the liquid controls the scrubbing. Removals of more than 90% are suspect since the column should not be capable of greater performance unless the flow rates are altered.

The experimental scrubbing system is shown in Figure 10-1. It consists of two similar packed columns, the first of which humidifies the air supply to simulate typical rendering plant effluent conditions. Odorants are metered into the air stream after humidification. Odor generators consist either of a lecture bottle of the gaseous odorants or saturators. A saturator is a fritted glass bubbler.

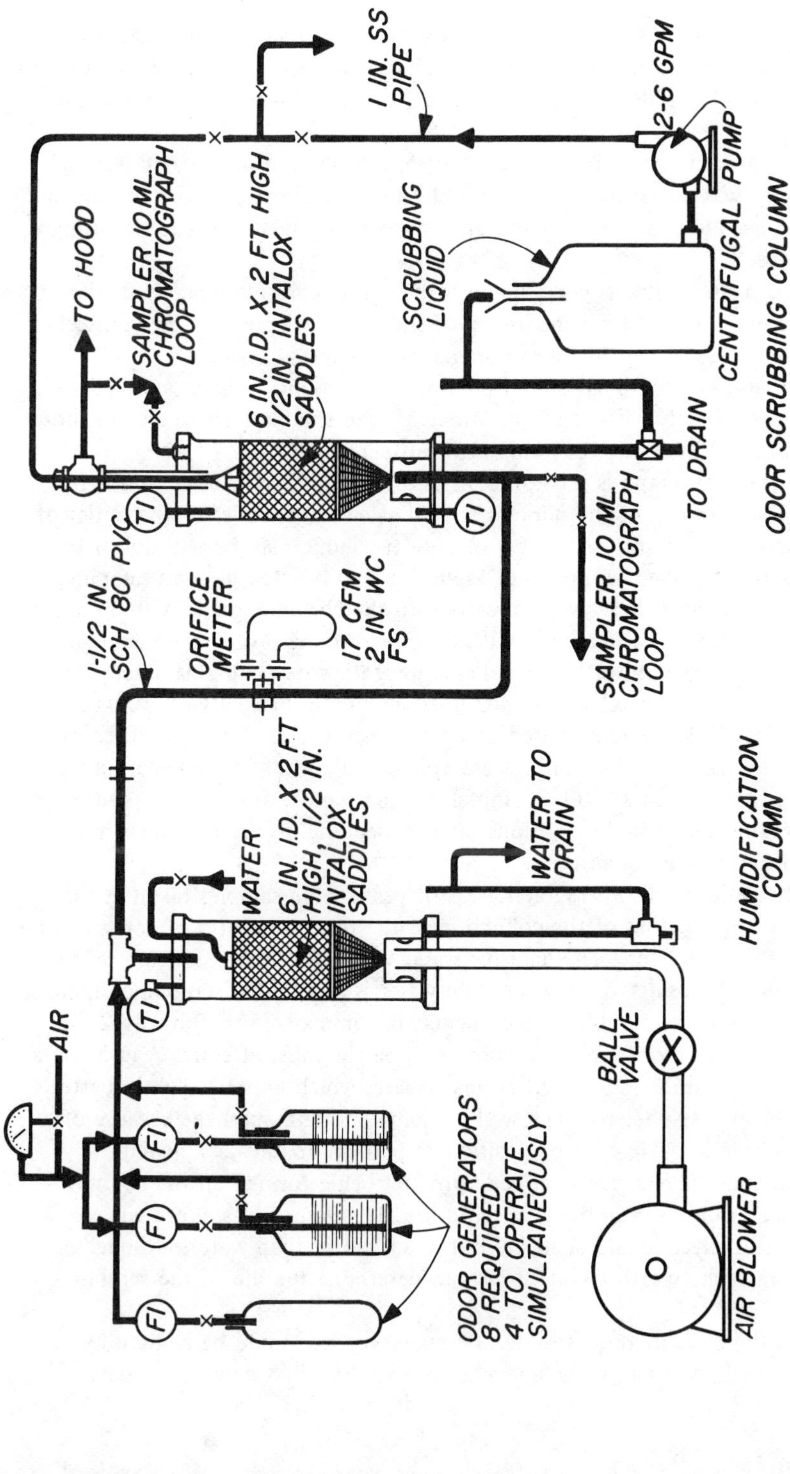

Figure 10-1. Laboratory odor scrubbing system.

Air is metered through this bubbler at a rate to supply the desired concentration after it is diluted with the main air stream. The main air flow is supplied by a blower. The flow is throttled and metered by a standard orifice.

Exhaustion tests were not made, since information on reagent eventual capacity was obtained from the bubbler tests. Only in the case of potassium permanganate was there any evidence of exhaustion of the reagent during a 1-hr test.

The main air line is constructed of polyvinyl chloride pipe, while the scrubbing liquid line is constructed of stainless steel pipe and polyethylene tubing. Usually two to four odorants were tested at once.

Some other combinations of odorants were found to react and had to be run separately. The concentrations used in the tests are similar to those encountered in strong rendering plant effluents.

A chromatograph is incorporated in the system in a manner which permits the sampling of the odorant stream at either the inlet or the outlet of the scrubber column. A sample of known volume may be withdrawn from the odorant stream by means of aspiration and isolated in a gas sampling loop, from which it is swept directly into the chromatograph with an adequate volume of carrier gas. A Perkin-Elmer Model 881 dual-column gas chromatograph with differential flame ionization detector was used. The columns are 1/8-in. x 10-ft Teflon-lined aluminum packed with 100/120 mesh SP-1000. Helium is used as a carrier gas at a flow rate of 60 cc/min. Inlet and detector temperatures are approximately 200°C and the sample loop is maintained at 100°C. Initial column temperature is 60°C and is programmed to 180°C at various rates, depending on the particular mixture of odorants being monitored.

The ratio of chromatograph areas of peaks from samples taken at the top and the bottom of the column gives the reduction ratio of concentration due to the scrubber. The data for water were of interest to assure reproducibility of results and also because water is a potential scrubbing liquid.

Results of the scrubber experiments are summarized in Table 10-2. Sodium hypochlorite, NaOCl, appears to be the most effective reagent on a broad spectrum of odorants. Permanganate, which appears slightly better in bubbler tests, behaved less well on some odorants, perhaps because of a slower reaction rate. Decomposition of permanganate was apparent, a brown stain of manganese dioxide product being formed on the column packing. Sodium bisulfite is specific to aldehydes, as was expected. Sodium hydroxide and acids are not much better than water, although they might be worth trying in tests to determine the life of the scrubbing solutions.

On the basis of these results, the main reagent should be sodium hypochlorite. If two scrubbing stages are needed, the first could be water,

Table 10-12. Scrubber Removal, %

	Valeraldehyde	Dipropyl Sulfide	Amyl Alcohol	Trimethyl-amine	Butyric Acid	Butanedione	Tert-Butyl amine	Heptadiene	Dimethyl Disulfide
Water	30	0	80-90	80-90			>90	0	10
NaOCl, 1%	10	>90	80	>90				20	50
H_2O_2, 3%	>90	0	75				>90	0	slight
Potassium permanganate, 3%	30	10-25	40-80	>90				25	20-75
Sodium bisulfite, 5%	>90	10	75						
HCl, 5%	0	0	80	>90					
Sodium hydroxide, 5%	10-30	0	0-60	0	>90	>90	>90		

followed by hypochlorite, provided that the spent water can be disposed of without causing an odor problem. Or the first liquid could be hypochlorite, followed by either permanganate, hydrogen peroxide, a second stage of hypochlorite, or an acid. An acid is effective on amines and it increases the solubility of alcohols.

The bubbler test results (in Table 10-11) gave higher odor removals than the corresponding scrubber tests. This was expected since the scrubber is limited by mass transfer as well as by chemical reaction rate. The fraction removed by a scrubbing tower is determined by its height, and it is possible from these data to design a scrubbing tower to achieve higher reduction rates.

ADDITIONAL STUDIES IN PROGRESS

The experimental work is being continued by conducting scrubber tests with the laboratory scrubber using actual plant air onsite at a rendering plant. This work will result in selection of the optimum scrubbing liquids, as well as more reliable scrubber design data and revised calculations of the cost of odor removal by scrubbing.

CHAPTER 11

ODOR CONTROL IN THE PHARMACEUTICAL INDUSTRY

George A. Herr

Eli Lilly & Co.
Lafayette, Ind. 47902

The so-called process industry is the most diverse and complicated of all industries. Within this family group, the pharmaceutical industry stands out as the one that has the largest product mix and the greatest range of process activities. This industry not only produces a very complex product mix—it also produces waste streams that defy adequate description.

A few years ago a new type of corporate entity appeared on the American industrial scene—the conglomerate. Diversification by acquisition, integration, or what have you became a goal and, eventually, a way of life. This required the procurement of expertise in allied and in completely foreign fields.

In a sense, the large pharmaceutical companies have employed techniques and procedures spanning several process industries for years. When the age of diversification arrived, it was natural that many of these companies could introduce new products that were entirely foreign to their normal marketing area. The process technology for the manufacture of these new products was well within the range of their expertise.

The net result of this crossover into other market areas is that there are now no large pharmaceutical companies limited solely to the manufacture of pharmaceuticals.

The typical large pharmaceutical firm today is a research-oriented, multi-plant, international organization whose product mix embraces entries into the human and animal health care fields and who makes fine chemicals and odors and fragrances. It may also manufacture agricultural chemicals, cosmetics, golf balls, industrial chemicals, packaging materials, etc.

Although the discussion which follows divorces itself from those activities that produce nonpharmaceutical products, it must be pointed out that these other activities can also produce odors. This fact becomes important when a single plant is engaged in process activities in other fields as well as process activities in the pharmaceutical field, *i.e.,* other processes which also produce odors. When odor-bearing substances are combined, one of the following things may occur:

1. The odors interact with one another, are synergistic, and produce an odor that is worse than the sum of its parts.
2. The odors may interact with one another and only the strongest odor will be detected.
3. There is no interaction; both odors still exist, and each is detectable.
4. The odors may cancel each other out.

This same set of facts also applies to odors that emanate from neighboring plants.

The prescription pharmaceutical industry is made up of 600 to 700 different firms. Fewer than 100 of these do any significant amount of research and development work, and 90% of all the active medicaments are made by 10% of these firms.

PHARMACEUTICAL PROCESS METHODS

Pharmaceuticals may be manufactured by several different process methods or by combinations of such methods: extractive; organic chemical synthesis; antibiotics manufacturing (fermentation and purification); antibiotic modification; and biological.

Perhaps to this list should be added the oldest method of drug preparation, *i.e.,* the grinding and formulating of medicinal herbs and other substances. This is a miniscule part of the industry, if not totally absent. Furthermore, there is nothing to be found in the literature on the antipollution efforts of medicine men, witch doctors, or fakirs.

Extractive

The use of the extractive processes was the first improvement over the use of the total drug-containing substance to treat a patient. These processes may be used to extract the active medicinal out of the leaf, bark, root, tissue, or wherever it might be found, then to concentrate and purify the active substance and, somewhere along the line, convert it into a form which is easily assimilated by the body. The source of the material could be plant or animal. This method is used to isolate several important drug substances: insulin, morphine, codeine, vinca alkaloids, belladonna and digitalis, to name a few.

Organic Chemical Synthesis

A high proportion of the drugs offered for sale today are the products of chemical synthesis. The reaction processes used to make them include most of the reactions found in the synthetic organic chemical industry. Also employed are many formidable, exotic reactions whose magnitude of cost and difficulty render them impractical for making industrial chemicals.

From a comparative pollution standpoint, the waste streams from this type of pharmaceutical manufacturing are more numerous and varied than from an operation of similar size in the synthetic organic chemical industry. The ratio of product produced to raw materials used is much smaller than for a like ratio in the manufacture of synthetic organic chemicals. The pollution problems created in this area of pharmaceutical manufacturing are quite formidable. A few of the classes of medicines made by this method are barbiturates, anesthetics, analgesics, tranquilizers, diuretics, vitamins, and bactericides.

Antibiotics Manufacturing (Fermentation and Purification)

Activities in this sphere of manufacturing activity embrace fermentation and product isolation followed by purification. Fermentation is an important production process within the pharmaceutical industry. Because of the high initial capital costs for equipment and the waste problems created by this manufacturing procedure, only a few companies are active in the field. Penicillin, streptomycin, erythromycin, and terramycin are some of the products made by fermentation. This mode of manufacture produces a very high waste-to-product ratio. The size of the waste load from this type of operation is quite formidable but is, for the most part, biodegradable.

Chemically Modified Antibiotics

Many of the newest life-saving drugs are made by this route. It consists of taking an antibiotic produced by fermentation, stripping it of its side chains by chemical cleavage, and reshaping the molecule through chemical reaction. The active substances dealt with are fragile molecules that are heat and pH labile. The wastes generated are more difficult to deal with than those of fermentation and organic chemical synthesis combined. The products include the so-called syncillins and the cephalosporins.

Biological

This method of pharmaceutical production makes use of reactions that take place within living animals, within organs removed from living animals, within embryonated chicken and duck eggs, and also in synthetic media.

The products may be vaccines to immunize man or other animals against such things as smallpox, measles, polio, and distemper. Other products are toxoids, used to protect man from tetanus and diptheria, or bacterial vaccines to combat typhoid and cholera.

ODOR PRODUCTION

The odors that are by-products of the processes used to produce pharmaceuticals are as varied and numerous as the processes themselves. The odors to be dealt with range all the way from odors similar to those found in nature to odors that are by-products from the harshest chemical reactions. Every operation has some characteristic odor associated with it. Fortunately, we need only be concerned with those that are offensive, highly concentrated, or toxic. Some odors are offensive in any concentration; others are pleasant when dilute but, when present in moderate to high concentrations, can be quite offensive. If an odor denotes the presence of a toxic material, the odor control strategy must be one that will completely eliminate the toxicant.

The following material describes the types of odors likely to emanate from these five different process methods of preparing pharmaceuticals.

Extractive

The odors encountered as by-products from animal extractive processes closely resemble those secured from food processing and rendering plants. Odors are produced as a result of the breakdown of carbohydrates and proteins in the absence of sufficient oxygen. When carbohydrates undergo anaerobic breakdown, a sour odor is produced. Proteins contain both nitrogen and sulfur and, under anaerobic conditions, form mercaptans, fatty acids, amines, and many other odorous substances. Fats can produce foul odors under both aerobic and anaerobic conditions through hydrolysis and oxidation reactions, which result in the formation of acids and aldehydes.

The fresh and spent sources of these natural products are handled in such a way as to minimize degradation and the production of foul odors. The same conditions that would promote destruction of fat and protein would also permit destruction of the desired active ingredient. While awaiting process, the raw materials are kept frozen. In process, the conditions are carefully controlled to minimize degradation. Frequently, agents are added during process to retard or prevent destructive bioprocesses. The agents protect both the product and the host material. After process, the waste tissue is treated in such a manner as to render all the fats and to recover the protein. These by-products thus become marketable commodities.

In the extraction processes which result in the isolation of an active substance from a plant source, the chief source of odor would be the vegetable substance itself or the solvent used in the extraction process. The source material is kept cool and dry until it is used. The process conditions are mild enough to protect the active ingredient as well as the source. The spent vegetable matter is disposed of by incineration, burial, etc., before it has a chance to decompose and cause further odor problems.

Another important consideration in the processing of animal or vegetable tissue is housekeeping. If this gets lax, many problems will arise. The odors that result from these primary problems are of relatively small concern compared to the problems themselves.

Synthetic Organic Medicinals

A wide range of odor types results from the manufacture of organic medicinals. Many of these may be described as very concentrated, very offensive, and easily identified as to source. If not controlled, these types of odors attract immediate public attention. Fortunately, the mode of manufacture employed in the synthesis of most organic medicinals–*i.e.,* batch manufacture in closed vessels–permits the containment of most odors produced during the reaction part of the process. The odors that escape from reaction vessels are minimal and are directed to the atmosphere through vent pipes. The vents may also be routed to scrubbers, fume incinerators, adsorbers, or other odor control devices.

The odors that escape from transfer, filtration, and drying operations are difficult to contain. They are usually removed from the manufacturing site by spot ventilation or area exhaust. The odor now is oftentimes quite dilute, compared with the odor that emanated from the reaction vessel vent, and in low concentration is much more difficult to deal with. While the same method may be used to destroy the odor, whether it is concentrated or dilute, economics will usually dictate that the dilute odor be treated in a manner different from that used to treat the concentrated odor.

The odors from this method of pharmaceutical manufacturing can be described in chemical terms as aldehydic, aromatic, phenolic, ammoniacal, acidic, etc. They may also be described by using such abstract terms as ethereal, balsamic, ambrosial, repulsive, nauseating, etc. Or, following still another system, they may be described with words that convey cogent meanings even to the layman: flowery, fruity, resinous, etc. Odors to fit any of these descriptions may be found within this manufacturing method.

Fermentation and Purification

Most of the odors encountered in the manufacture of an antibiotic occur during the fermentation process. The odors secured from purification are mostly solvent in nature with an underlayment of fermentation odors and are fairly easy to deal with. The odors from fermentation are very similar to many naturally occurring odors. The overall characteristic odor can be described as moldy or musty. (Within the trade, the mass of microbial cells that produce the antibiotic are sometimes referred to as "the mold.") This moldy odor is present in all antibiotic fermentations, but frequently acts as only a background for other odors. The odors from fermentation range all the way from innocuous to nauseous. The main difference between many naturally occurring odors and those by-product fermentation odors is concentration. As these odors leave a fermenter, they are much more concentrated than those found in nature. These odors are detectable in minute concentrations and have the ability to travel great distances.

A fermentation process is carried out in the presence of large amounts of excess air. The reaction media are continually air stripped so that any by-product odors, pleasant or otherwise, are constantly wafted toward the nostrils of a highly critical and odor-conscious public. Any method to control or eliminate unpleasant odors from this source must also be able to cope with tremendous quantities of hot moist air.

Antibiotic Modifications

As stated before, this is the newest variation for the preparation of pharmaceuticals. In this method, more than in any of the other four methods, the designated process must be strictly followed in order to secure any product at all. Any process alteration whatever could destroy the yield entirely. In this mode of manufacture, the odors and odor problems closely resemble those secured from synthetic organic medicinal manufacture. They can consist of solvent odors, acid-type odors, amine odors, and odors due to the nature of some of the reagents used. There can also be carryover odors from the fermentation processes.

Biological Manufacture

The raw materials are frequently contained within the bodies of living animals. Animals ranging from monkeys to horses have been used to generate biological substances. If a living animal serves as the reaction vessel and a large number of them are needed, the odor problems that arise can be similar to those in a zoo or barnyard. These are best combatted through prompt removal of waste and good housekeeping.

In the procedures involving isolated organs, eggs, or a synthetic media, the processes all take place within closed systems. Proper reaction conditions and good housekeeping are mandatory in order to produce the desired product without infecting the operating personnel. The same operating principles that promote good yields and protect personnel also prevent the production of bad odors.

ADDITIONAL ODOR SOURCES

Animal Testing

Most large pharmaceutical firms make large research efforts. An important part of the research activity is the animal testing function, where new drug substances are tested for activity and toxicity before they are ever tried in man. In addition, many drugs require a battery of animal tests on every routine batch or lot of material produced, before it is released for sale.

A large variety of animal species are used. The odors from such a function must be carefully controlled on a continuing basis or they can become overwhelming. A vital part of any such function is an incinerator or crematorium in which the bodies or organs of dead animals may be consumed in an odorless fashion at high temperatures.

Waste Treatment

Large pharmaceutical firms also have large waste disposal problems. Many of these firms treat their biodegradable wastes in a microbiological waste treatment plant. These plants are susceptible to upset, which means malodor production. Even under normal operating conditions, these plants can give off odors that are objectionable. Many of the substances that caused odor problems in production have now moved to the waste treatment plant. There is no reason to believe that they will not also cause odor problems in their new habitat–and they do.

If a company has a microbiological waste treatment plant, that company should also be prepared to deal with the types of odors unit processes in such a plant can give off. Some of these are odors from activated sludge units, trickling filters, anaerobic digesters, sludge lagoons, sludge driers, and sludge incinerators.

The odors from an out-of-balance or overloaded waste treatment plant are without a doubt among the worst odors to be found anywhere. If they emanate from a supposedly aerobic process, the same problem exists here as with fermentation odors; that is, they are contained in large volumes of warm moist air.

TREATMENT OF ODORS

Due to the wide range of odor types, sources, and concentrations experienced within the industry, the methods used to combat them must also be numerous and varied.

In treating an odor, the degree of difficulty frequently varies inversely with the concentration. The more concentrated an odor, the easier it is to treat.

Generally speaking, regardless of the degree of repellency, the odors that are the most difficult to treat are those that have area or multiple sources. Unit operations carried out in the open would fall into this category, as well as such sources as sludge lagoons and sludge drying beds. Odors that emanate from processes such as fermentation, which employ large volumes of air, as well as certain processes within a waste treatment plant that require an abundance of air, are extremely difficult and expensive to treat.

The methods of odor control fall into the following categories:

1. Process modification
2. Equipment modification
3. Dispersion
4. Masking
5. Chemical reaction
6. Thermal oxidation (flame and catalytic)
7. Adsorption

Process Modification

This is a very desirable way to eliminate or control an odor; however, this method is not widely practiced in the pharmaceutical industry. Frequently the process in question is used to produce a life-saving drug or a drug that will prolong life or alleviate suffering. The important thing is to get the right product in sufficient quantity to supply the market. Most times, in order to accomplish this, the process must remain inviolate and the pollution problems are dealt with "at the end of the pipe."

Equipment Modification

Equipment modification is most often used in conjunction with some other method of control. The equipment used to run an odorous process is completely enclosed. All piping is of welded construction; the reactors are all closed, and are opened only after careful decontamination. Materials are transferred by sealed pump whenever possible. All reactor vents, spot ventilation, and area ventilation are ducted together to one common piece of control equipment.

In controlling fermentation odors the process air is frequently exhausted from the fermenter into a duct system. The system can be used to pick up the air from all the fermenters in the plant and from other odorous processes that use large quantities of air. The duct serves as a means of gathering the foul air and directing it to a point of treatment.

Dispersion

Dispersion can, under certain circumstances, do an adequate job where the idea is just to get by. This technique makes maximum use of local meteorological factors—wind speed, topography, elevation, temperature, volume, and velocity of the gases containing the odorous substance. The use of tall stacks has been found effective in reducing the ground-level concentrations of air contaminants.

Many odors produced in the pharmaceutical chemical industry, while objectionable, are harmless and can only be detected in high concentrations. Odors of this type need only be exhausted from the roof of the manufacturing building to achieve the desired dispersion. However, in an area subject to many inversions it is a method which can deliver the odor, completely undispersed, back to ground level, with embarrassing results.

This is a method that has been used widely in the past to solve all sorts of gaseous particulate pollution problems. It is the embodiment of the theory "The solution to pollution is dilution."

Masking

This method can most succinctly be described as one that makes the olfactory sense less sensitive to an odor, or an odor more acceptable to the olfactory sense. Odors may be masked or neutralized. In the first case, the disagreeable odor is overpowered by a stronger and more agreeable odor. In this instance, even though the masking odor may be more "agreeable" to us, the receiving public may contain individuals who do not agree with us, and we may have created a greater problem. Even as beauty is in the eye of the beholder, so is stench in the nose of the smeller.

The masking agents are usually organic aromatic compounds, natural or synthetic, such as heliotropin, vanillin, eugenols, benzyl acetate, and phenylethyl alcohol.

The methods used to apply a masking agent are many and varied. The mask may be fogged into the base of a stack. The agent may sometimes be introduced into the raw material before processing. If the odor emanates from an area source such as a lagoon, the agent may be pumped directly in, sprayed on the surface, or introduced into the surrounding atmosphere by fogging nozzles. The agent may also be introduced into a scrubbing solution through which the air stream containing the odor is passed. Other pollutants are removed but the odor rides right through. The air stream strips the masking agent from the scrubbing solution and it escapes to the atmosphere in combination with the odor, where it performs its masking function.

Neutralization or counteraction is a better method, and much work has been done in this field. Neutralizing agents are offered which will counteract the following odor types:

1. Phenolic
2. Amine
3. Mercaptan
4. Organic fatty acids
5. Aldehydes
6. Solvent odors
7. Aromatics

The methods used to apply the neutralizing agent are the same as those used to apply the masking agents.

These techniques of odor control are frequently employed because capital costs for equipment are low; however, the cost of the masking agent can become a significant factor. The technique, therefore, falls short of being a long-term effective solution to an odor problem.

Chemical Reaction

This method is most widely used in containing odors from organic chemical synthesis. Many of the raw materials are strongly basic or acidic substances. The reactions carried out can produce by-product odorous substances with strong acid or base characteristics. Some of these materials can be highly toxic. Therefore it is doubly important that they be controlled.

The control methods are varied and depend on the nature of the odor substance. In some instances an "acceptor" or neutralizing agent may be added directly to the reactor. Most cases, however, require post-reactor treatment.

Odor streams that escape from vents or other point sources, as well as the exhaust from a ventilating system, may be scrubbed with a stream containing an absorbing or neutralizing agent. This may be done in a packed or plate tower, in a water or steam ejector, or by any other method that will bring the scrubbing liquid into intimate contact with odor-containing air. If the acidic or basic odor substances are water soluble, the simplest scrubbing liquid to use is water. If water consumption is a problem, these scrubbing systems can be of the recirculating type and the pH of the solution can be controlled by the addition of selected acids or bases. The capital and operating costs for such systems can vary widely, depending upon the particular application. Generally speaking, the capital costs are low to moderate and the operating costs are low.

For odor substances that do not have basic or acidic properties, water solutions of strong oxidizing agents may be used in place of water alone. This is an old method of odor control that is becoming more attractive. Because of the cost of the oxidizing agent, it is mandatory that the scrubbing liquid be recirculated to secure maximum use of the agent. Some common agents are potassium permanganate, sodium hypochlorite, and gaseous chlorine. There are also proprietary solutions on the market that may be regenerated by electrolysis.

These systems have to be tailored to fit a particular situation. The agent and the odor must be mated. If the right combination can be worked out, much can be said for the method. If not, something else must be tried. Wet oxidation systems are characterized by moderate capital costs of purchase and installation, and are followed by moderate to high operating costs.

Another oxidizing agent that is used widely in odor control is ozone. This method requires the use of an ozone generator, a reaction chamber, a system of ducts to gather and convey the odor-bearing air to the chamber, and, finally, a means for dispersing the treated air to the atmosphere.

This method does not work on all odor substances. In many instances the ozonides formed on the ozone itself may have an odor that is more disagreeable than the one that is under attack. The concentration of ozone entering and leaving the system must be carefully controlled because of the toxicity of ozone. Capital costs for the installation of such a system are high while the operating costs are low to moderate.

Thermal Oxidation

Flame Incineration

Flame incineration may be used when an odor substance is combustible or can be modified by high heat. It is characterized by very high removal efficiencies over a wide range of odor substances. In most instances the concentration of combustibles in the gas stream is well below the lower combustible limit, so an auxiliary fuel must be used. Temperatures as low as 800°F may be used to destroy a malodor, but frequently a temperature range of 1,400 to 1,500°F is required for complete destruction.

Purchase and installation of fume incinerators require large capital outlays. The operating costs are high. On large installations, a heat exchanger may be used to recover some of this waste heat. The recovery methods center upon the principle of using the hot odor-free exhaust air to preheat the entering odor-bearing air.

Coal and oil-fired boilers double in some pharmaceutical houses as fume incinerators and steam generators. This method greatly reduces the capital expenditures, and the operating expenses are minimal. This method is most commonly used to incinerate fermentation or waste treatment odors.

Further economies can be achieved by removing all entrained moisture from the incoming air. One drawback to this method is that the odor-bearing air or the products of combustion therefrom contain compounds that are corrosive to the steam generating system.

Catalytic Combustion

This is a variation of flame incineration, the basic difference being that a metallic catalyst is used to promote combustion at lower temperatures. The process requires that the combustible pollutant be present in concentrations below the lower explosive level. Since the catalyst may be poisoned fairly easily and, at best, progressively decreases in activity with use, it is important that the relationships between temperature, time, and activity be thoroughly understood in order to maximize the efficiency of the unit.

As with flame incinerators, auxiliary fuel must be used. Heat exchangers are employed to recover heat values—as with flame incinerators. The capital costs for purchase and installation of these units are high while the operating costs are moderate.

Adsorption

Adsorption is another method for removing odorous substances from an air stream. This process depends upon the attractive forces between a solid surface and a gaseous molecule. These forces can be either chemical or physical in nature. If they are chemical, the process is probably not reversible; if physical, the adsorbent may be used over and over again after going through a regenerative cycle.

Although activated carbon is the most common adsorbent, many other materials, such as activated alumina, bauxite, molecular sieves, silica gel, clay and ion exchange resins can be used.

If the odor is due to a solvent or some other compound with a monetary value, and is not contaminated with other substances, it may be recovered through the regenerative process and be reused. If this is the case, the system may pay for itself in a short period of time.

If the odor is in low concentration, is readily adsorbable on carbon, and can be removed therefrom easily, this still may be the method of choice, regardless of the economic value of the odor substance. This method is most commonly used in dealing with odors that are contained in large volumes of fairly dry air. The adsorption equipment is often used in conjunction with spot ventilating or area exhaust systems. While the initial capital costs for equipment and adsorbent can be high, the cost for operating such a system is usually moderate or low.

The selection of the proper piece of equipment or method for solving an odor problem may depend upon many things. First, the method chosen has to be capable of doing the desired job, *i.e.,* destroying that specific odor or reducing it to an acceptable level. Then one has to determine what an acceptable level of destruction is: 100%–or reduction to a level that will produce an acceptable, minimal number of odor complaints. The method chosen must be in harmony with the surroundings. If dispersion is the method of choice and the plant is near an airport, there will be restrictions as to how high the stack can be.

When considering flame or catalytic oxidation, the units must be located in areas that are free of combustible vapors and fuel must be available.

To help conserve energy, and in those cases where we have a choice of methods, the method requiring the least amount of energy should be chosen.

There is a danger of substituting one objectionable odor for another. As long as an odor is an odor, it will be objectionable to some segment of the populace. If toxic substances such as ozone or chlorine are used to destroy an odor, the system must be foolproof. It does not make sense to replace a substitute that may only be objectionable with one that is toxic.

The economics of comparable methods should be studied. The cost of odor removal has to be an important consideration, if not the overriding factor, in the final choice of method.

CHAPTER 12

ODOR EMISSION SOURCES IN THE CHEMICAL AND PETROLEUM INDUSTRIES

Yousuf H. Habib

Foster Wheeler Energy Corp.
Livingston, N.J. 07039

Chemical and petroleum industries are major sources of odor, along with sanitary processes, animal feedlots, and rendering plants. The types of odorants are many and varied. The olfactory thresholds of these odorants are expressed in parts per billion and are capable of causing a nuisance at very dilute levels. Table 12-1 lists some of the odorants that can originate from chemical and petroleum industries, together with their characteristic odors.

Table 12-2 ranks some of these odorants in order of their odor intensity.

POINT AND MULTIPLE SOURCES

One can start to appreciate the extent of the problem when dealing with a multitude of these compounds and their extremely low olfactory perception level. More often than not, sources of these odors will be complex—a series of compounds contributing simultaneously. Another characteristic of the chemical industry is its trend to concentrate where energy and/or raw materials are available. Such an industry grouping is a significant factor in terms of the resulting odor pollution. The simultaneous production of varied products and compounds in a concentrated area results in complex emissions that are almost impossible to trace back to their source.

Leonardos *et al.* studied the potential of creating odors downstream from a point source. They found that a leakage of 0.1 lb/hr of butyric acid would result in a concentration 1.5 times the threshold limit of 0.001

Table 12-1. Major Odorants and Their Characteristic Odors: Chemical/Petroleum Plants

Acetaldehyde	CH_3CHO	Pungent
Acrolein	CH_2CHCHO	Burning fat
Allyl disulfide	$(CH_2CHCH_2S)_2$	Garlic
Allyl mercaptan	CH_2CHCH_2SH	Garlic
Ammonia	NH_3	Pungent
Benzyl chloride	$C_6H_5CH_2Cl$	Aromatic
Benzyl mercaptan	$C_6H_5CH_2SH$	Unpleasant
Butyric acid	$CH_3CH_2CH_2COOH$	Rancid
Camphor		Aromatic
Carbon disulfide	CS_2	Rotten
Chlorine	Cl_2	Pungent
Diethylsulfide	$C_2H_5C_2H_5S$	Garlic
Dimethylamine	CH_3CH_3NH	Fishy
Dimethylsulfide	CH_3CH_3S	Decayed cabbage
Ethyl mercaptan	C_2H_5SH	Decayed cabbage
Ethylseleno mercaptan	C_2H_5SeH	Foul
Formaldehyde	$HCHO$	Pungent
Hydrogen sulfide	H_2S	Rotten eggs
Methyl mercaptan	CH_3SH	Decayed cabbage
Phenol	C_6H_5OH	Emphyreumatic
Propyl mercaptan	C_3H_7SH	Unpleasant
Pyridine		Emphyreumatic
Sulfur dioxide	SO_2	Pungent
Trichloroethylene	$CHClCCl_2$	Aromatic
Trimethylamine	$CH_3CH_3CH_3N$	Fishy
Valeric acid	$CH_3CH_2CH_2CH_2COOH$	Body odor

ppm one mile away with the wind blowing at eight miles per hour.[1] If there are multiple sources, as is usually the case in chemical and petroleum industries, the combination of more than one odorant results in a much more pronounced odor level than a single point source.

CHEMICAL STRUCTURE AND ODOR INTENSITY

To get a better insight into the source of odorant emissions from chemical plants and refineries, knowledge of several general correlations of odor perception to the chemical structure is helpful in predicting the effects of chemical changes on the intensity of odors. Moncrieff has made

Table 12-2. Odor Rankings

Odorant	Ranking
Alcohols	1
Ammonia	1
Ethers	1
Halogens	1
Hydrocarbons	1
Esters	2
Ketones	2
Phenols	2
Sulfides	2
Sulfur dioxide	2
Aldehydes	3
Unsaturated ketones, ethers, alcohols	3
Amines	4
Hydrogen sulfide	4
Mercaptans	4
Selenium compounds	4

an extensive survey of these correlations.[2] Some of the more important generalizations are summarized below.

1. Amines will smell pungent (ammoniacal smell) in concentrated form but will smell fishy (characteristic amine smell) in dilute solution.
2. Sweet-smelling alkyl halides give out stronger smells as one moves from chlorine to iodine. Stronger odor also results if more than one halogen atom is present.
3. Side chains play a significant role. The closer the side chain to a polar group, the stronger the odor. Similarly, unsaturation will also increase odor strength–more so if it exists closer to a polar group. Branching in a paraffin series may also increase odor intensity.
4. The olfactory influence of the ring will predominate over the individual olfactory functions of sulfur, nitrogen, and oxygen. If the heterocyclic ring is reduced, sulfur, nitrogen, and oxygen will regain their olfactory effects.
5. For the homologous series, the ascending order will increase the odor effect to a maximum, which will then fall off as the increasing chain length reduces the volatility of the compounds. Polymerization will also reduce (and in some cases destroy) odor.

ODOR SOURCES

The chemical industry can be classified as inorganic and organic. Individual industries vary from fertilizer plants, tanneries, dye and rayon manufacture, plastics and resins manufacture, chemical processing plants, soap and detergents production, oils and fats manufacture, and specialty chemicals production.

The petroleum industry includes refineries and petrochemical plants. Refineries contribute hydrocarbons, oxides of sulfur, aldehydes, organic acids, and ammonia from a varied array of sources. Catalyst regenerators, air and steam blowing units, tank vents, barometric condensers, flare systems, process heaters, blowdown systems, pumps and valves, and storage vessels are the most important potential sources of odorants from the petroleum industry.

Table 12-3 lists some of the major sources of odors in the chemical and petroleum industries and their major odorants.

Table 12-3. Major Sources of Odorants in Chemical and Petroleum Industries

Refineries	Blowdown and blowdown recovery systems	SO_2, H_2S, NH_3, HC, organic acids, aldehydes, mercaptans
	Catalytic crackers	SO_2, NH_3NH, aldehydes
	Fluid catalytic units	SO_2, NH_3, aldehydes, HC
	Process heaters	SO_2, NH_3, H_2S, aldehydes, HC
	Storage vessels	Hydrocarbons
Inorganic Chemical Industry	Fertilizers, phosphate production	NH_3, aldehydes, SO_2
	Phosphoric acid	SO_2, H_2S, aldehydes and other odorants
	Soda ash and sodium hydroxide	NH_3
	Sulfuric acid, nitric acid, lime	NH_3, aldehydes, SO_2
Organic Chemical Industry	Organic chemicals	Mercaptans, NH_3, HC, SO_2, aldehydes, organic acids, and other odorants
	Paint industry	NH_3, HC, SO_2, aldehydes, organic acids, and other odorants
	Plastics	NH_3, HC, SO_2, aldehydes, organic acids, and other odorants
	Rubber	NH_3, HC, SO_2, aldehydes, organic acids
	Soap, detergents	NH_3, HC, SO_2, aldehydes, organic acids, and other odorants
	Textiles	NH_3, HC, SO_2, aldehydes, organic acids, and other odorants

Paints and Varnishes

A wide variety of natural and synthetic materials are employed in this industry. An equally wide range of odors may occur, according to the types of raw materials and solvents employed. Commonly used natural oils are linseed, soybean, tall oil, castor oil, tung and fish oils. Varying consistencies are attained through heat application accompanied by oxidation, polymerization, or esterification. Additives range from glycerine, pentaerythritol, and phthalic anhydride to various kinds of resins.

Varnish cooking is carried out either in open kettles or, more preferably, in completely enclosed stationary kettles to modify natural or synthetic oils or resins. Heat processing is used to modify these oils and resins to film processing vehicles. The final product will be a fast-drying chemical which will form a stable film. Emissions from varnish cookers have a potential for noxious odor. Acrolein is one of the major odorants emitted during the cooking of oils. Most of the varnish cooking odor emissions start being given off around 350°F. At this temperature, the majority of the solids/liquids used begin to vaporize. As with most other operations, the emission rate and the extent of malodors will vary with the temperature levels of heating, as well as with the type of additives used, the amount of blowing air utilized, the types of starting chemicals employed, and the cooking period required.

Some of the typical odorants from the lacquering and enameling industry include methyl methacrylate (sickening sweet), acrolein (acrid), allyl sulfide (garlic), hydrogen sulfide (rotten eggs), N-butyl mercaptan (skunk), and aldehydes (pungent). Tall oil esterification with pentaerythritol and glycerine results in emissions of butyl mercaptan, hydrogen sulfide, allyl sulfide, and thiophene. Sulfur in tall oil is generally responsible for hydrogen sulfide emissions during the heating phase of esterification, increasing with increasing heating temperature. It is at higher temperatures also that N-butyl mercaptan is formed.

Unsaturated aldehydes may be produced during the manufacture of oleoresinous varnishes as well as alkyd resins. Oleoresinous varnish cooking will also produce phenols, terpenes, and acroleins. Unsaturated aldehydes also are emitted as a result of oil breakdown and additive decomposition. Acrylic aldehydes are formed as decomposition products during the heat bodying of natural oils, along with fatty acids, acrolein, and glycerine. In gum running, terpenes and its oils and fatty acids are formed.

Solvents that are used and lost during grinding and thinning processes are a cause of highly odorous emissions. Water emulsion and water-soluble coating techniques use raw materials that are responsible for more odor problems. For example, the overpowering sweet odor of methyl methacrylate is due to this monomer, which is used as a raw material in the manufacture of latex coatings.

Rubber Industry

In rubber manufacturing, compounding and vulcanizing give rise to various highly odorous emissions which are usually incinerated to get rid of the odors. Rubber manufacture involves the use of adhesives. Whenever protein adhesives are used, their degreasing, hydrolyzing, and drying will entail a potential for malodorants. Incineration is the best way to destroy these odors, although other forms of oxidation, as well as hypochlorite scrubbing, have been utilized.

Detergent Industry

Odor-causing processes from the manufacture of detergents include oleum sulfonation and liquid sulfur trioxide sulfonation of organic bases. During the liquid SO_3 sulfonation, a completely closed system will give out only very silute odors. Liquid SO_3 is combined with the alkylate to react according to $RH + SO_3 \rightarrow RSO_2OH$. If the alkylate is not dilute enough, odor-causing long-chain olefins will be formed on the addition of liquid SO_3. Mild odors will occur in a closed system only when the final sulfonic product is discharged to its storage place.

In oleum sulfonation, oleum of 20 to 30% strength is used to attach the sulfonic acid group to the hydrocarbon group carbon, according to $RH + H_2SO_4 \rightarrow RSO_2OH + H_2O$. The alkylate and oleum are mixed, reacted, cooled, and diluted. Spent acid is separated and sulfonic acid is either pumped to neutralization and storage or is vacuum deodorized and then neutralized.

Long alkyl chains, if present in the alkylate, may cause dealkylation, resulting in long-chain olefins as by-products. These long-chain olefins are malodorous and a potential nuisance. Vacuum deodorizing is used occasionally to scrub out the odorous compounds.

One of the sulfoalkylation processes also has potential for odorous emissions. Sulfoalkylation entails complex sulfonation of a simple sulfonate. It involves catalytic reaction of fatty acids with isothionate as the sulfating agent. Isothionate itself is a sodium derivative of hydroxyethane sulfonic acid. Beta-sulfoesters are formed from this type of reaction. The vacuum system used in this process employs barometric condensers which discharge to a hot well. This hot well forms one of the sources of odorous emissions, particularly fatty acids. Another potential odor source is the gaseous emission from the reactor recirculation system. Noncondensibles in this emission cause odors. The final product holding tank is also a source of malodorants when water vapor is flahsed off from the tank.

Synthetic Textiles Industry

Viscose rayon production entails odorous emissions. In viscose production, treatment of the cellulose xanthogenate, dissolved by an acid in coagulating baths of soda lye, solidifies the thread. Carbon disulfide and hydrogen sulfide are given off and are the causes of odors.

Spinning operations of synthetic polymers give rise to malodorants. Glass fiber spinning, proofing, and crease-resistance operations all involve odiferous emissions of a phenolic nature, as well as those associated with solvents such as formaldehyde. Conversion of natural fibers and synthetic organic polymers to textiles in general involves emissions of odorous solvents like acetone, methanol, carbon disulfide, methylene chloride, etc. These solvents are concentrated by adsorption and, when recovery is not required, are destroyed by combustion.

Sulfur Plants

In order to reduce sulfur compound emissions from oil refineries, off gases are treated to reduce these sulfur compounds to elemental sulfur. The origin of these sulfur-bearing compounds can be traced to the sulfur content of the crude oil. During refining this sulfur is mostly converted to hydrogen sulfide, forming an integral part of the waste gases. Since these waste gases have a heating value content, it may be desirable to burn them. However, incineration will result in an SO_2 emission–another undesirable pollutant. These gases are therefore–prior to being used as a fuel–treated to remove most of the hydrogen sulfide.

Amines are generally used to treat the H_2S of the waste gas. The resultant waste gas can either be used as refinery fuel or burned as a flare. Historically, Claus-type sulfur reduction units have been employed to oxidize part of the H_2S to SO_2 and then react the SO_2 with the remaining H_2S to form sulfur: $H_2S + \frac{1}{2}SO_2 \rightarrow S + H_2O$. Small fractions of highly undesirable H_2S in the acid gas will pass unreacted and create a potential for toxic and odorous emissions. A Claus-type plant will therefore be followed by an incinerator to oxidize the H_2S, which in turn will give rise to SO_2. Additional treatment is then necessary to deal with the SO_2.

Resins Industry

The majority of odor sources in plastics-resins plants emanate from solvents emissions. Storage and thinning of resins, as well as the main resin reaction, all result in solvent emissions. In polyurethane resins manufacture, noncondensibles may prove a significant cause of odor problems. Excess tolylenediisocyanate may emit from finished polyurethane resin and cause an odor nuisance. Storage of phenolic and amino resins can result in characteristic aldehyde odor. Any leakage may also cause these odors.

In the manufacture of oil-modified alkyds and polyester resins, discharges from barometric condensers and reaction kettles are a potential source of solvent emissions, as well as oil cooking odors. For example, during its treatment with glycerine and phthalic anhydride, a drying oil is heated. This results in emission of acrolein and other odorous compounds. The intensity of these odors increases with the increase in heating temperature.

Leakages from polyvinylchloride and polystyrene facilities respectively give out characteristic vinyl chloride and styrene odors in addition to solvent odors. A polyvinyl acetate facility will give out characteristic vinyl acetate odor from the final product storage vessels as well as from barometric condensers.

Other Chemical Industries

A miscellaneous array of odor sources exists in many other chemical industries. For example, glass manufacturing may have a potential for hydrogen sulfide, as does stove clay production. Manufacture of barium chloride from barium sulfite—and of sodium sulfides as well as sulfur dyes—can result in malodorants emitted from unit processes and operations.

Production of phosphoric acid is another potential source of malodors. Phosphorus is burned to pentoxide, which is then reacted with water to yield phosphoric acid. The raw acid contains heavy metals, which are generally precipitated as sulfides by adding a slight excess of sodium sulfide, hydrogen sulfide, or sodium hyposulfide. Precipitation is necessary to assure high-grade phosphoric acid. Excess hydrogen sulfide may be removed by air blowing. When arsenic and antimony-type heavy metals are precipitated in this fashion, hydrogen sulfide is released. Usually the emitted hydrogen sulfide is sent to a phosphorus burning tower where it is burned off to sulfur dioxide.

Production of neoprene, ammonia, aldehydes, and other organic chemicals also involve emissions of unpleasant odors. The fertilizer industry also yields malodors from several sources. For example, the manufacture of urea, formaldehyde, phosphate, and ammonia can give out unpleasant odorous emissions.

Petroleum Industry

Petroleum refineries involve a large variety of processing systems. Petrochemical plants manufacture a wide variety of chemicals with petroleum as the base derivative. Despite the fact that the unit processes are geared to closed-loop operations, odors are produced and form the basis for a nuisance complaint.

Typical processing systems that create odors may be catalytic cracking and reforming units, lube oil treatment, asphalt production, and sulfur recovery plants. Typical causes may be gas leakages and crude oil and distillate vapors. The types of odors vary from sulfurous, ammoniacal, amino, and aldehydic to those of mercaptans.

The major portion of odor emissions originates from storage vessels and some process equipment. As an illustration, aldehydes usually originate from catalyst regenerators and compressor engines, ammonia from catalyst regenerators, mercaptans from gas deodorizing, and oxides of sulfur from incinerators, flares, boilers, and heaters. Hydrogen sulfide may be the largest odorous element in refinery operations. It can originate from heaters, sumps, storage vessels, pumping and loading operations, compressors, absorbers, and dehydrators. Other miscellaneous odors can come from various sources like air or stream blowing, barometric condensers, and process tanks.

Catalytic Regenerators

Oil refineries make extensive use of a wide range of catalysts. By the time these catalysts are spent and need to be regenerated, they have become carbon contaminated. This carbon is burned off. Flue gases from the regenerating process contain impurities which include hydrocarbons, carbon monoxide, ammonia, and organic acids, as well as oxides of sulfur and aldehydes. Usually these flue gases are passed through a mechanical collector to a CO boiler.

Pumps

A refinery uses pumps and compressors throughout its operations. Leakage is the major and perhaps the only source of odorous emissions from pumps. Application of pumps in a refinery is wide from the point of view of process application as well as materials of construction. Leakages that usually cause problems exist between the contact of the moving shaft and the stationary casing. Mechanical seals may help control some leakages. Oil-pressurized packing glands can also be effective. Mercaptans and hydrogen sulfide leakages nevertheless cause a nuisance problem.

Storage Vessels

Huge quantities of petroleum products (liquids as well as gases) have to be stored in petroleum production facilities. Various kinds of vessels are used. They can be pressurized vessels or open tanks, have floating or fixed roofs, or simply be reservoirs or pits.

Vessel vents are the usual source of malodors. Vapor recovery or flare systems may be applied to control these emissions. Storage vessels, however, emit more hydrocarbons and mists than malodorants. Loading operations of these vessels may create odorous emissions. The normal process of evaporation is another cause for these odors. The floating type of roof or a pressurized vessel, however, helps reduce odorous emissions.

During scheduled cleaning and maintenance of these vessels, purging is done to allow maintenance personnel to enter the tanks. Such purging can, however, cause a nuisance due to malodors. Malodorants vary in their concentrations, and thus in undesirability, with the type of material stored. Also, whenever a stored liquid needs agitation, the concentration of odorants and thus their nuisance potential will be greater.

Cooling towers are still another source of air contaminants from refineries. The heated water from various refinery processes must be cooled. This cooling water is usually contaminated to a small degree with odorous elements. Leakage from malfunctioning equipment or from a heat exchanger and contact with process streams usually result in contaminant carryover. When some of this water evaporates from the cooling towers, odorous elements are released.

Catalytic Cracking and Regeneration

During cracking, hydrogen sulfide is produced in the heavier fractions and mercaptans are produced in the lighter gasoline fractions. The extent of conversion depends on the amount of sulfur in the crude oil. Spent catalysts are then regenerated and give off noxious fumes, including sulfurous gases and ammonia. Oxidation is the usual control adopted for these odorous emissions.

Air Blowing

Sometimes air is used to agitate petroleum products. When it is vented, it carries with it malodorous compounds. This is particularly true in asphalt oxidation. Incineration is an effective way to control these odors.

More than 90% of the asphalt used in the U.S. is recovered from crude oil. When a high grade of asphalt is desired, plastic properties are imparted to it by blowing air onto the crude residue at high temperatures. This operation removes the gas-oil fraction and renders the asphalt paving-grade. Odorous materials are among the contaminants evolved during asphalt air blowing. Asphalt blowing emissions, therefore, require proper treatment to avoid a nuisance complaint.

Blowdown Systems

These systems are employed in a refinery to handle waste refinery gases, especially during emergencies, as when a compressor fails. Some refineries may also employ blowdown recovery systems if the blowdown material has to be recovered because of either economic or ecologic reasons. Flare systems are used in refineries to burn off blowdown materials.

A blowdown system typically consists of relief valves, safety valves, manual bypass valves, knock-out vessels, headers, and storage tanks. Whenever an emergency like an overpressurized vessel occurs, large amounts of odorous gases, mists, and hydrocarbons are vented through the blowdown system. It is therefore preferable to have a blowdown recovery system or a system of flares to burn off such odorous compounds as sulfides, mercaptans, and ammonia.

Effluent Treatment Systems

Large quantities of wastewater result from refinery operations and must be treated. These effluents are also a source for malodorous elements.

Mercaptans, ammonia, and hydrogen sulfide can produce objectionable odors through open tanks, channels, and leaks in sewer systems. To minimize these odorants, source treatment is the most effective method. Modification of the unit operation or process could eliminate the source of the problem. Sulfides, mercaptans, and ammonia can otherwise be removed from the waste stream by live steam stripping. The noncondensible vapors from the top can be sent to recovery units or to incineration.

Chlorine may also be used to precipitate sulfides. However, H_2S would be formed again if the effluent later contacted an acid stream. Hydrogen sulfide would also be re-formed if thiosulfates were bacterially reduced. Thus conversion of H_2S and mercaptans to thiosulfates through oxidation would prove ineffective if the possibility exists for a bacterial attack on the thiosulfates. Mercaptans and some other odorous materials can also be removed by caustic. However, the spent caustic may create an odor. Chemical treatments like neutralization and neutralization combined with oxidation have to be used to render the spent caustic odorless. Treatments convert compounds to thiosulfates, sulfates, and sulfites. In some cases, despite neutralization, effluent gases may have to be incinerated to destroy odorous compounds.

ODOR CONTROL METHODS

The most common control techniques are process change, catalytic/thermal combustion, diffusion, counteraction, absorption, and adsorption. In many applications, especially in refineries, a combination of two methods may be required for adequate control. For example, absorption of soluble odorous compounds followed by catalytic combustion of nonsoluble/noncondensible odorous compounds.

Sometimes a simple process modification or proper operation of a process may be sufficient to control odor emission. Use of mechanical seals in refinery compressors, proper operation of refinery heaters, and floating roofs for refinery storage vessels are examples of minor adjustments. Masking agents or odor counteractants are also used to attack specific problems. Dispersion is employed to bring concentration levels below detection. However, dispersion will largely depend on original concentrations, stack heights, and general wind conditions.

In textile industries the most common odor source is organic solvents. These may be absorbed in alkaline solutions or adsorbed on carbon. The concentrated odorant may then be combusted to destroy the odor.

In the petroleum industry, absorption and combustion are the control methods most often used. Flare systems are installed to take care of odorous emissions, but more importantly to handle emergency situations. Incineration is used most effectively for exhaust gases from vulcanizing rubber. The paint industry, similar to the textile industry, may emit vaporized solvents which can be controlled by absorption followed by adsorption or combustion.

Aldehydes are effectively treated by absorption in bisulfite solutions. Mercaptans have been absorbed effectively by sodium hydroxide. Carbon adsorption, as well as active MnO_2 on sawdust, have been found to be good methods for hydrogen sulfide removal, although absorption in amine solution has been used most often. Normally, H_2S absorption will have to be followed by incineration or oxidation. In case H_2S as well as NH_3 are to be removed, water scrubbing yields ammonium sulfide, which may be acid/base stripped followed by incineration of concentrated H_2S and NH_3.

When sulfides and amines from catalytic cracking operations are removed by combustion to oxides of sulfur and nitrogen, an incomplete combustion generally leads to aldehyde formation, which also can cause odors. Some products from petroleum operations are normally refined by washing followed by caustic neutralization. Flare combustion has to be used to oxidize mercaptans and hydrogen sulfide to avoid generation of obnoxious odors from the spent soda solution. Despite this, some gaseous components still are not oxidized completely, in which case nascent oxygen is added to assure complete combustion and prevent odor emission.

The chemical industry uses such a variety of processes and raw materials that each case should be dealt with on an individual basis. The specific type of control method must be determined by the nature of the chemical process as well as by the characteristics of the odorant emitted.

REFERENCES

1. Leonardos, G., D. Kendall, and N. Bernard. *J. Air Pollut. Control Assoc.*, **19 (2)**, 91-95 (1969).
2. Montcrieff, R. W. *The Chemical Senses*, 2nd ed. (London, 1951), quoted by R. D. Hoak, *Franklin Institute Symposium* (1958), p. 2.

CHAPTER 13

ENGINEERING ANALYSIS AND ODOR CONTROL

Robert M. Bethea

Institute for Environmental Technology
Texas Tech University
Lubbock, Texas 79409

The selection of suitable odor control methods depends largely on the odor-producing process, the odoriferous substances present, and the other substances in the waste gas stream. The general odor abatement methods may be classified as follows: process change, physical containment (which includes leak prevention and housekeeping), dispersion by both ventilation and diffusion (with the aid of tall stacks), chemical conversion (which includes flame, thermal, and catalytic combustion and chemical reaction or oxidation), odorant capture by either wet scrubbing or adsorption, and masking or counteraction.

PROCESS CHANGES

The best way to control odors is by preventing their release. This can be accomplished by either of two routes: (1) a change in the basic process which produces the odors or (2) better maintenance. Among the process changes which should be considered are changes in the temperature, pressure, and volume of the gas produced. Since evaporation and release of volatile components is directly dependent upon temperature, any process change which will reduce the temperature of the effluent will result in a decrease in the total odor concentration.

Another way to prevent leaks from gas handling equipment is to change from a positive pressure system to a negative pressure system. If this can be done, all leaks will be inboard. Although some additional amount of gas must be handled, this is preferable to the continued release of fugitive odorous emissions.

Other process changes which should be considered involve process chemistry and raw materials. By judicious or fortuitous selection of feedstocks, the potential for odor formation and release may be greatly reduced. On the physical side, changes in processing procedures may result in a decrease in odor emissions. These can be as simple as raw material segregation (as is customarily practiced in the rendering industry) or as complex as the change from evaporative concentration to freeze drying in the fruit juice industry to retain flavor and aroma in the product rather than lose these volatile components as odorous emissions.

COMBUSTION TECHNIQUES

Complete combustion has been generally accepted as the best way to deodorize foul-smelling gases. It may not, however, be the most economical method. The economic feasibility of odor control by air oxidation is being degraded by the emergence of a national energy crisis. Continued use of natural gas and other low-sulfur fuels for combustion processes is expected to decrease as the supplies diminish (and thus become more expensive). A significant problem associated with controlling odors by any of the various combustion techniques is that partial oxidation may actually increase the odor problem by the formation of more odoriferous components. An example is the formation of butyraldehyde or butyric acid from the incomplete oxidation of butanol.

Thermal Oxidation

There are three combustion techniques. These are (1) thermal oxidation (also called afterburning or direct-flame incineration), (2) direct combustion (or flaring), and (3) catalytic oxidation.

The first of these, thermal oxidation is perhaps the most certain as well as the most flexible combustion technique for controlling odors. In this method the odoriferous gas is preheated, often over an auxiliary fuel-fired burner, and passed into a combustion chamber. This odor control method requires very careful regulation of combustion conditions if unacceptable emissions are to be avoided. In particular, the burner should be designed so that the odor-containing gases pass directly through the flame, thus minimizing the possibility of odor escape.

Of particular importance to the effectiveness of this combustion process are three primary factors: (1) residence time in the preheat and combustion zones, (2) temperature, and (3) the amount of turbulence created. These three factors are obviously interdependent. For example, the time required for complete combustion of the odoriferous material is directly affected by the operating temperature in the unit and the turbulence provided in the afterburner design to insure adequate mixing of the waste gas with the combustion air. The processes which affect combustion in controlled flames of this type are physical convection and diffusion. They govern the rate of reactant mixing and product removal from the hot zone and heat transfer by thermal convection and radiation. These affect the three primary factors listed above, in that they control the degree of mixing and the combustion or oxidative reaction rate.

This method of odor control is well suited for controlling emissions of organic aerosols, mists, or droplets. It is the recommended technique when the odorant concentration is below the lower explosive limit (LEL). In some cases it may be necessary to dilute the gas stream with air to achieve this low concentration for safety and insurance purposes.

Thermal oxidizers (afterburners) generally operate in the range of 1,200 to 1,600°F and require a minimum residence time of 0.3 sec in the oxidizing zone. Although this is often the most expensive combustion-based odor control process, it is admirably suited for emissions containing combustible particulates, or if heavy metals or catalytic poisons are in the gas stream. Operating costs can be offset to some extent if the gas is already considerably above ambient temperature and if the heat from the combustion process can be recovered for preheating the incoming air stream.

Another way to minimize energy costs in thermal oxidation is to direct the low-volume, high-concentration odor emissions to a boiler firebox. This method is attractive since additional fuel is seldom required. Users of this approach, however, have found that the combustion system must be well designed and constructed to prevent both odor blowback and corrosion in the boiler tubes. If the requirement for steam and energy of these boilers is variable, some provision must be made to control the odorous emissions when the boilers are at low-fire conditions during slack periods in plant operation.

Direct Combustion

This is the most effective technique for controlling high concentrations of odors in low volumes of air. Flaring the gas in this manner, is however, wasteful of its fuel content. For economy, this odor control method is restricted to combustible gases. A common requirement is that the waste gas supply 50% of the fuel value of the combustion mixture. An auxiliary fuel, such as natural gas, is often needed to initiate and maintain a stable flame, especially if the odoriferous process gas stream is subject to fluctuations in flow rate. Careful design of air-collecting hoods and ducting systems is necessary to reduce, if not eliminate, dilution by ambient air so as to minimize any supplementary fuel costs.

Catalytic Oxidation

Combustion by this flameless oxidation process occurs at 600 to 800°F. A minimum contact time of 0.3 sec is recommended. The decrease in temperature, as compared with other air oxidation or incineration techniques, results in a significantly lower energy requirement. The economics of the process still require that a large portion of the heat of combustion be recovered and used to preheat the incoming odor-containing air streams.

The use of precious metals is necessary for adequate effectiveness. Since these materials are easily blinded, their use is restricted to gases free of particulates and any combustible material having a solid residue. The catalysts are easily poisoned by heavy metals such as mercury and lead. The presence of halogens or SO_2 generally suppresses catalyst effectiveness. The design of an odor control system using catalytic oxidation must include provisions for washing the catalyst beds, for their regeneration, and for preheating the gas to vaporize any liquid or solid aerosol such as fatty particulates, polycyclic organic matter, etc., before it reaches the catalyst surface.

Odor control by catalytic oxidation has the advantage of operating at lower temperatures and thus with lower fuel costs than thermal incineration. Neither process is economical for heating large volumes of essentially zero-BTU air. For the most efficient and economical control of odors by this incineration technique, the emissions from the different odor sources should be treated separately and not mixed with ventilation air. Catalytic combustion offers virtually complete odor elimination–if sufficient catalyst area and contact time are provided–so long as the temperature remains above the catalyst ignition point. This technique is used most often when the odorant concentration is considerably below the LEL.

Some of the problems associated with air oxidation as a general technique for odor control are those of maintaining the proper temperature in the combustion zone, the loss of possibly valuable material, and the problems associated with heat recovery. In addition, if catalytic combustion is the method of choice, there are the problems of catalyst cleaning, regeneration, and poisoning to be considered.

The applications for air oxidation as a control technique for odoriferous emissions are many and are distributed throughout all aspects of industry. Any of the three combustion methods discussed above can be expected to work for organic vapors. Possibly the best choice for polycyclic organic matter would be thermal oxidation. Odors arising from the production of phenol, phthalic anhydride, and their derivatives are effectively controlled by catalytic oxidation. Emissions from soap and detergent making can be controlled by thermal oxidation, which is quite suitable for the control of odors resulting from rubber compounding and vulcanization. Either catalytic or thermal oxidation can be used for the odors coming from the production and curing of cores used in foundry operations and for the odors associated with the production of intermediate resins and finished plastics. Either method may also be used for the control of odors from the production and coating of glass fibers. Odorous emissions from textile-finishing operations are probably best handled by thermal oxidation techniques.

WATER WASHING

Odor removal by washing the gas with water is somewhat limited as a control technique: too many of the components already identified in various odorous streams are essentially insoluble in water. The substitution of a different liquid, in which the odoriferous materials are soluble, converts the simple gas washer to an absorber. In either case, contaminant removal is achieved through the mechanism of solubility and is thus entirely physical in nature.

REACTIVE SCRUBBING

Let us differentiate reactive scrubbing from washing and absorption in the following manner. Reactive scrubbing involves the removal of odorous materials by neutralization, oxidation, or other chemical reaction. In other words, reactive scrubbing is primarily chemical in nature. Of course, the capture of the contaminant must, as with simple absorption, occur by the physical process of solubilization.

Odor removal by any liquid-gas process is a function of the residence time of the odor-laden air in the control unit, the available contact areas, the solubility of the odorant in the liquid phase, and the concentration of the odor in the inlet gas stream.

Mechanism

Washing, absorption, and reactive scrubbing involve gas-liquid diffusion for removal of odoriferous components from the vapor phase and the removal of particulates by inertial impaction and interception in the scrubber. The efficiency of these devices for odor removal depends on the equilibrium between the solvent and the odorant, and is thus temperature-dependent. The soluble components can be absorbed by the liquid phase. Vapors of insoluble materials can often be condensed and subsequently removed by the liquid phase. This odor control technique, which handles large gas volumes economically, is probably more likely to be selected than any of the air oxidation methods if the gas has a high moisture content. Complete odor removal, of course, depends upon the degree of completion of the reactions involved and is facilitated by precipitations—provided the resulting solids can be handled by the device of choice.

Characteristics

Gas absorbers and reactive scrubbers are quite flexible in application, dependable, and can be highly effective. They have a large capacity at a reasonable pressure drop (depending upon internal configuration). These devices are easy to install, suitable for controlling both gases and particulates, and usually have favorable configuration and space requirements. The only disadvantage is the limited degree to which modifications can be performed once the system has been installed. Some applications of gas absorbers are the following: organic sulfur compounds, nitrogen bases, organic acids, oxygenated hydrocarbons, reactive gases, and halogenated compounds.

The selection of the chemicals for wet scrubber service depends entirely upon the characteristics of the odorants to be removed. Table 13-1 is a brief compilation of different types of chemical oxidants for a fairly representative variety of service requirements.

Table 13-1. Absorber Reagents

$KMnO_4$	General, rendering, vaporous polycyclic organic matter
Cl_2	Phenolics, rendering, sewage
NaOCl	Protein adhesives
$(C_2H_5OH)_2NH$	H_2S
Na_2SO_3	Aldehydes
NaOH	Cl_2, pesticide and herbicide manufacturing
$Ca(OH)_2$	Paper sizing and finishing
H_2SO_4	Nitrogen bases

Design Criteria

A great many criteria must be satisfied for the proper design of any wet scrubber, the most important of which is the total odor emission involved. This is the product of the air flow rate and the odor concentration. Also to be considered are the required mass transfer for satisfactory performance and the potential for increasing capacity with no increase in the physical dimensions of the scrubber. Another important factor is the effect of pressure drop on efficiency as it affects both the mass transfer and particulate removal rates. The gas handling capacity and the contact time required in the scrubber will determine the energy required for its operation. Changing the type of scrubber will have a great effect on pressure drop characteristics. Two variables often overlooked in the design of wet scrubber systems are the effective concentration of the control reactant and pH. Associated with these two criteria is the necessity for providing by-product recovery, sludge removal, and reagent makeup. Other factors which must be considered in the design of a wet scrubber are the materials of construction (with regard to corrosion resistance) and the reliability of the entire system, which includes any required control instrumentation.

Design Checklist

The design of a wet scrubber system is composed of 10 separate items: (1) the prime movers for both phases, (2) selection of corrosion- and erosion-resistant materials, (3) emission collection and containment subsystems, (4) inclusion of a combustor for explosive gases, (5) a pre-scrubber for particulate removal, (6) addition of a gas cooler for heat recovery or vapor condensation as a preconditioning step to the scrubbing process, and (7) the scrubber itself (which may be selected from one of the many different types available today: venturi, packed bed, fluidized or jiggling bed, tray or plate tower, spray chamber, injector, cross-flow, and centrifigal or cyclonic scrubbers).

For best operation, the scrubber (8) should be followed by a demisting section to prevent carryover of any entrained liquids in the gas phase. In line with recent OSHA requirements, sound control or attenuation (9) should be provided if possible. The last necessary portion of any well-designed scrubber system is (10) treatment of the effluent so that it does not become a water pollution problem.

It is absolutely unacceptable to merely exchange an air pollution for a water pollution problem. It is entirely possible that the water effluents from other plants may contain reagents which will reverse the scrubbing reactions one has used, thus causing the release of odorants in highly concentrated form downstream.

Scrubber Configuration Choice

The selection of the particular scrubber configuration required for a given odor control problem is not necessarily simple. Almost any type of scrubber will be effective if the odorants are easily absorbed in the scrubbing fluid. In the case of slow reaction or low odorant solubility, either the packed bed or plate tower should be selected as it will give adequate residence time for mass transfer. If solids are also present in the odor-laden gas stream, the selected scrubber must be able to handle them as well. For these, spray chambers or injectors might be an adequate solution. They can be used as prescrubbers or preconditioners for fluidized beds and cross-flow-type scrubbers.

Cross-flow scrubbers operate at lower pressure drops than countercurrent packed beds for the same application. The flooding potential is considerably reduced in the cross-flow design. They are suitable, as static packed beds are not, for handling gas streams with high particulate burdens. Cross-flow scrubbers generally have lower operating costs. Other than these differences, the operational criteria are much the same. However, the design procedures are vastly different. In the packed bed design, concentrations are considered in only one direction. The design procedures required for cross-flow scrubbers are much more complex in that the concentrations change through the scrubber in a horizontal direction and vertically within the packed sections.

ADSORPTION

Adsorption can be an effective and economical control method for emissions with low concentrations of odorous compounds. Adsorption is a physical process in which molecules from the gas phase are captured by and retained upon the surface of a solid (adsorbent). The process is diffusion-limited and is dependent upon the active surface area available. The mechanism is principally dependent on the van der Waal's forces of attraction between the adsorbent and the contaminant to be removed from the gas stream (adsorbate). As expected, the efficiency of operation is governed by the thermodynamic equilibrium which exists between the adsorbate and the adsorbent.

Applications

The primary use of adsorption is the concentration of highly dilute odors from enclosed atmospheres for subsequent destruction or recovery. This includes the elimination of solvents and vapors from stack effluents. Subambient temperature operation is sometimes required for adsorption to be effective. This is especially true if ammonia, hydrogen sulfide, and formaldehyde are to be controlled. Although these are special cases, the exothermic heat of adsorption must be removed in order to keep the bed temperature low enough (120°F is a reasonable upper limit) for effective capture and retention of the odoriferous components in the gas phase.

Activated carbons have a wide application range even in humid atmospheres. The selectivity of silica gels is somewhat narrower. Another class of adsorptive materials is the reactive adsorbents, such as potassium permanganate and manganese dioxide.

Design Variables

The efficiency of the adsorption technique depends on the temperature and pressure of the bulk-gas phase, the contact time allowed, absorbent replacement or regeneration frequency, bed depth, the characteristics of the adsorbent (specificity for the odorous gases, surface to volume ratio), and the following properties of the odoriferous gases: degree of unsaturation; molecular size, configuration, and polarity; and vapor pressure at bed temperature. One of the most important variables is retention, as opposed to breakthrough. Breakthrough can occur in two ways: (1) by the desorption of some lightly held components caused by the presence in the vapor state of more strongly adsorbed materials or (2) when the capacity of the adsorptive bed has been reached with regard to a particular component.

Adsorptive systems are normally installed as parallel beds, so that one bed can be on-stream while the other bed is being regenerated. Regeneration is normally accomplished by steam stripping; boil-off at elevated temperatures followed by incineration; vacuum desorption followed by condensation for recovery of valuable components; or a combination of these techniques. An appreciable water pollution problem can be created by steam stripping if proper preventive measures are not taken.

Adsorption has been used for odor control in many ways: controlling the spray dryer emissions associated with soap making; controlling odors from the manufacture of protein adhesives and pesticides; after the evaporators and coolers in fermentation processes; in combination with wet scrubbers to control odors coming from the production of paints and varnishes; in controlling odors arising from the manufacture of flavorings and perfuming agents; and for solvent recovery in the manufacture of synthetic fibers.

Some of the problems of adsorption are batch-to-batch variation in adsorptive capacity, low capacity for some odorous substances, relatively short adsorbent life without regeneration, and blinding or occlusion of active sites by any particulates in the gas stream. These problems have prevented this technique from achieving popularity in certain segments of industry.

CONTROL METHOD EVALUATION CRITERIA

Now that we have discussed the basic techniques for odor control and their design and operational criteria, it is pertinent to turn to the criteria which will be used in evaluating the tradeoffs between these different systems. The evaluation criteria can be divided into three principal portions: (1) the technical feasibility of the device or system, (2) the economic feasibility of the particular odor control technique, and (3) the industries and sources to which any particular technique can and will be applied.

Technical Feasibility

The technical feasibility of any particular method depends upon the general status in its development and the performance of the control devices. The general status must include consideration of the operational principles, the types of units which are available commercially (if they are not available, consideration must be given to the extent of testing which has been completed and whether this testing has been accomplished at least in pilot plant or semi-works scale), and the number of similar units that have already been placed in commercial operation. With this information and with some feedback from users of the devices in question, rational decisions regarding selection become much simpler.

The other aspect of technical feasibility is device performance. Characteristics of each candidate device which must be considered are efficiency, operational flexibility, whether stream preconditioning is required, whether the device can handle cyclic or fluctuating loads, the suitability of the device to handle multiple pollutants, the energy requirements for both the gas and liquid phases, the material and chemical requirements for odorant control and corrosion prevention, reliability and maintainability, and the control requirements of the device itself.

Economic Feasibility

The economic feasibility of a particular candidate control device to meet a given set of requirements requires comparison of both the initial and the operating costs of such a system. The initial costs include the capital costs of the control device, the costs of auxiliary equipment such pumps, fans, and water pollution control subsystems, installation cost, the costs associated with land acquisition and area modification, and the costs of any required modifications to the existing processes. The operating costs are associated with energy, maintenance, by-product disposal, reagents, effluent purification, the cost of the capital, and operator training.

Application Potential

The evaluation criteria conclude with examination of the industries and sources within those industries to which the candidate control techniques will be applied. The three crucial aspects of this portion of the evaluation are (1) is the candidate technique suitable for old as well as new plants? (2) what are the required interfacings; and (3) what will be the ultimate impact on the characteristics of the emissions from the source?

We can sum up our evaluations of the candidate technique or process by performing a cost/benefit analysis. The goals of such an analysis are to show that implementation of the control technique will reduce the odor emissions, will lower the cost of air pollution control, and will give a reasonable if not adequate approach to achieving the performance standards.

REFERENCES

Bethea, R. M. "Solutions for Feedlot Odor Control Problems," *J. Air Poll. Control Assoc.*, **22**, 765-771 (1972).

Bethea, R. M., B. N. Murthy, and D. F. Carey. "Control Technology for Rendering Plant Odors," *Environ. Sci. Technol.*, 7, 504-510 (1973).

Chass, R. L., V. V. Kanter, and J. H. Elliott. "Contribution of Solvents to Air Pollution and Methods of Controlling Their Emissions," *J. Air Poll. Control Assoc.*, **13**, 64-72 (1963).

Cleary, G. "Some Considerations in the Evaluation and Control of Odors," paper EN 34C, pp. 876-883, in *Proceedings of the Second International Clean Air Congress*, H. M. Englund and W. T. Beery, eds. (New York: Academic Press, 1971).

Elkin, H. F. and R. A. Constable. "Source/Control of Air Emissions," *Hydrocarbon Processing*, **51** (10), 113-116 (1972).

Elliott, T. C. "Odor Control: Its Time Is Coming," *Power*, **115** (6), S12-S13 (1971).

Faulkner, W. D., W. G. Schuliger, and J. E. Urbanic. "Odor Control Methods Using Granular Activated Carbon," paper presented at 74th National Meeting, Am. Inst. Chem. Engrs., New Orleans (1973).

Huch, R., H. Beine, and W. Brocke. "Studies and Developments on the Reduction of Environmental Odor Pollution," paper EN 28A, pp. 779-783, in *Proceedings of the Second International Clean Air Congress*, H. M. Englund and W. T. Beery, eds. (New York: Academic Press, 1971).

Hudson, J. L., E. H. Johnson, D. F. S. Natusch, and R. L. Solomon. "Hydrogen Sulfide Adsorption by Manganese Dioxide and Activated Carbon," paper presented at 74th National Meeting, Am. Inst. Chem. Engrs., New Orleans (1973).

Lundgren, D. A., L. W. Rees, and L. D. Lehman. "Odor Control by Chemical Oxidation: Cost, Efficiency, and Basis for Selection," paper 72-116, 65th Annual Meeting, Air. Poll. Control Assoc., Miami Beach (1972).

Murthy, B. N. "Odor Control Research," Div. of Process Control Engineering, Natl. Air Poll. Contr. Admin., DHEW, Cincinnati (1969).

Onnen, J. H. "Wet Scrubbers Tackle Pollution," *Environ. Sci. Technol.*, **6**, 994-998 (1972).

Sheppard, S. V. "Particulate Collection with Packed Cross-Flow Scrubbers," *J. Air Poll. Control Assoc.*, **22**, 278-281 (1972).

Sullivan, R. J. "Preliminary Air Pollution Survey of Odorous Compounds," Report APTD 69-42, Natl. Air Poll. Contr. Admin., DHEW, Raleigh, N.C. (1969).

Turk, A. "Industrial Odor Control and Its Problems," *Chem. Eng.*, **76**, 70-80 (Nov. 3, 1969).

Turk, A., R. C. Haring, and R. W. Okey. "Odor Control Technology," *Environ. Sci. Technol.*, **6**, 602-607 (1972).

Walsh, M. A. and J. R. Katzer. "Catalytic Oxidation of Phenol in Dilute Concentration in Air," paper presented at 74th National Meeting, Am. Inst. Chem. Engrs., New Orleans (1973).

Wicke, M. "Collection Efficiency and Operation Behavior of Wet Scrubbers," paper EN 16H, pp. 713-718, in *Proceedings of the Second International Clean Air Congress,* H. M. Englund and W. T. Beery, eds. (New York: Academic Press, 1971).

Yocom, J. E. and R. A. Duffee. "Controlling Industrial Odors," *Chem. Eng.*, **77** (13), 160-168 (1970).

CHAPTER 14

COMBUSTION METHODS FOR ODOR CONTROL

Mahesh V. Bhatia

Engelhard Minerals and Chemicals Corp.
East Newark, N.J. 07029

Paul N. Cheremisinoff

New Jersey Institute of Technology
Newark, N.J. 07102

ODOR CONTROL BY COMBUSTION

Combustion is one of the effective means to control odor. Incinerators and afterburners are used for combustion of malodorous waste stream fumes. By combustion, hydrocarbon fumes are oxidized to water and carbon dioxide. Additionally, nitrogen in waste streams and air used for combustion is converted into NO_x, sulfur forms sulfur dioxide, and chlorides emit hydrochloric acid. It may therefore be essential to have a secondary system to remove objectionable SO_2, HCl, etc., by scrubbing or other means. When waste gas streams contain large concentrations of chlorides, phosphates, etc., combustion is not practical.

Combustion can be employed when malodorous fumes are toxic, have a low concentration of combustibles, and have no heat-recovery value. In fact, this may be the only acceptable method, since fumes do not lose toxicity after masking or scrubbing, and the problem is only transferred from a gas to a liquid phase. Combustion offers a total destruction method. Exhaust fumes, even if they contain only a relatively small concentration of malodorant components, may create problems as great as highly concentrated fumes and require odor abatement systems. This is due to the fact that intensity (I) of odor increases slowly with concentration (C) above the threshold intensity, and is defined by the Weber-Fechner law:

$$I = K \ln C \quad (1)$$

where

- I = intensity of odor
- C = concentration of malodorant component
- K = a constant for that component

Oxidation or combustion must be complete since partial oxidation will cause additional problems. Figure 14-1 shows that "intermediates," obtained after partial oxidation, may have worse odors than some original compounds.

TYPES OF INCINERATORS

There are three types of incinerations: (1) direct flame, (2) thermal, and (3) catalytic.

Direct Flame

Direct-flame incinerators are used when the fumes are combustible and need no additional air for combustion. Waste stream fumes should have at least 100 Btu/cu ft heating value. Combustion takes place in the presence of the flame. Heating value can be determined by Dulong's formula:[2]

$$\text{Gross Btu/lb} = 145.4C + 620\left[H - \frac{O}{8}\right] + 41S$$

where C, H, O, and S are weight percentages in gases of carbon, hydrogen, oxygen, and sulfur respectively. The term (H - O/8) represents hydrogen in fuel that is combined with oxygen as moisture or in some other form. If fumes are below the lower flammable limit, it may be essential to add small amounts of fuel. The fume to air ratio must lie between the lower and upper flammable limits. Refer to Table 14-1 for values of lower and upper flammable limits of some of the fume constituents.[3]

Generally, high temperatures up to approximately 2,500°F are used in direct-flame incinerators. Since the temperature is high, heat-recovery units are often used with the combustion units, as indicated in Figures 14-2 and 14-3.

Thermal Incinerators

In many instances fumes are dilute, *i.e.*, the quantity of combustibles is very small and the heating value is very low—approximately 1 to 15 Btu/cu ft, compared to 100 Btu/cu ft for direct-flame incinerators. Therefore, it is not possible for fumes to sustain combustion without additional fuel. It also follows that the quantity of fuel required is higher for thermal incinerators.

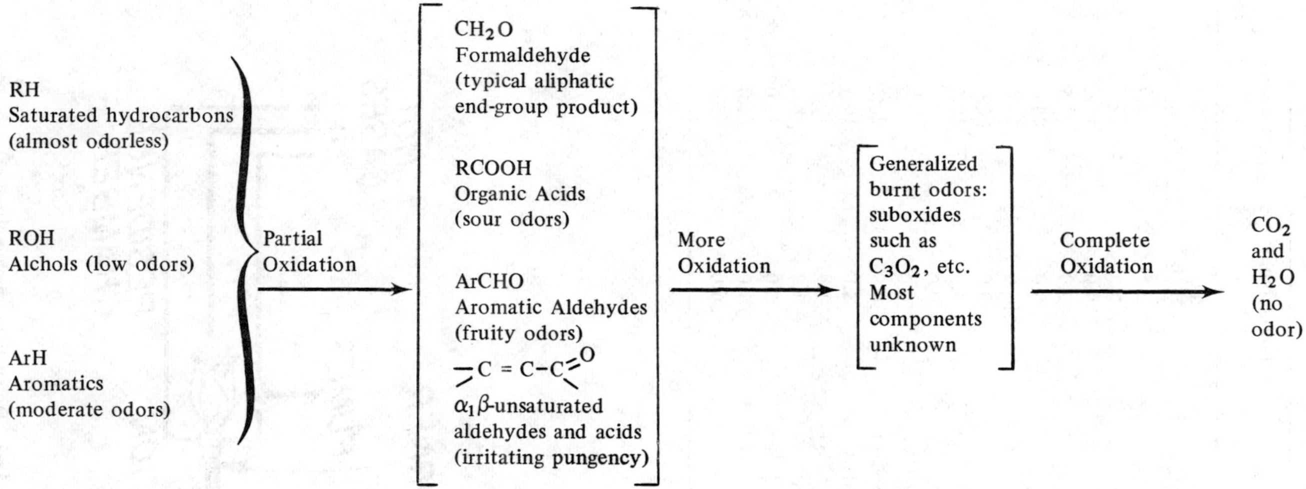

Figure 14-1. Partial oxidation can cause odors even worse than the original problem.[1]

Table 14-1. Flammable Concentrations of Gases and Vapors in Air

	Lower Limit %	Upper Limit %
Petrol vapor	1.5	6.0
Benzene (C_6H_6)	2.7	6.3
Ethane (C_2H_6)	2.5	5.0
Methane (CH_4)	5.5	14.5
Alcohol (C_2H_5OH)	4.0	14.0
Diethyl ether [$(C_2H_5)_2O$]	2.9	7.5
Acetylene (C_2H_2)	3.0	73.0
Carbon monoxide (CO)	15.0	73.4
Hydrogen (H_2)	10.0	66.0
Hydrogen Sulfide (H_2S)	4.5	45.5
Chlorine Dioxide (ClO_2)	10.0	
Ozone (O_3) in oxygen (O_2)	10.0	
Monomethylamine (CH_3NH_2)	4.95	20.75
Dimethylamine [$(CH_3)_2NH$]	2.8	14.4
Monoethylamine ($C_2H_5NH_2$)	1.8	10.1
Diethylamine $(C_2H_5)_2NH$	1.2	8.0
Natural gas	5.0	12.0
Town gas	7.0	21.0
Blast furnace gas	36.0	65.0

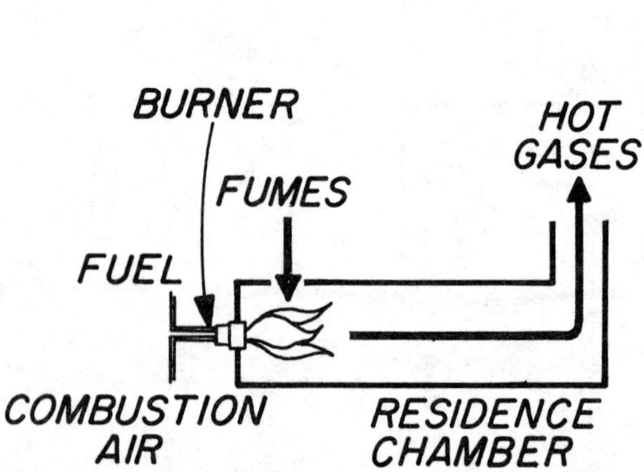

Figure 14-2. Direct flame incinerator.

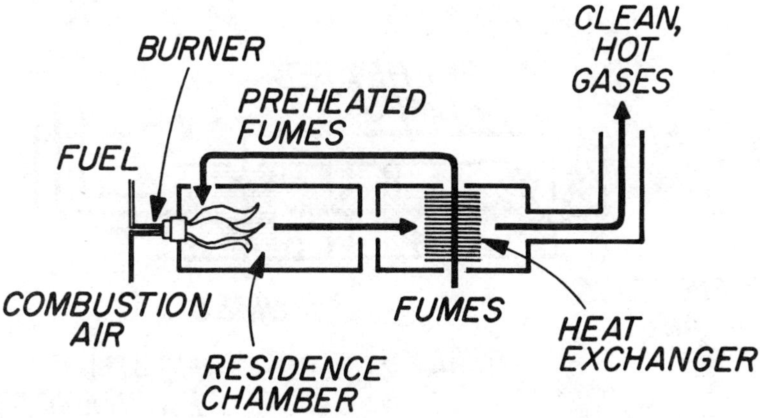

Figure 14-3. Thermal incinerator with heat recovery.

In thermal incineration the major portion of the fumes is injected immediately next to the burner flame. Part of the fumes are bypassed and then mixed with the combustion mixture. Sometimes part of the fumes is mixed with the auxiliary fuel stream. Thermal incineration temperatures range from 1,100 to 1,500°F.

Catalytic Oxidizers

A catalytic oxidizer flow schematic is shown in Figure 14-4. In the presence of a catalyst, air oxidation requires much lower combustion temperatures. Reaction does not take place in the presence of flame; however, a flame is used to preheat the fumes. The preheat burner supplies heat to raise the temperature required for catalytic oxidation. Generally, the temperature upstream of the catalyst is in the range 600 to 900°F, and 800 to 1,100°F downstream of the catalyst. The increase across the catalyst is due to an exothermic catalytic oxidation reaction. Waste streams containing elements or metals damaging to catalyst activity must be avoided.

PRINCIPLES OF COMBUSTION

Before one designs an incinerator, it is necessary to understand combustion principles. Significant features are considered here, but for a more thorough discussion the reader is referred to more detailed works.[4]

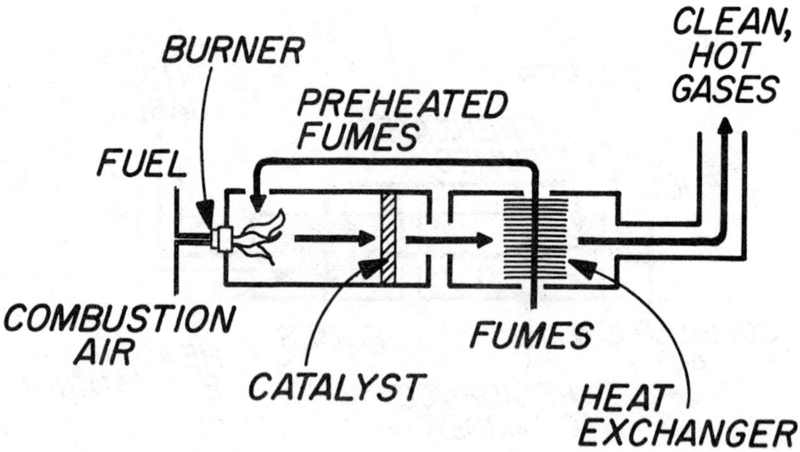

Figure 14-4. Catalytic oxidizer with heat recovery.

During the combustion process, fumes are air oxidized to carbon dioxide and water. We may consider oxidation of a malodorous component as an example. Most of the fumes are dilute; *i.e.,* concentration of a malodorous component is extremely low in the waste stream, and the reaction can be considered to be first-order with respect to concentration. The rate of reaction also depends on temperature. Hence

$$\text{Rate of reaction} = W\ (\text{lb/cu ft/hr}) = W(C) \times W(T) \quad (2)$$

from Arrhenius equation

$$W(T) = k_o e^{-E/RT} \quad (3)$$

where

k_o is a constant
E = activation energy Btu/mole or cal/mole
R = gas constant Btu/°R · lb mole or cal/°K · g mole.

Now consider an adiabatic well-insulated incinerator, as shown in Figure 14-5. As combustion takes place, heat is liberated. The amount of heat liberated is proportional to the rate of reaction. In fact, graphs plotted for rate versus temperature and rate versus heat liberated can be superimposed on each other by changing ordinate scales.

If one further considers that generated heat is carried away by reaction products leaving the incinerator and assumes that perfect mixing takes

Table 14-2. Influence of Temperature on Reaction Rate[4]

Temperature °K	Values of $e^{-E/RT}$			
	For E = 20,000 cal/mole		For E = 40,000 cal/mole	
500	2×10^{-9}	1	4×10^{-18}	1
1,000	4×10^{-5}	2×10^{4}	2×10^{-9}	5×10^{8}
2,000	6×10^{-3}	3×10^{6}	4×10^{-5}	1×10^{13}

Indicates the effect of temperature on rate of reaction. As can be seen, for activation energy of 20,000 cal/mole, reaction rate is increased 20,000 times when temperature is changed from 500 to 1,000°K.

place so that the concentration of fumes leaving the incinerator is the same as the concentration inside the incinerator, the incinerator is a "back mixed" type.

Let W be the rate of reaction of pounds of substance per unit time per unit volume. Q_1 is heat liberated in Btu/hr, q is heat value in Btu/lb, Cp is specific heat in Btu/lb °R, V is the volume of incinerator, and λ_R is residence time in hours. Then

$$Q_1 = W\,q\,v = \text{Btu/m} \tag{4}$$

$$= k_1 \frac{(C_i - C)}{\lambda_R}\,q = k_1 e^{-E/RT}\,c\,q \tag{5}$$

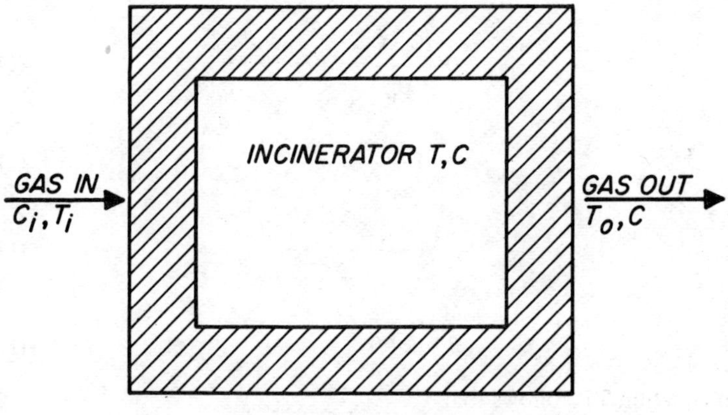

Figure 14-5. Adiabatic incinerator.

For a first-order reaction, k_1, the rate constant, has dimensions hr^{-1}. Therefore

$$k_1 = 1/\lambda_k \tag{6}$$

$$Q_1 = \frac{qe^{-E/RT} C}{\lambda_k} \tag{7}$$

Heat carried away by the gas stream per hour is:

$$Q_{II} = C_p(T - T_i)/\lambda_R \tag{8}$$

At steady state, the heat liberated must be equal to the heat carried away by the gas stream.

$$\frac{Ce^{-E/RT} q}{\lambda_R} = \frac{C_p (T - T_i)}{\lambda_R} \tag{9}$$

Let

$$\lambda_{RK} = \lambda_R/\lambda_K \tag{10}$$

which is dimensionless or reduced residence time.

As will be expected, longer residence time (higher λ_R) or a higher reaction rate constant (lower λ_K) would mean higher conversion of malodorous compounds to their ultimate combustion products. λ_{RK} is therefore an important parameter in incinerator design. The effectiveness of an incinerator in odor abatement can be expressed in terms of the concentration of the malodorous component of the outlet of an incinerator.

Let ϕ be effectiveness, defined by

$$\phi = \frac{C_i - C}{C_i} \tag{11}$$

$$= 1 - \frac{C}{C_i} \tag{11a}$$

$$= 1 - \bar{C} \tag{11b}$$

where

$$\theta = RT/E \tag{12}$$

From equation 5 it follows that

$$\phi = \lambda_{RK}\, e^{-1/\theta} (1 - \phi) \tag{13}$$

or

$$\phi = \left(\frac{1}{1 + \frac{e^{1/\theta}}{\lambda_{RK}}} \right) \tag{13a}$$

Hence Q_I can be plotted as θ versus ϕ, using λ_{RK} as a parameter. The heat liberated will be zero when no conversion is taking place and waste streams are at the preheater temperature only.

A similar analysis can be made for the heat carried away by the stream (Q_{II}):

$$\phi = 1 - \frac{C}{C_i} = 1 - \frac{Cq}{C_i q} \tag{11c}$$

$$Q_{II} = \frac{C_p}{\lambda_R} (T - T_i) \tag{9a}$$

If

$$m = RqC_i/EC_p \tag{14}$$

then

$$\phi = \frac{1}{m} (\theta - \theta_i) \tag{15}$$

The heat removal line of ϕ versus θ has a constant slope of 1/m. At $\theta = \theta_i$, $\phi = 0$. For the maximum value of $\theta = \theta_{max}$, $\phi = 1$. Therefore,

$$\theta_{max} = m + \theta_i \tag{16}$$

Thus, the highest temperature attained depends on the preheat temperature, concentration of the combustible in the fumes entering the incinerator, heat value and specific heat. From equation 16 it follows that

$$T_{max} = \frac{qC_i}{C_p} + T_i \tag{17}$$

This T_{max} is known as *flame temperature.*

At a steady state, these two graphs will meet each other (Figure 14-6). Points of intersection 1, 2, and 3 are steady-state conditions.

Consider an infinitesimal change $d\theta$ at $\theta = \theta_1$. At temperatures $\theta_1 + d\theta$, heat carried away is more than heat liberated. Hence the system will tend to cool down, and the temperature will drop back to θ_1. Similarly, at $\theta_1 - d\theta$, heat liberated exceeds heat carried away, and the temperature will rise to θ_1. Therefore, whenever the temperature changes around $\theta = \theta_1$,

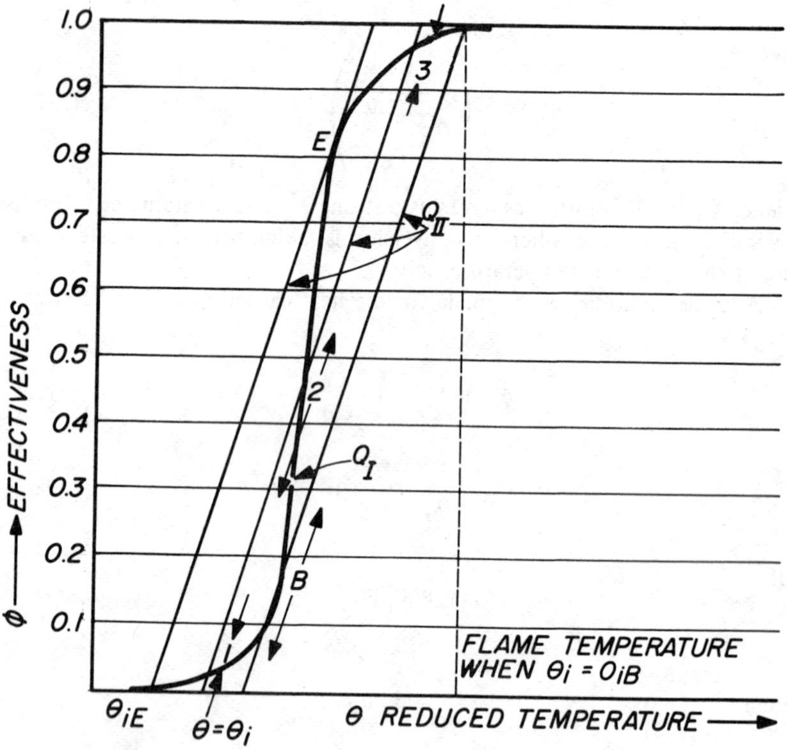

Figure 14-6. Graph: effectiveness versus temperature.

the system will return to its steady state. Similarly, a change in either direction around θ_3 will return to a steady-state condition, represented by point 3 in Figure 14-6.

The steady-state condition represented by point 2 is, however, not similar to the other two steady states as discussed above. At $\theta = \theta_2 - d\theta$, the heat carried away is more than the heat liberated, and the waste stream will be cooled–until point 1 is reached. Thus, steady state 2 is really unstable.

As gases are preheated to a higher temperature, Q_{II} line is shifted to the right. When waste stream gases are preheated to θ_{iB}, Q_I and Q_{II} meet each other tangentially, at point B. Here any change in either direction, however small, will increase the fume temperature. This phenomenon is called *thermal explosion.*

When the preheated gas temperature is at $\theta = \theta_i$, θ_2 is called *ignition temperature.* If preheated gas is at $\theta = \theta_{iB}$, θ_B is called *ignition temperature.* The lower steady-state region is called the *oxidation region*, and the upper steady-state region is known as the *combustion region.*

As the inlet temperature is reduced, the Q_{II} line is shifted to the left. At inlet temperature θ_{iE}, Q_{II} and Q_I intersect each other tangentially at E. Here the phenomenon is just opposite to thermal explosion. Slight changes in either direction will cool the gases, and the incinerator may come down to the oxidation region steady state or become extinguished. At point E, *extinction* takes place.

As can be seen from Figures 7A, 7B, and 7C, ignition temperatures are related to actual conditions. The change in the heat removal line takes

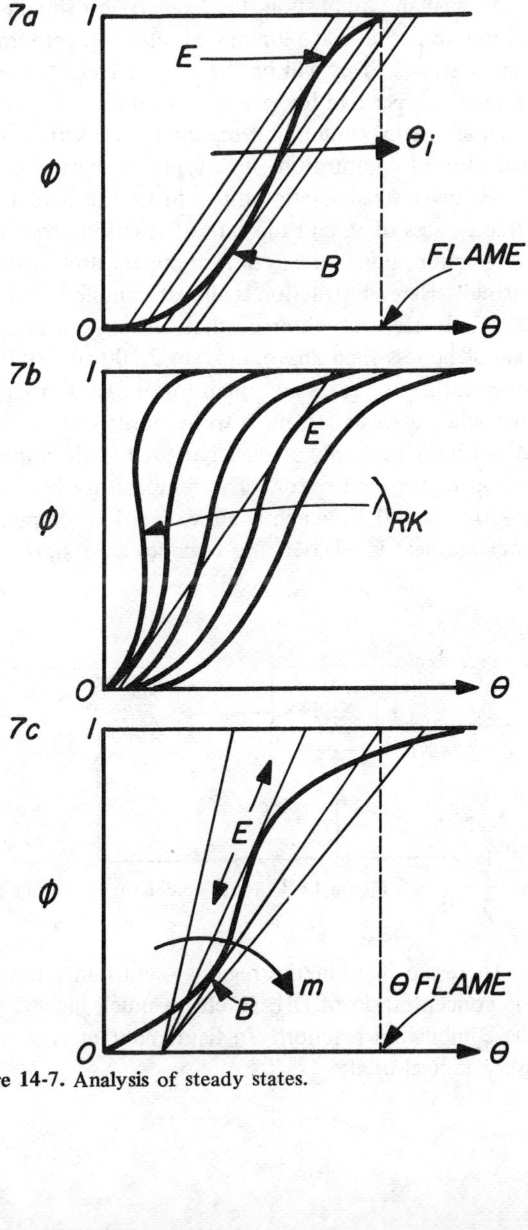

7A. Change of Inlet Temperature θ

7B. Change of Residence Time λ_{RK}

7C. Change of Heating Value of Mixture

Figure 14-7. Analysis of steady states.

place due to changes in concentration of the combustible component inlet preheat temperature. Heat liberation is changed by residence time λ_{RK}.

When concentration (C) of a malodorous component is very small, the heat removal line Q_{II} is almost vertical, and graphically it is difficult to find the ignition temperature. In such cases ignition temperature is taken as θ at $\phi = 0.9$. Quantitative relationships have been derived to estimate ignition or extinction temperatures, represented by the tangential intersection of Q_I and Q_{II}.

To sustain combustion, heat generation should be at least 10 Kcal/mole greater than the heat required to raise the gas temperature to ignition temperatures. This will be the driving force for combustion of unburnt regions. Upper and lower explosive limits or flammability limits are defined as the maximum and minimum concentrations of fuel in air which will support combustion. For typical values refer to Table 14-1.

At lower steady-state temperatures, the oxidation region, conversion or effectiveness of ϕ, and the rate of reaction are much lower. For complete combustion, upper steady-state temperatures in the combustion region are desired. Here, destruction is almost complete and the rate of reaction is faster. It has been observed that, in most cases, stable flame temperatures cannot be less than approximately 2,100 to 2,400°F. However, ignition temperatures of gases are much lower (1,000 to 1,200°F). Hence it follows that when a waste stream is to be burned off, it should not be completely mixed with air; only a small part of the air is mixed with the fumes and passed to the incinerator. The temperature here is raised to approximately 1,400 to 1,500°F, which is above ignition temperature. Therefore, the available heat is released and organics are oxidized.

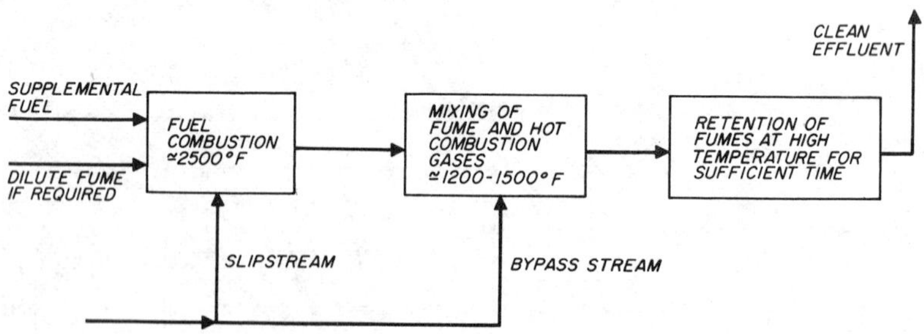

Figure 14-8. Flow schematic of thermal incineration.

There are two kinetics regions—cool flame and hot flame. In hot flame the concentration of OH, H, etc. is much higher. This helps in accelerating the combustion reaction. In flame contact two radicals are produced for every radical used:

$$H + O_2 \rightarrow OH + O$$

$$O + H_2O \rightarrow OH + OH$$

and

$$O + H_2 \rightarrow OH + H$$

$$H + H_2O \rightarrow H_2 + OH$$

Therefore, one may logically think that flame contact is essential for effective combustion in the incinerator. While achieving flame contact, the waste stream may quench the flame and combustion will be only partial. For this reason, too much emphasis should not be placed on flame contact.

DESIGN PRINCIPLES

Having reviewed combustion principles, a greater appreciation of incinerator design may be obtained. As indicated, the concentration of combustibles in fumes is generally very small. Therefore, thermal incinerators or catalytic oxidizers are used in odor abatement.

Considering the design principles of thermal incinerators, there are three major parameters: residence time, temperature, and turbulence or mixing. The last is perhaps the most important, and yet, too often neglected. Many incinerators are ineffective due to poor mixing. Time and temperature are relatively easy to adjust. All three parameters are interrelated to each other, and variation in one can affect changes in others.

Temperature

As indicated in the discussion under combustion principles, the temperature required for complete combustion should be near the upper steady state–in the combustion region and above the ignition temperature. Further, from Table 14-2 (p. 221) and Figure 14-9 it can be seen that within small temperature ranges, rates of combustion reactions jump from almost negligible to very high values. Incinerators should be operated within these temperature ranges.

Generally, for greater than 99% destruction, operating temperatures are between 1,200 and 1,500°F. For 50 to 90% odor destruction, the required temperatures are 1,000 and 1,200°F. For 90 to 99% odor destruction, incinerators are operated between 1,100 and 1,300°F. It may be noted that carbon monoxide, which is a partial oxidation product in the combustion region of hydrocarbons, is difficult to oxidize and higher temperatures are required for complete destruction. Similarly, methane, toluene, cellosolve, etc., are stable. For this reason, higher temperatures are required to

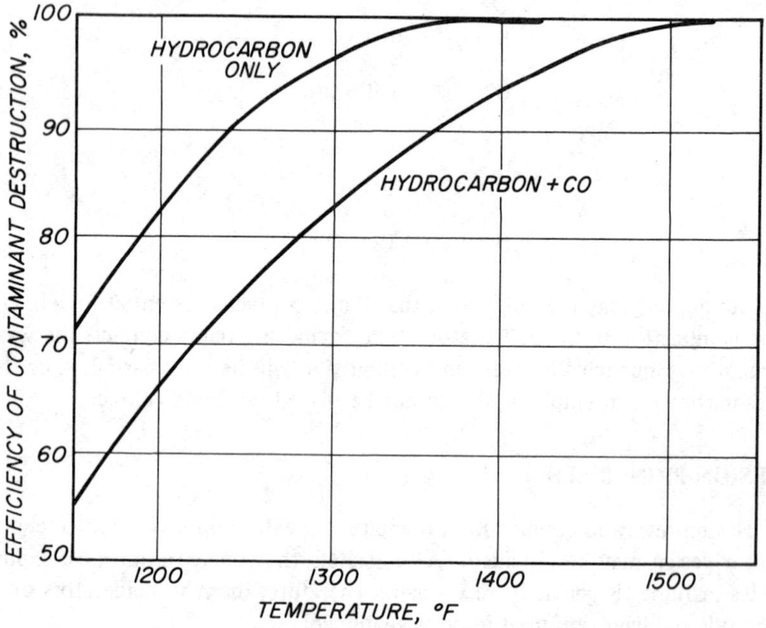

Figure 14-9. Typical effect of operating temperature on effectiveness of thermal incinerator for destruction of hydrocarbons and CO.[5]

incinerate these offensive odorous compounds. Table 14-3 gives auto-ignition temperatures of some of the major components in malodorous waste streams.[6,7,8]

Residence Time

Time is closely interrelated to temperature. Higher temperature means higher velocity, and hence lower residence time. At 1,200°F, T = 460 + 1,200 = 1,600°R. At standard conditions, T = 520°R (or 60°F). At 1,200°F, the actual volume at atmospheric pressure is approximately three times larger than in standard conditions. Therefore, residence time is approximately one-third of that at standard conditions.

Figures 14-10 and 14-11 further indicate effects of time. Remember an increase in λ_{RK} shifts the steady-state point to the left (Figure 14-7B), or toward a lower temperature if effectiveness, ϕ, is constant. In general, the residence time required for odor destruction (50 to 99% +) is 0.3 to 0.5 seconds.

Table 14-3. Auto-Ignition Temperatures of Organic Compounds[6–8]

Compounds	Auto-Ignition Temperature, °F	Compounds	Auto-Ignition Temperature, °F
Acetone	1,000	Hydrogen	1,065
Ammonia	1,200	Hydrogen cyanide	1,000
Benzene	1,076	Hydrogen sulfide	500
Butadiene	840	Kerosene	490
Butyl alcohol	693	Linseed oil	820
Carbon disulfide	257	Maleic anhydride	890
Carbon monoxide	1,128	Methane	1,170
Castor oil	840	Methyl alcohol	878
Chlorbenzene	1,245	Dichloromethane	1,185
Cresol	1,038	Methyl ethyl ketone	960
Cyclohexane	514	Mineral spirits	475
Dibutylpthalate	760	Petroleum naphtha	475
Ethyl ether	366	Nitrobenzene	924
Methyl ether	662	Oleic acid	685
Ethane	950	Phenol	1,319
Ethyl acetate	907	Phthalic anhydride	1,083
Ethyl alcohol	738	Propane	898
Ethyl benzene	870	Propylene	940
Ethyl chloride	965	Stoddard solvent	444
Ethylene dichloride	775	Styrene	915
Ethylene glycol	775	Sulfur	450
Ethylene oxide	804	Tung oil	855
Furfural	739	Toluene	1,026
Furfural alcohol	915	Turpentine	488
Glycerin	739	Vinyl acetate	800
		Xylene	925

Mixing or Turbulence

Figure 14-8 indicated the entire waste stream is not mixed with auxiliary air. Part of the stream, generally 50%, is bypassed and only 50% is mixed with air. This mixture of air and slip stream is passed into the afterburner. The temperature of this initial mixture has to be very near the stable flame temperatures (2,400 to 2,500°F). Then this hot mixture is mixed with the rest of the fumes. These two streams should be perfectly mixed. The secret of successful incinerator performance lies in achieving proper mixing. This mixture must be retained at high temperatures for the time required for complete combustion reaction or total destruction of malodorous components does not occur.

The total residence time in the incinerator is a fraction of a second. This includes the time required for initial heating of the slip stream mixture, mixing

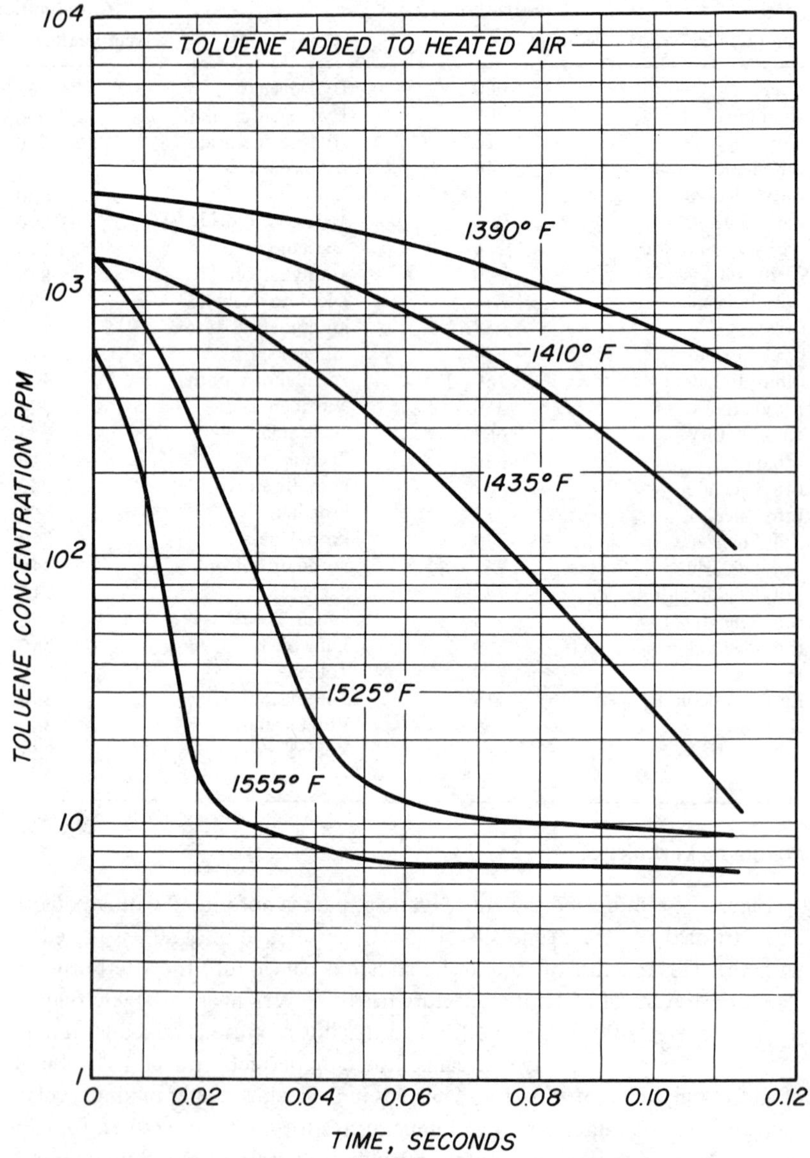

Figure 14-10. Toluene oxidation kinetics.[5]

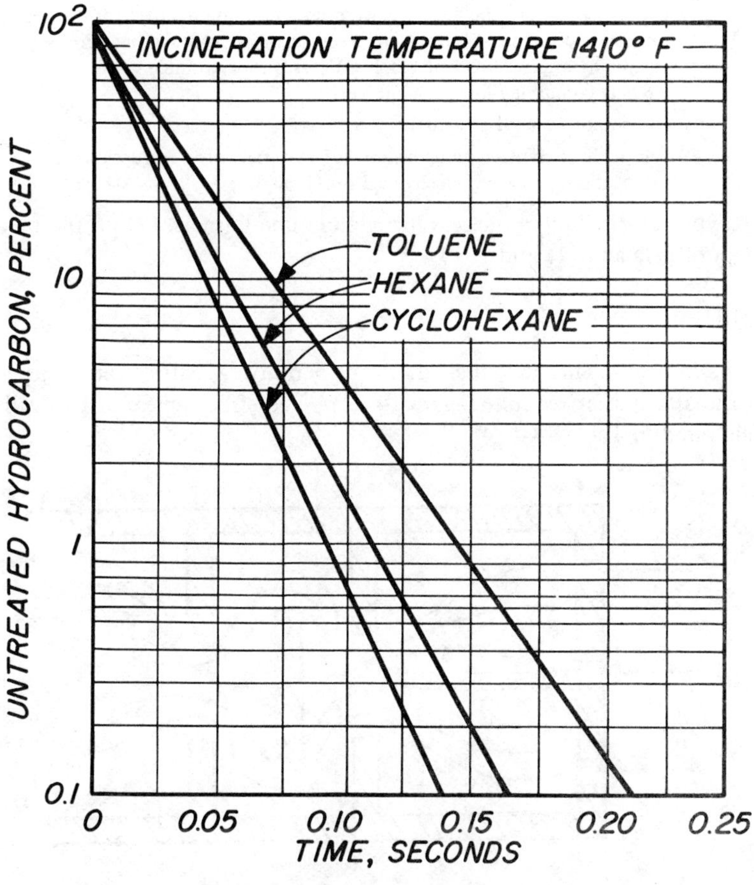

Figure 14-11. Hydrocarbon oxidation rates in absence of flame.[5]

it with the bypassed stream, and then retaining it for the time required for complete combustion. Proper selection of burners, incinerators, internals, and size can achieve the goal of complete mixing, but additional fuel may be required.

A logical approach is to divide the flame into many small surrounding jets or lines of flames. This is done by using distributed burners. There are, however, some limitations in using distributed burners, which include

1. Fouling–waste streams may contain some noncombustibles. Also condensibles from fumes may accumulate near cooler areas, and burners may become fouled.

2. Outside air cannot be additionally used and contain sufficient (16%) oxygen if distributed burners are used. Coffee roasting and wire enameling are examples of batch processes that produce waste streams containing low oxygen.
3. Oil cannot be used as a fuel.
4. Due to heat transfer, temperatures of materials of constructions rise substantially, creating mechanical problems.

Types of distributed burners include (1) line burners, (2) profile plates, (3) multijets and (4) grids.

Baffles

As in heat exchangers, pressure drops permitting, baffles can be provided in the afterburner section (Figure 14-12). The effectiveness of baffles is illustrated in Table 14-4.[5]

Figure 14-12. Baffles increase turbulence.

Table 14-4. Effectiveness of Baffles. Multijet burner - 1400°F. Combustion Chamber Temperature 0.3 seconds residence time

	Without Baffle	With Baffle
Inlet		
Hydrocarbons	900-1200 ppm	1300-1400 ppm
Carbon Monoxide	10 ppm	30 ppm
Outlet		
Hydrocarbons	250-700 ppm	100-110 ppm
Carbon Monoxide	20-100 ppm	-10 ppm
Maximum-Minimum temperature found in traverse across combustion chamber	120-240°C	- 30°F

Other important parameters in achieving proper mixing are the location and size of burners, and hence the velocity of gases flowing through. Burners must not be in dead zones or dead spots; they must be placed where the flow is fully developed and turbulent. Velocity of gases is generally 10 to 30 ft/sec. This velocity is sufficient for molecular mixing. As a rule of thumb,

$$\frac{(\mathrm{SCFM})\mathrm{D}}{\mathrm{A}} > 120 \qquad (18)$$

where

SCFM = standard flow, cu ft/min
A = area, sq ft
D = equivalent diameter, ft

Catalytic Oxidation

A catalytic afterburner is schematically shown in Figure 14-4. Contaminated fumes are preheated in an incinerator to 500 to 900°F, which is required for catalytic oxidation, and preheated fumes are passed over a catalyst bed or matrix. Oxygen must be supplied evenly to all parts of the catalyst. This type of incinerator is operated at much lower temperatures than thermal or flame incinerators. The fuel requirement is only 20 to 40% of that for thermal incinerators. Refer to Table 14-5 for temperature comparisons between thermal and catalytic oxidizers.

Figure 14-13. Honeycomb catalyst matrix.
(courtesy Engelhard Minerals & Chemicals Corp., Engelhard Ind. Div Systems Dept.)

Table 14-5. Comparison of Temperatures Required to Convert Combustibles to CO_2 and H_2O[7]

Combustible	Ignition Temperature, °F: Thermal	Ignition Temperature, °F: Catalytic	Difference °F	Catalyst/ Thermal %
Benzene	1076	575	501	53.4
Toluene	1026	575	451	56.0
Xylene	925	575	350	62.2
Ethanol	738	575	163	77.9
MIBK	858	660	198	76.9
MED	960	660	300	68.8
Methane	1170	932	238	79.9
Carbon monoxide	1128	500	628	44.3
Hydrogen	1065	250	815	23.5
Propane	898	500	398	55.7

Noble metals such as platinum, palladium, and rhenium, dispersed over finely divided porous ceramic substrates, are typical catalyst systems used for this purpose. Catalytic oxidizers are much smaller and lighter than thermal incinerators. The cost of catalysts must, however, be considered. For a smaller stream capacity (up to 12,000 cfm), costs of both types of incineration are almost equal. In cases of dilute concentrations of organic pollutants that have high odor, catalytic units have lower capital costs as well as lower operating costs, compared to thermal incineration.

For most fume abatement applications, supports used for catalysts are in the form of open ceramic honeycombs. Honeycomb catalyst support provides a low pressure drop and maximum surface area. Because the cross-section of the honeycomb is sized to enhance catalytic activity, the residence time required for combustion is reduced. As a result, a small volume of catalyst can handle a large flow. Another attractive feature is that the honeycomb catalyst can be mounted in various configurations, such as vertical or horizontal, to fit into existing ovens and meet plant space requirements. The matrix is within a rigid stainless steel frame to facilitate handling and allow rapid replacement and inspection.[9]

If waste streams contain certain components or elements damaging to catalyst activity, they must not pass across catalyst surfaces. These are listed below per their classifications.

1. Fouling agents–alumina and silica dust, silicones, iron oxides, particulates. (They deposit over catalyst active centers and block contact with oxidants.)

Figure 14-14. Catalytic oxidizer. (courtesy Engelhard Minerals & Chemicals Corp., Engelhard Ind. Div. Systems Dept.)

Table 14-6. Characterization of Combustibles Emission Sources[5]

Source	No. of Sources Approx. 1970 Emissions mm lb[c] of Organics	Ease of After burner Control[d]	Known Emissions: Hydrocarbons: Aromatics	Non-aromatics	Organic Halides	Organic Acids	Alcohols	Aldehydes	Nitrogenous Organics	Esters	Ketones	Ethers	Sulfurous Organics or H_2S	Odor or Other	HI Index of Emission's Offensiveness[e]
SOLVENT EMITTERS	14524		X	X	X		X	X	X	X	X	X		Odor	3.7
Adhesives		2	X	X			X			X	X				3.7
Degreasing metal parts		1			X										6.2
Dry cleaning		2	X	X	X									Odor	4.1
Fiberglass curing		3		X										Odor	2.7
Herbicides		1	X	X											3.5
Rubber products fabrication		3	X	X	X						X			Odor	3.5
Surface coating mfg.		3	X	X			X			X	X	X		Odor	3.6
Surface coating (industrial)		3	X	X			X		X	X	X	X		Odor	4.3
Surface coating (trade)		1	X	X			X			X	X	X			3.6
Surface coating removal		1	X	X	X					X					2.8
Graphic arts		3	X	X			X			X	X	X		Odor	3.6
FOOD AND BYPRODUCTS				X		X		X	X				X	Odor	–
Breweries	185 plants													Odor	–
Charbroiling restaurants	a	3												Odor,Smoke	–
Coffee roasting	1.3mm lb	3	X	X		X		X						Smoke	4.6
Deep fat frying	a	3		X										Smoke	2.7
Fish canning	320 plants	2							X					Odor	8.5.
Meat and bonemeal drying	a	3												Odor	–
Meat packing plants	2700 plants	2												Odor	–
Poultry dressing plants	840 plants	2												Odor	–
Rendering plants	268 plants	2							X				X	Odor,Ammonia	8.4
Smokehouse	1.2mm lb	3	X	X		X		X			X				4.5
Vegetable oil extraction	338 plants	3		X											2.7

COMBUSTION PROCESSES				X		X			X						Odor,Smoke	
Autobody incinerators	3 mm lb	3		X		X			X						Smoke	4.7
Brake shoe reclamation	2 mm lb	3		X		X			X							4.7
Copper wire insulation burnoff	2 mm lb	3		X		X			X							4.7
Fuel oil burning	364 mm lb	b		X					X						SO_2	4.5
Refuse incineration	3505 mm lb	3		X		X			X						Odor, Smoke	4.7
Secondary nonferrous metal recovery	403 plants	3													Smoke	–
Steel drum reconditioning	8 mm lb	3		X		X			X						Smoke	4.7
GENERAL INDUSTRIAL PROCESSES			X	X	X	X		X	X	X	X	X	X	X	Odor	
Carbon black manufacture	1449 mm lb	3		X											Combustible particulates	2.7
Charcoal manufacture	426 mm lb	3	X	X		X		X								4.0
Chemical plants	194 mm lb	1	X	X	X	X	X	X	X	X	X	X	X	X	Odor	4.9
Hardboard manufacture	a	3												X		2.7
Gray iron foundries	1061 plants	2													CO	–
Metallurgical coke manufacture	289 mm lb	2	X	X					X						Ammonia,Odor	3.5
Oil refineries	2788 mm lb	2	X	X										X	Odor	4.4
Paper manufacture (pulp mill)	279 plants	1												X	Odor	8.3
Pharmaceutical manufacture	875 plants	3													Odor	–
Plastic fabrication	5000 plants	1	X	X											Odor	3.5
Soap manufacture	668 plants	3													Odor	–
MISCELLANEOUS SOURCES			X	X		X			X						Odor	–
Automobiles	32060 mm lb	b	X	X		X			X						CO,NO_X	4.6
Gasoline marketing	1852 mm lb	b	X	X											–	3.5
Oil tempering of metals	a			X											Smoke	2.7
Sewage systems	a														Odor	–

[a]Information not available
[b]Emissions have been included even though process is not amenable to control
[c]mm is used to designate million pounds
[d]A rank of 3, 2, or 1 was assigned to each source according to whether it was controllable with considerable difficulty, control was impossible, or possible only with extreme difficulty, respectively.
[e]The index can range from zero for emission with no offensive qualities to ten for an emission with maximum offensiveness.

2. Suppressants–halogens, both as elements and in compounds, and sulfur compounds. (Chemical reaction with catalysts is reversible.)
3. Poisons–phosphorus, bismuth, mercury, lead, tin, zinc, arsenic, antimony.

Combustion takes place on the surface of catalysts. It is necessary that the gas stream entering the catalytic bed is uniformly heated. Hence a preheat section is used as a mixing chamber in the catalytic oxidizer.

If sufficient quantities of combustibles are present in the waste stream, the need of an auxiliary heat source is minimized. The temperature of the catalyst should not exceed 1,472°F for the practical life of the catalyst.

APPLICATIONS

Table 14-6 indicates possible applications for odor incineration. Additionally, the following case histories provide more specific data on successful applications. Catalytic oxidizers are being used to burn off hydrocarbons and odor-causing pollutants in the form of exhaust from diesel engines and automobiles (Figures 14-15, 14-16, 14-17).

Pulp and Paper

The kraft process is a common method of pulping in the paper industry. Almost 60% of the paper manufactured in the U.S. is made by this method. Much concern has been expressed about malodorous fumes evolved in this process. Major sources of malodorous fumes in kraft processes are stack gases from recovery furnaces, direct-contact evaporators, noncondensibles from digester blow tanks, and multi-effect evaporators.

Hydrogen sulfide, when reacted with the methoxy group of lignin, forms organic sulfur compounds which are odorous. Also, hydrogen sulfide gas is generated in this process, and has a well-known, disagreeable smell and toxicity. Combustion of these fumes has been achieved on a satisfactory level. Sulfur dioxide emissions, after combustion of sulfur compounds, have been reported at acceptable levels.

In the direct-contact gas evaporator, black liquor feed is evaporated, while concentrating and evaporated fumes contain odorous sulfides. Even if paper mills use oxidized black liquor to reduce hydrogen sulfide emissions, objectionable organic sulfides are still present. Direct-flame incinerators may be used for odor abatement. Some of the noncondensibles, although much less in volume than stack gases, also cause odor problems. These may be in sufficiently high concentrations to exceed lower explosive limits, and hence must be diluted before combustion. Sometimes lime kilns are used to mix these noncondensibles with combustion air.

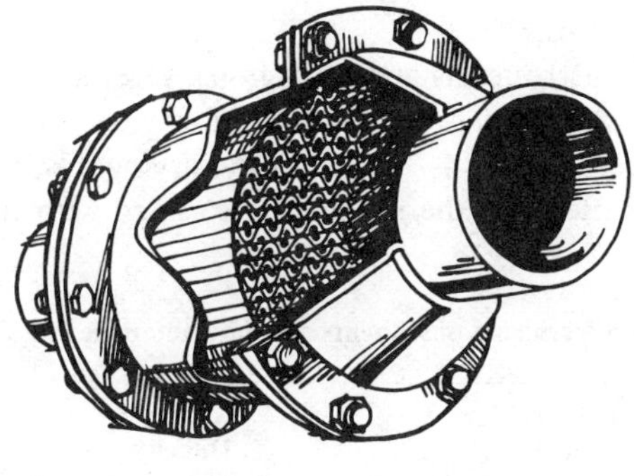

Figure 14-15. Typical catalytic oxidizer unit for auto exhaust emission control. (courtesy Engelhard Industries)

Figure 14-16. Catalytic converters are used in auto exhaust systems. (courtesy Engelhard Minerals & Chemicals Corp., Engelhard Ind. Div.) PTX®–Engelhard Minerals & Chemicals Corp.

Varnish Cookers[8]

Of the methods proposed for air pollution control—condensation, absorption, adsorption, chemical oxidation, electrical precipitation, combustion, and flocculation—only scrubbing and combustion have proved practical. Scrubbing with water may remove some air contaminants, but it is unable to eliminate odors and noncondensibles. Combustion, therefore, is more popular in odor abatement from varnish cookers. Thermal as well as catalytic oxidizers have been widely and successfully used in controlling varnish cooker emissions.

In the varnish industry resins and drying oils, together or separately, are heated in kettles, during which vaporization of solvents, oils, and other fractions takes place. Table 14-7 indicates the drying oils, resins, thinners, and other reactants used and the problems associated with these materials.

Table 14-7. Common Varnish Constituents and Characteristic Air Contaminants[8]

Class	Drying Oils, Film-Forming Materials	Resins	Diluents, Thinners	Other Reactants
Ingredients	Safflower oil Linseed oil	Natural shellac, rosin	Turpentine Petroleum solvents	Fatty acids Glycerol
	Tung oil	Dammar, copal, kauri	Coal tar solvents	Pentaerythritol
	Soya oil Oiticica oil Castor oil (dehyd.)	Synthetic phenolic alkyd or glyptal	Alcohols	Phthalic anhydride Maleic anhydride Fumaric acid
	Tall oil Fish oils	Epoxy Coumarone-indene Melamine		Styrene Cyclopentadiene
Materials identified in air contaminants	Acrolein Free fatty acids Glycerol Sulfur compounds	Terpene oils Phenol Formaldehyde Styrene Monomer	Terpenes Hydrocarbons	All reactants may be lost to some degree
Contaminant class	Odors Irritants Liquid particles	Solid particles Liquid particles Odors Irritants	Odors Organic vapors	Odors Liquid particles Solid particles Irritants

Spray towers are provided between hoods and the combustor to avoid flashbacks and separation of condensibles (Figure 14-17). Ignition temperatures of most of the components in varnish cooker emissions are given in Table 14-8.

Oxidation of most of the products takes place at a low temperature. By the time these materials come to the incinerators, they may be already

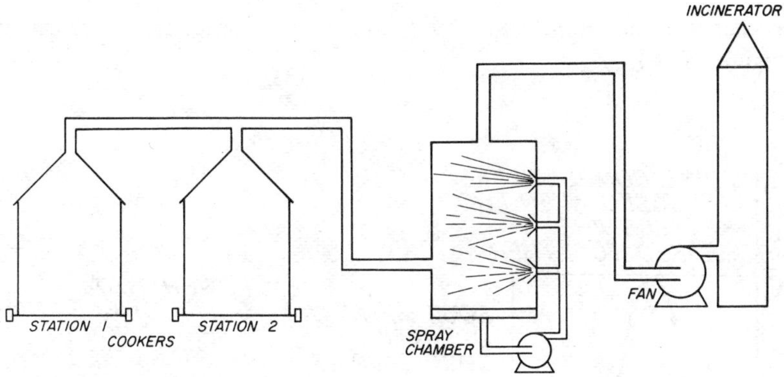

Figure 14-17. Varnish cooking flow schematic with air pollution control equipment.

Table 14-8. Ignition Temperatures of Selected Materials Used in Varnishes[8,10]

Material	Auto-Ignition Temperature, °F	Ignition Temperature of Mist, °F
Stoddard Solvent	444	–
V M & P Naphtha	450	–
Mineral Spirits	475	–
Linseed Oil	820	1150
Castor Oil	840	–
Tung Oil	855	–
Maleic Anhydride	890	–
Toluene	1026	–
Phthalic Anhydride	1083	–
Refined Linseed Oil	–	1130
Linseed and Fish Oil (50-50 mix.)	–	1140
Z-3 Linseed & Z-3 Castor Oil	–	1120
Alkyd, Tall, Pentaerythritol	–	1280
Varnish	–	980-1160
Phenol	1319	–

partially oxidized to aldehydes, acids, and peroxides. These "intermediates" are more objectionable than the initial fumes from cookers, and must be oxidized completely to H_2O and CO_2.

It can be seen From Figure 14-17 that several cooking stations can be served by one combustion chamber. By this method, peaks in combustibles from different cookers can be staggered to achieve a lower cumulative peak, and hence economy in fuel cost and protection of active centers of catalysts (Figure 14-18).

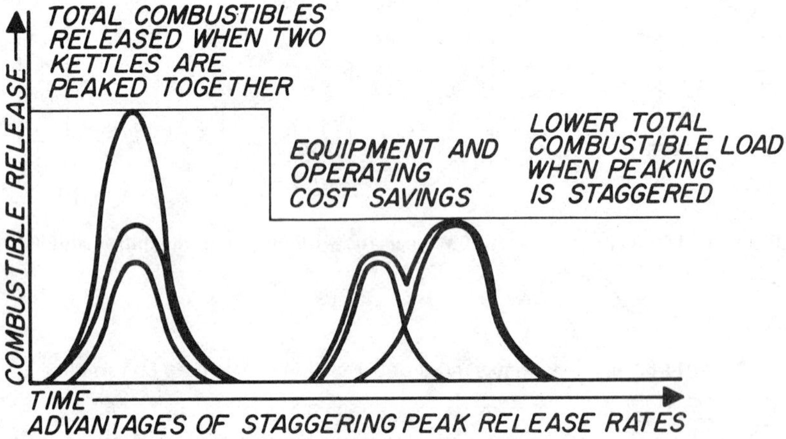

Figure 14-18. One incinerator can be used for more stations to achieve economy.

Coffee and Chicory Roasting[11]

In small amounts, coffee odor may be termed pleasant, but in larger quantities it has been a cause of malodor. Coffee contains proteins, fats, sugars, dextrine, cellulose, caffeine, organic acids, etc. During the roasting process, some of these compounds volatilize, decompose, and partially oxidize, producing a bad odor. It has been reported that ammonia and its organic compounds, acids, aldehydes, ketones, phenols, sulfur compounds, and furfuryl mercaptans are responsible for the odor.

Both thermal and catalytic afterburners have been used to control these odors. The results of analysis on tests are shown in Tables 14-9, 14-10, 14-11, and 14-12, given by Sullivan *et al.*[11]

During testing, it was observed that temperatures up to 1,200°F were required for thermal incinerators, and temperatures up to 700°F for catalytic oxidizers. Activity of the catalyst was sufficiently long, and deposits could be removed by simple washing.

Drying Operations

Drying operations have been known to cause fumes. If these fumes contain constituents having odors, they necessitate an odor abatement program. Many manufacturing processes have drying, baking, or curing routes.

Table 14-9. Peak Concentrations of Hydrocarbons and Aldehydes From Chicory Roast and Direct-Fired Afterburner[11]

Afterburner Condition	Hydrocarbons ppm as n-Hexane	Aldehydes ppm as HCHO	Remarks
Not operating	600	132	Heavy plume, acrid smell
Outlet at 720°F	150	45	No plume, acrid smell
Outlet at 900°F	35	15	Slightly acrid smell
Outlet at 1050°F	10	6	Odor removed

Table 14-10. Concentrations of Organics Expressed as Hexane at Various Operating Temperatures on Catalytic Afterburner Treating Coffee and Chicory Roast Gases Measured at Peak of Cycle[11]

Temperature °F	Organics as Hexane Parts Per Million	
	Coffee Roast	Chicory Roast
Untreated	140	600
700	40	90
800	35	60
900	30	40

Table 14-11. Concentrations of Various Constituents Tested Following Coffee Roast and Catalytic Afterburner Operating at 710°F[11]

Constituent	From Roaster	After Preheat	After Catalyst
Particulates mg per cu m averages	208-396	78-198	3-21
Total organics as hexane ppm peaks	143	84	40
Aldehydes as HCHO ppm			
averages		7.5	6.8
peaks		13.2	13.8
Mercaptans ppm	1.0		Nil
Organic acids as CH_3COOH mg per cu m	380.0		118.0
Phenols as C_6H_5OH mg per cu m	7.0		Nil
Odor	very acrid	acrid	slightly acrid but acceptable

Table 14-12. Concentrations of Various Constituents Following Chicory Roast and Catalytic Afterburner Operating at 900°F[11]

Constituent	From Roaster	Between Flame and Catalyst	After Catalyst
Particulates mg per cu m averages	178	28	12
Total organics as hexane ppm	400-600	45-55	20-40
Aldehydes as HCHO averages	27.5	13.4	1.6
Phenols as C_6H_5OH mg per cu m	7.5	Nil	Nil
Organic acids as CH_3COOH mg per cu m	425	104	52

Hardboard plants temper boards by dipping them in oxidizing oils made of linseed oil, petroleum products, polymers, and a drier such as iron tallate or iron naphthenate. When these boards are subsequently baked at 300°F, malodorous fumes are evolved. The experience of one company has been that an incinerator with heat recovery was a satisfactory route to eliminate odor and visible contaminants.[12]

The foam rubber used for carpet backing involves curing light mineral-oil-containing rubber in forced convection air ovens. Since oil aerosols are much more stable, even high-pressure water-jet scrubbing cannot eliminate odorous mists from the curing oven fumes. These fumes have a tendency to settle down. Incinerator is perhaps the only effective way of control. If lubricating and compounding ingredients do not contain catalyst poisons, catalytic oxidizers can also be used. A flow schematic of a complete abatement system is shown in Figure 14-19 and performance test data are given in Table 14-13.

Wire Enameling

Figure 14-20 shows a wire enameling process. Bare wire is coated with resins and varnishes. These are mixtures of volatile solvents and solids. During the drying process, fumes from these solvents are evolved. Most of the time condensibles from these vapors accumulate in the stack and cause a fire hazard, in which case scrubbing is not recommended. Table 14-14 indicates the performance of incinerators used for odor control in a typical wire enameling industry. When solvents do not contain catalyst poisons, catalytic oxidizers have been used.[14]

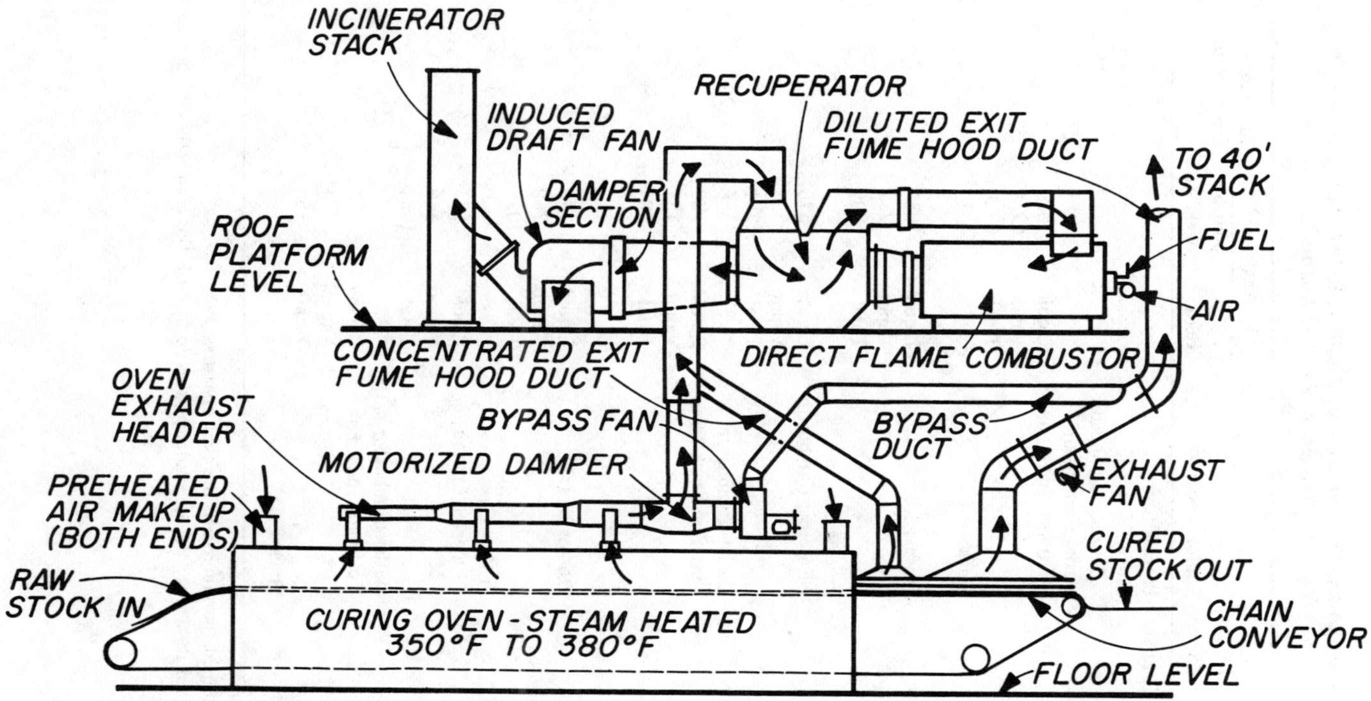

Figure 14-19. Schematic flow diagram of system for odor abatement for rubber curing.

Table 14-13. Performance Test Results in Rubber Curing Process Using Incinerator for Odor Control[13]

Run	Station Location	HC Analysis** Item	HC Analysis** ppm by Weight	Incineration Temperature °F	Flow lb/hr
Blank Run	Process Riser				26,000
	Process Gas Entering Combustor				
	Outlet Combustor			1114	
	Upstream Stack Fan	Liquid	0.5		
		Vapor	69.0		
		Total	69.5		
	Stack Exit				28,200
First Operating Run	Process Riser	Liquid	1077		29,800
		Vapor	232		
		Total	1309		
	Process Gas Entering Combustor	Liquid	576		
		Vapor	725		
		Total	1301		
	Outlet Combustor			1120	
	Upstream Stack Fan	Liquid	20		
		Vapor	187		
		Total	207		
	Stack Exit				31,000
Second Operating Run	Process Riser	Liquid	795		29,800*
		Vapor	260		
		Total	1055		
	Process Gas Entering Combustor	Liquid	635		
		Vapor	446		
		Total	1081		
	Outlet Combustor			1190	
	Upstream Stack Fan	Liquid	35		
		Vapor	54		
		Total	89		
	Stack Exit				31,000*

* Assumed the same as in the first operating run.

**Liquid results are the sum total picked up in both the condenser and the aerosol filter.

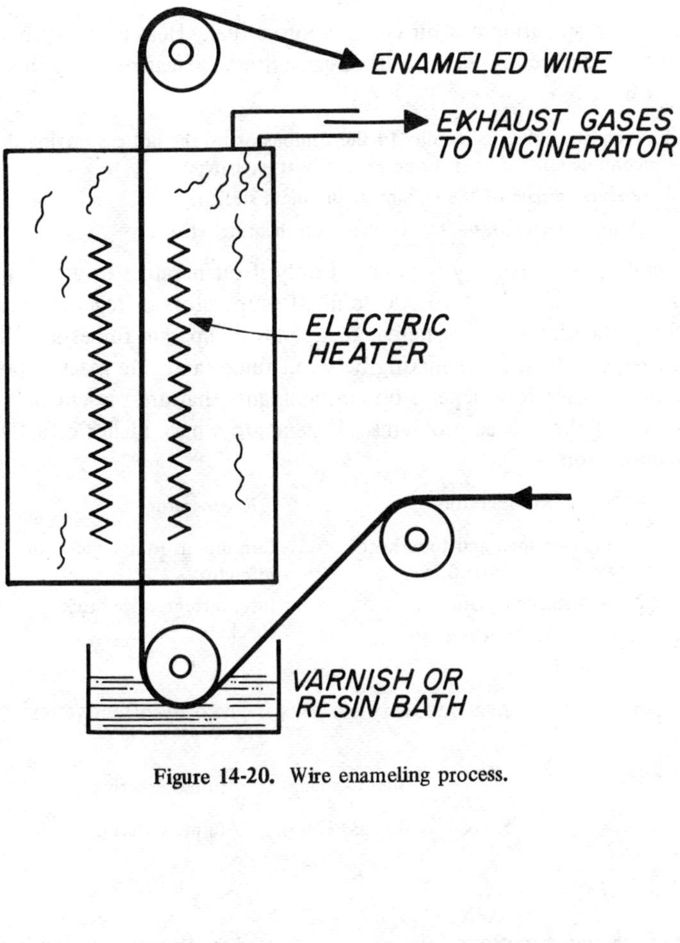

Figure 14-20. Wire enameling process.

Table 14-14. Summary of Odor Panel Tests in Wire Enameling[14]

Approximate Incineration Temperature °F	Odor Strength of Inlet Average Odor Units/scf	Odor Strength of Outlet Average Odor Units/scf	Effect of Thermal Incineration on Odor Strength
1400	2600	70	97% reduction
1200	2500	350	86% reduction
1000	1300	2100	Increased

HEAT RECOVERY

The major operating cost in combustion is fuel. Heat recovery is therefore an important consideration in design criteria. Heat recovery can be achieved by

1. Preheating the gases going to the combustor by the hot gas leaving the combustion chamber, using gas-gas heat exchangers.
2. Recycling part of the stream to the main stream.
3. Power regeneration—steam generation, heating, etc.

Generally, heat recovery is practical only if incinerators or afterburners have a capacity of at least 10,000 scfm. Recuperators or regenerators are used for preheating gases. Whereas recuperators separate fumes and flue gases in regenerators, some mixing between fumes and flue gases is possible. Selection will therefore depend on contaminants that are present in the gas stream, and if they cause problems. Regenerators have higher effectiveness than recuperators.

Recuperators	**Regenerators**
1. Tubular shell and tube heat exchanger, cross flow	1. Continuous rotary metal or refractory
2. Flat stacked plate	2. Intermittent valve-brick checkers
3. Corrugated surface	
4. Finned type	

Performance of these heat recovery systems is given by effectiveness η, defined as:

$$\eta = \frac{Q}{Q_{max}} = \frac{T_{\text{fumes leaving}} - T_{\text{fumes entering}}}{T_{\text{flue gas entering}} - T_{\text{fumes entering}}}$$

BURNERS

There are two functions for the burners used in afterburners or incinerators: (1) to supply the heat source, and (2) to accelerate the oxidation rate, since the concentration of free radicals is much greater in the flame.

The range of input rates within which a burner operates is termed the "turndown" ratio. This is defined as a ratio of maximum heat input to minimum heat input within which a burner can operate satisfactorily. Maximum heat load is limited by the "blow-off" phenomenon. If the velocity of the fuel-gas mixture exceeds the flame velocity, the flame will be extinguished (blow-off). On the other hand, "flashback" is the limitation on minimum heat input; this occurs when the flame velocity exceeds the gas-air mixture velocity. Since velocity through an orifice is proportional to the square root of the pressure drop across the orifice, the turndown can also be defined by the square root of ratio maximum to minimum pressure drop across the burner orifice.

As was mentioned earlier, there are two classifications of burners for the mixing of fumes—distributed and discrete-type burners. Normally, however, the main classifications are based on the type of fuel used: (1) gas burners, (2) oil burners, or (3) gas-oil combination burners. From a practical standpoint, most gaseous burners are the distributed type, whereas most of the oil burners are discrete.

Gas Burners

Gas burners can be classified based upon the mixing of gas and air: premixed or nozzle type. In the premixed type of gas burner, air and gas are mixed upstream of the burner prior to going to the burner. Based upon the pattern of ports, premixing types can be either (1) small port or ported manifold premixed burners (distributed), or (2) long port pressure-type premixed burners.

In small ports, a number of ports may surround the main port in line, ribbon, grid, doughnut, or spoke and wheel patterns. If the flame from one port is temporarily extinguished, adjacent ports will reignite it. High heat is liberated in a small area, so materials of construction must be considered. This type, therefore, is not used for high heat loads.

Figure 14-21. Small port Premix Burner for direct preheating of make-up air in a duct. The venturi-like stainless steel shrouds assure rapid mixing with secondary air from the duct while maintaining flame stability over a wide turndown range. (Courtesy North American Mfg. Co.)

Long port or pressure-type premixed burners have only one main port. It produces a short, intense flame within the burner tile. The burner tile is diverging, conical, or cylindrical in shape. This creates a recirculating zone around the burner for combustibles and free radicals. It helps in stabilizing combustion within a wide turndown ratio.

Based upon the type of mixing, premixed burners are classified as (1) inspirator, (2) aspirator, or (3) fan.

Inspirator and aspirator types use venturis for mixing. In the inspirator premixed type, gas is passed through the venturi and air is sucked in, depending on the flow of gas, by expanding gas energy. Gas loses substantial pressure while passing through the venturi, and this helps in sucking air. The gas must therefore be available at sufficient pressure (3 to 50 psi). Since small amounts of gas are used to induce a substantial volume of air, turndown is limited.

The aspirator premixed type can be used even when gas is at low or atmospheric pressure. Air passes through a venturi and sucks gas, as air energy is expended at the throat of the venturi. Gas volume depends on air flow rate. To adjust the gas to air ratio, air flow can be adjusted by a valve.

In the fan premixed type, the gas enters directly to an air blower and the air ratio cannot be adjusted automatically. There is also some danger of flashback.

Nozzle-Type Gas Burners

This is also known as diffusion mixing. In this type of mixing, air and gas enter separately and are not mixed before they enter the burner port. As gas and air pass the nozzle orifice, rapid mixing takes place. Air is passed by a swirling movement, and thus the gas jet or jets are mixed rapidly with air near the ignition point.

In this type, air can be passed to the surrounding furnace gas if there is a possibility of higher concentrations of CO_2 sufficient to extinguish the flame. This will give stability to the flame. If longer flames are not required, radiation-type burners can be used. These burners heat their surrounding refractory tiles, which radiate heat to the furnace. Most incinerators use nozzle-type burners due to their greater turndown, stability, and the impossibility of flashback in using lower pressure blowers. They are adaptable to gas as well as oil.

Oil Burners

Since liquid is used as a fuel and must be vaporized, this type of burner is also called a vaporizing burner. Oil fuel undergoes the following sequence. (1) Oil is passed through the atomizer under pressure; (2) it is atomized by high- or low-pressure air or atomizer nozzles; (3) it comes out in conical liquid form and is broken into small droplets; (4) the droplets are heated by radiation from the flame and burner tile and by convection from hot gases. Vapors are formed around the droplets and the vapors are mixed

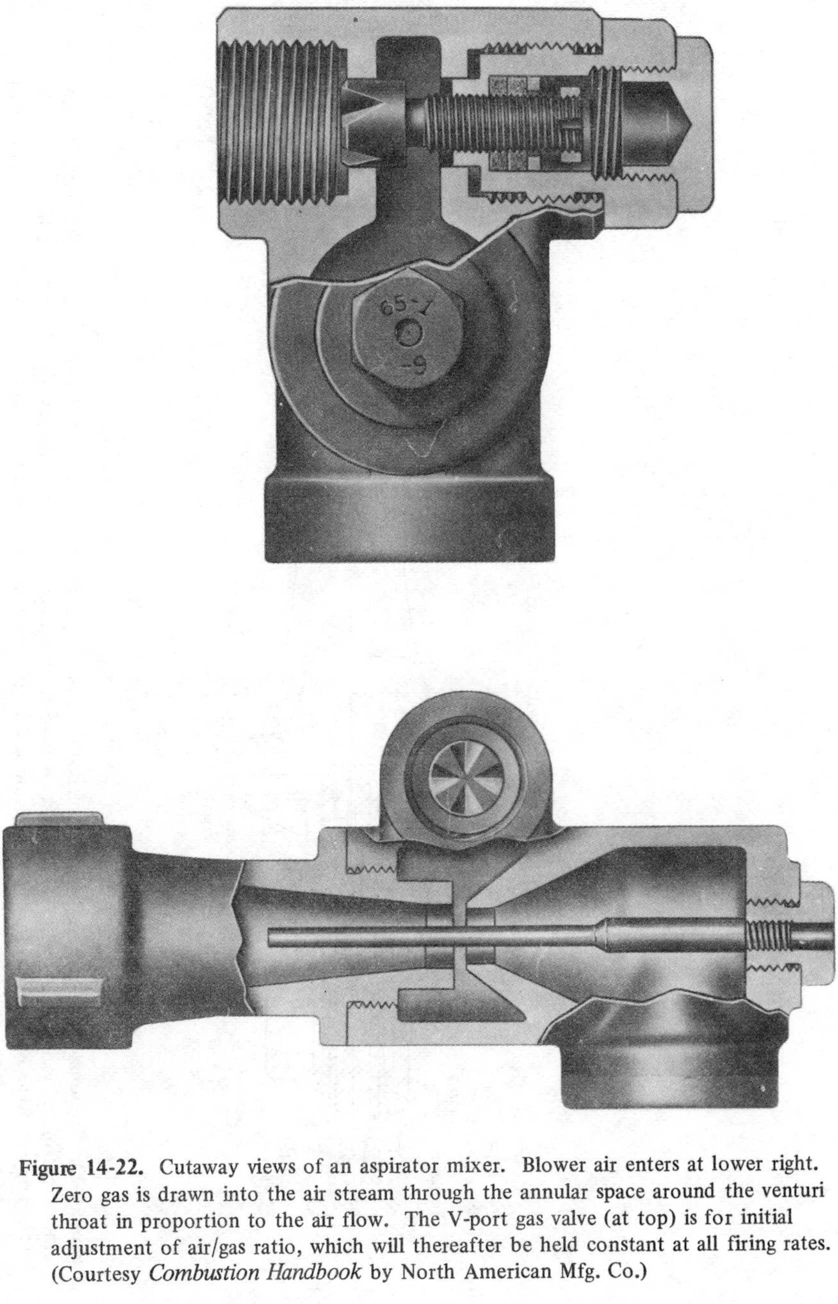

Figure 14-22. Cutaway views of an aspirator mixer. Blower air enters at lower right. Zero gas is drawn into the air stream through the annular space around the venturi throat in proportion to the air flow. The V-port gas valve (at top) is for initial adjustment of air/gas ratio, which will thereafter be held constant at all firing rates. (Courtesy *Combustion Handbook* by North American Mfg. Co.)

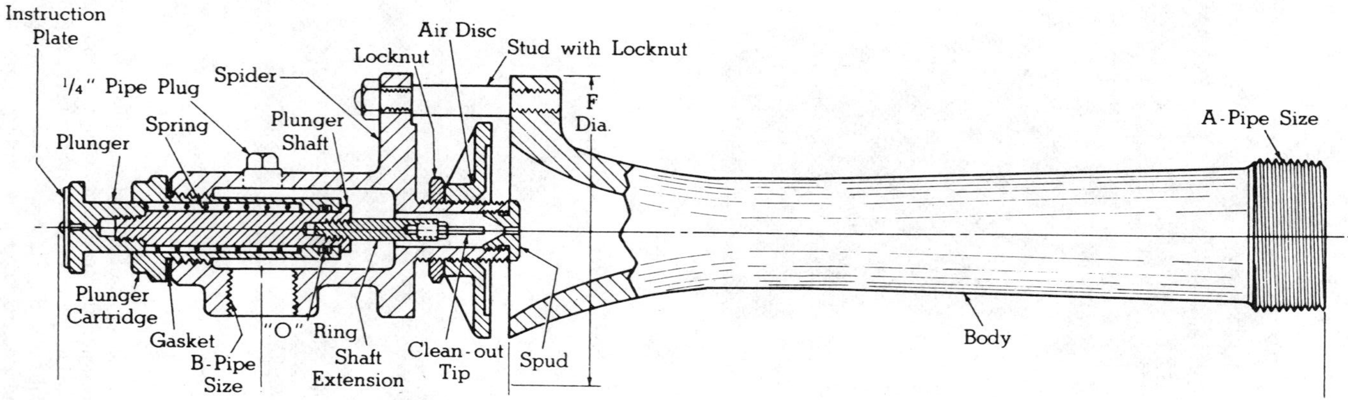

Figure 14-23. Sectional view of an inspirator mixer. High pressure gas discharging from the spud or gas orifice forms a high velocity jet in the throat of the venturi-like body. The gas jet entrains air in proportion to the gas flow, and the gas and air are mixed. The gradual enlargement of the body results in a re-conversion of the mixture velocity to static pressure. The position of the disc may be adjusted to produce the desired air-fuel ratio. (Courtesy *Combustion Handbook* by North American Mfg. Co.)

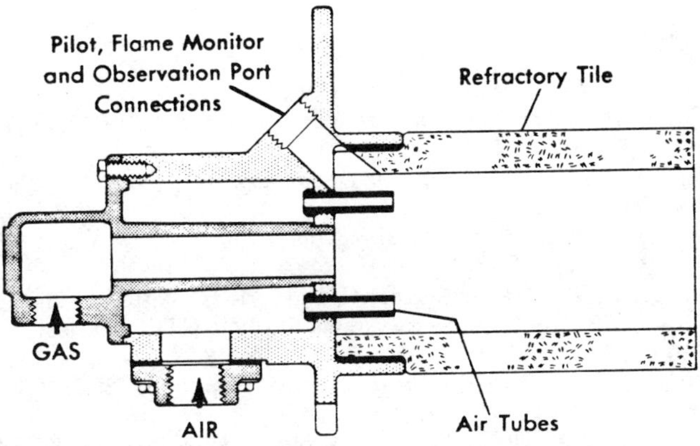

Figure 14-24. Sectional view of a nozzle mixing burner. Air and gas meet near the point of ignition. (Courtesy *Combustion Handbook,* North American Mfg. Co.)

with turbulent hot gases, ignited, and then evaporated. Or they may be chemically broken down and become ignited. If droplets leave any char, it may be completely burned off or carried away to the stack.

Combination Burners

It is practical to have burners which can use both oil and gas as fuel. This will give fuel flexibility to burners. Figure 14-25 explains the design of this type of burner.

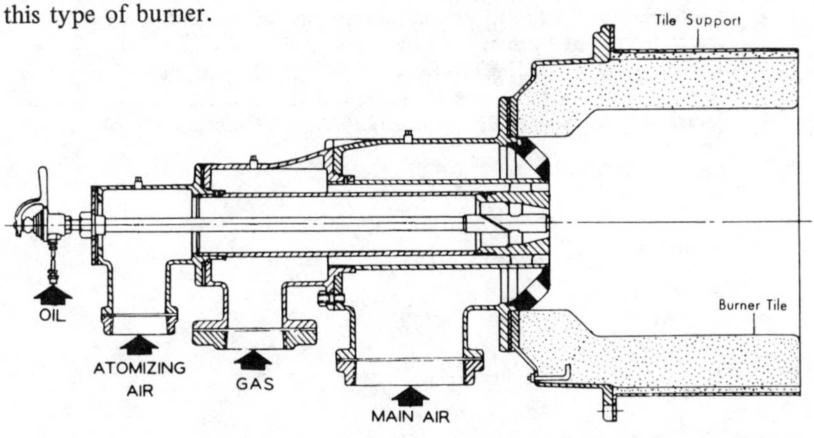

Figure 14-25. A 4-pipe ultra-stable combination burner capable of burning heavy oil or any gas. This burner can be operated over a wide range of fuel/air ratios at all firing rates in cold, tight combustion chambers. The gas tube and gas connection can be omitted and a shorter oil tube substituted to make a less expensive straight oil burner. The metal tile support shown should be used only when the burner is installed in a thin metal wall—not when installed in a refractory wall. (Courtesy *Combustion Handbook,* North American Mfg. Co.)

REFERENCES

1. Turk, A. *Chem. Eng.*, **77** (9), 199 (1970).
2. Corey, R. *Principles and Practices of Incineration.* (New York: Wiley-Interscience, 1969).
3. Summer, W. *Methods of Air Deodorization.* (New York: American Elsevier, 1963).
4. Vulis, L. *Thermal Regimes of Combustion.* (New York: McGraw-Hill, 1961).
5. Rolke, R. W., C. Hawthorne, C. Garbett, E. Slater, T. Phillips, and G. Towell. "Afterburner Systems Study," NTIS Publication PB-212-560 (1972).
6. Lund, H. *Industrial Pollution Control Handbook.* (New York: McGraw-Hill, 1971), p. 7-7.
7. Thomaides, L. *Pollution Engineering,* **3**, 32 (May-June 1971).
8. Mills, J., W. Hammond, R. Adrian. *J. Air Poll. Control Assoc.,* **10** (2), 161 (1960).
9. Engelhard Industries. "Deoxo Abatement Systems for Pollution Control of Organic Fumes," Bull. EM-9605-2.
10. Benforado, D. and G. Cooper. "The Application of Direct Flame Incineration as an Odor Control Process in Kraft Pulp Mills," presented at 22nd Engineering Conference Process System and Controls–Water and Air Pollution, TAPPI, Atlanta, Georgia (September 19-22, 1967).
11. Sullivan, J. L., F. Kafka, and L. Ferrar. *J. Air Poll. Control Assoc.,* **15** (12), 583-586 (1965).
12. Wehrle, A. *Plant Engineering* (November 1970).
13. Sandomirsky, A. G., D. M. Benforado, L. D. Grames, and O. E. Panletta. "Fume Control in Rubber Processing by Direct Flame Incineration," paper presented at 59th annual conference of Air Pollution Control Association, San Francisco (June 1966).
14. Benforado, D., and J. Waitkus. "Exploring the Applicability of Direct Flame Incineration," paper presented at the Annual Convention of Electrical Conductor Div. of the Wire Assn., Chicago (October 23, 1967).

CHAPTER 15

FLUID BED INCINERATION

Paul N. Cheremisinoff

New Jersey Institute of Technology
Newark, N.J. 07102

Fluid bed incinerators are finding increasing consideration and application in the combustion of solid and liquid wastes. The potential of fluid bed techniques over older and more conventional equipment shows possible improvements in economics and efficiency. Fluid bed incineration may be considered for higher thermal efficiency, odor control improvement, and particulate matter emission.

Previously, fluid bed reactors have been used in such process applications as catalytic cracking of petroleum, drying, and roasting. The quest for new techniques by researchers and practitioners in pollution control applications has led them to examine this method and its advantageous potentials and design flexibility. Fluid bed incinerators have been demonstrated in the combustion of many types of solid wastes, including municipal garbage, sewage sludge, and petrochemical and refinery wastes. They are applicable to almost any waste material that can be destroyed by combustion. One of the distinct advantages of this high thermal efficiency method of combustion is its ability to incinerate materials with high water content without using supplemental fuel.

WHAT IS A FLUID BED?

When a gas passes upward through a bed of finely divided solid particles, four general class conditions may exist, depending upon the characteristics of the particles, the geometry of the bed, and the velocity of the gas (Figure 15-1). These situations are (1) fixed bed, (2) dense phase fluidization, (3) two-phase fluidization, and (4) pneumatic transport.

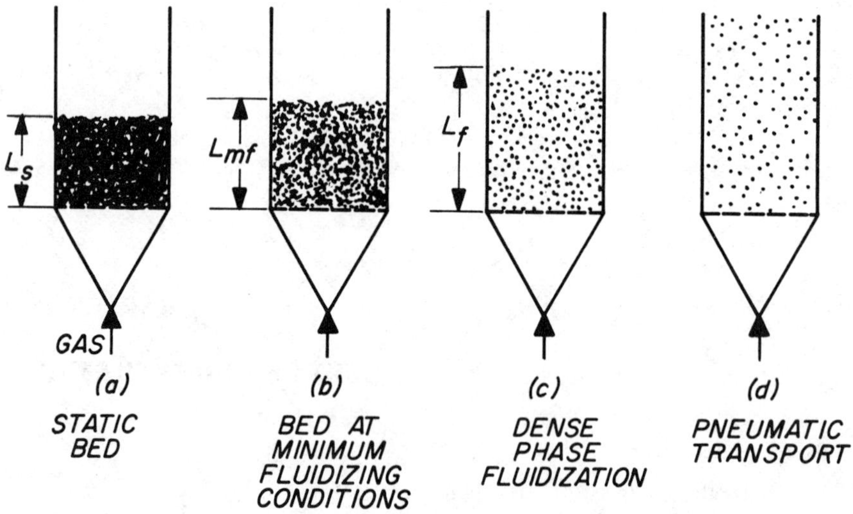

Figure 15-1. Fluid bed situations.

In a fixed bed at low velocity, the passage of a gas up through the bed causes a relatively low pressure drop. The solid particles remain essentially stationary and the gas merely percolates through the void spaces in the bed. The particles remain in place in the fixed bed as long as the pressure drop is less than the weight per unit area of the bed. This reaction is represented in graphical form in Figure 15-2. Any further increase of the gas velocity causes the bed to expand slightly, and the particles gain some freedom to move.

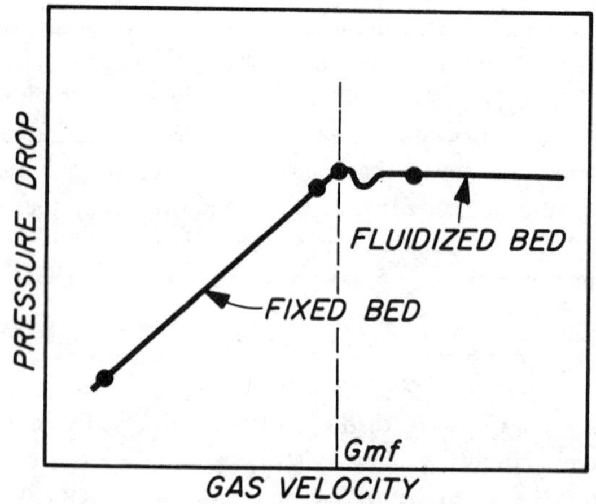

Figure 15-2. Variation of pressure drop with velocity in fixed and fluidized beds.

At a certain flow rate, indicated by G_{mf}, the pressure drop becomes equal to the weight of the bed per unit area and the particles are suspended in the gas stream. At this point the bed is said to be in the fluidized state. This gas-solid system has liquid-like characteristics. Intense and violent mixing between the solids and gas occurs. The results are completely uniform temperature, composition, and particle size distribution throughout the bed. The inception of fluidization, G_{mf}, is dependent on fluid and particle properties. The maximum gas velocity that can be allowed in a fluid bed is the terminal velocity, G_t, of the particle. These velocities are used in pneumatic transport. The limits of fluidizing velocity are $G_{mf} < G_f < G_t$. Depending on design, the actual gas velocity is chosen between the two extremes. It is because of this extensive combustion surface that the high thermal efficiency is possible in fluid bed reactors. At operating equilibrium, the fluid bed acts as a boiling liquid.

THE FLUID BED REACTOR

A fluid bed reactor or incineration system will consist of the following major components.

1. The reactor section is constructed of carbon steel and lined with insulation and a refractory to meet required incineration conditions.

2. The freeboard or disengagement zone is designed so that there is minimum solid particulate carryover. Exiting gas may be brought in contact with the feed and may be cooled by water content in the feed. It is the largest section of the reactor. Proper design of this section allows disengagement of solids from gases. This section may be designed with an increase in diameter to reduce exit gas velocity and ensure reduced dust loading.

3. The plenum chamber ensures proper air distribution below the orifice plate.

4. The distributor or orifice plate separates the plenum chamber from the fluid bed zone. The fluidizing and combustion air passes through this plate, which passes air but prevents the backflow of bed material into the plenum chamber.

5. Air supply (3 to 8 psig) for fluidization and combustion. The combustion system is usually forced draft. While a preheater may be desirable for start-up, secondary fuel may be provided for injection directly into the fluid bed and/or freeboard section.

6. Dust collection systems collect particulate matter from the exit gas. This may consist of both primary and secondary systems of low-pressure

drop cyclones and wet scrubbers to control solid particulates in the submicron range. The latter would, of course, merit high-energy venturi scrubbers for consideration.

7. Feed and discharge systems to suit the particular materials handled. These include such hardware as screw feeders, rotary valves, etc., depending on design and material requirements (solid or liquid type).

8. Instrumentation to automate the operation. Fluid bed systems lend themselves to a wide degree of instrumentation and imaginative engineering possibilities.

A schematic for a typical fluid bed incineration system is shown in Figure 15-3, with interrelation between the components.

FLUID BED INCINERATION EXAMPLES

Fluid bed incineration obviously has wide potential application in the combustion of both solid and liquid wastes. Solid wastes can be divided into two categories: (a) solid wastes such as municipal garbage, general refuse, and packing waste, and (b) semi-solid wastes such as domestic sewage sludge and highly viscous, almost non-flow-type combustibles. Liquid wastes may consist of such materials as hydrocarbon and chemical mixtures of liquids, clarifier effluents from pulp and paper plants, and petrochemical and oil refinery wastes.

Solid wastes can be fed to a fluid bed incinerator by either screw-type or ram feeders. The fluidizing medium can be generated ash or sand of 0.03-inch nominal-size diameter. Fluidizing velocities are typically 2 ft/sec at 1,400°F, which is in the combustion temperature range. The mass of a fluid bed medium is kept at the combustion temperature by initially heating the bed with an auxiliary fuel injected into the bed at the top or bottom of the reactor. Heating the bed from the top can be more economical from a capital cost standpoint since it eliminates more costly refractory lining of the plenum chamber. Combustion and fluidizing air is introduced from the bottom of the bed. These units are designed to have a burning capacity of 300,000 Btu/sq ft/hr.

Semi-solid wastes can be fed to the incinerator by a screw pump. Thermoplastic materials such as greases are most readily fed by first melting and then pumping them. In cases where auxiliary fuel is required, the main body design remains essentially as described.

In liquid wastes, feed materials for combustion are introduced at the top of the bed in the freeboard area. This enables sufficient residence time for heat exchange to take place between incoming wet feed and exiting combustion gases. Since these hot gases are generally thought of as totally

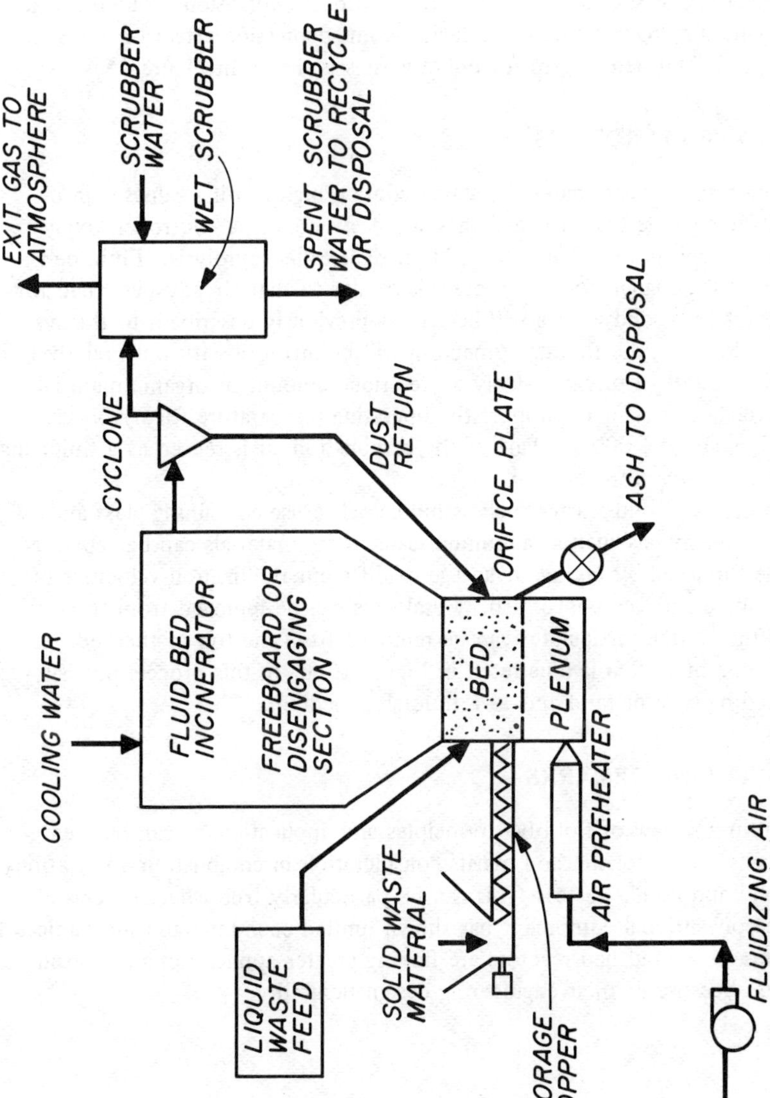

Figure 15-3. Typical fluid bed incineration system.

wasted, evaporation of water from incoming feed by direct heat exchange has the effect of thermal efficiency improvement. In designing a system for inorganic wastes, particular attention should be given to the chemical composition of the feed material. In addition to corrosion problems, trace elements may have a fluxing or fusion point depression effect on the ash. A typical fluid bed system for liquid wastes is shown in Figure 15-4.

FLUID BED PYROLYSIS

Some solid waste materials, which contain high heating values, can be used for pipeline gas, provided they are destructed in a controlled oxygen or inert atmosphere. This type of process is called pyrolysis. Fluid bed equipment designed for this process is similar to that already described for incineration. Solid wastes will be fed, as previously described, to the system. However, the fluidizing medium will be nitrogen with only relatively small amounts of oxygen. Only a prescribed amount of organic material is burned–sufficient to support the operating temperature for pyrolysis, which is about 1,000°F. Part of the combustion air is reused as a fluidizing gas to conserve nitrogen.

In cases of solid wastes such as municipal refuse containing glass and nonferrous metals such as aluminum cans, these materials can be separated in the fluid bed by taking advantage of different fluidization velocities of sand, metal, and carbon or ash. Metal cans can be removed from the bottom, the lighter particulates can be removed from the top of the bed. This type of system is illustrated in Figure 15-5. In this process hot, suspended bauxite or sand acts as a fluidizing medium.

CONCLUDING REMARKS

From a review of fluid bed principles and applications it can be seen that this process or method merits consideration in combustion applications of solid and liquid wastes. This can be particularly true where older and more conventional equipment has shown limited economic and mechanical advantages. Fluid bed reactors are finding greater application and consideration because of their engineering design flexibility.

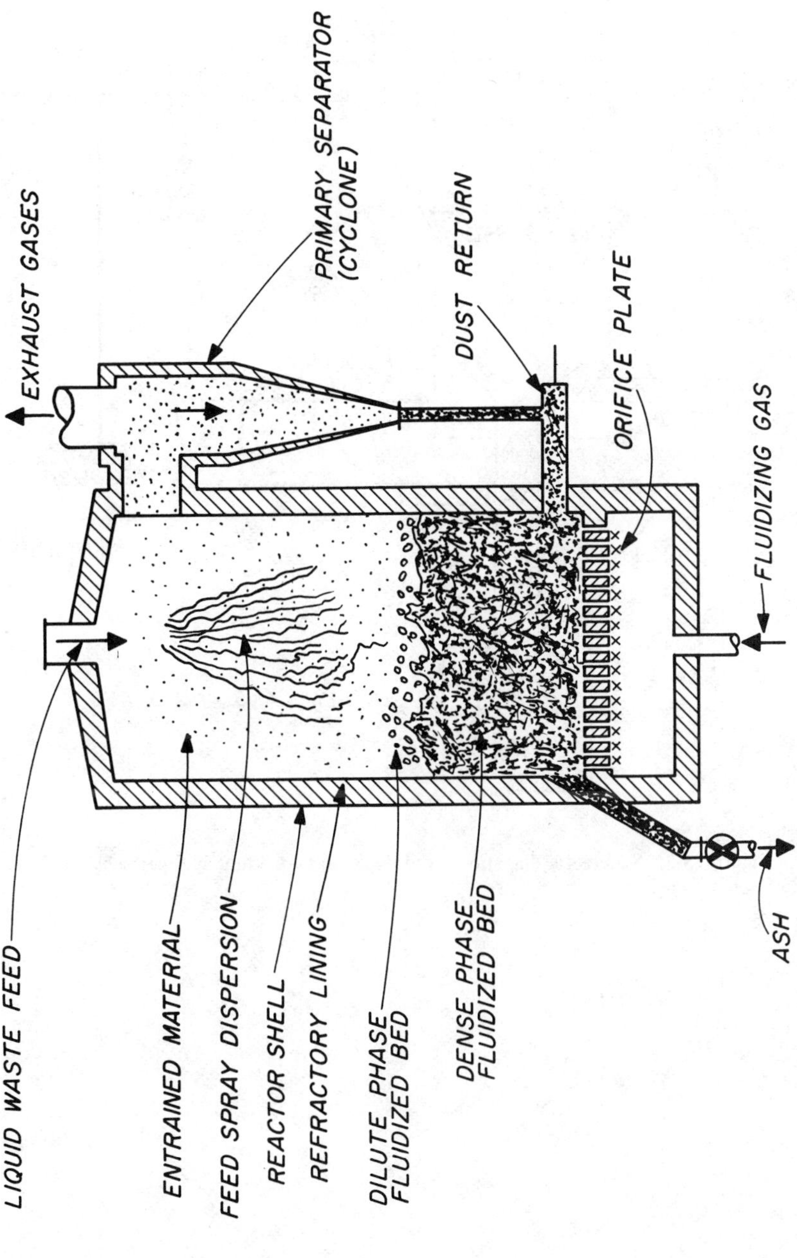

Figure 15-4. Liquid waste incineration system.

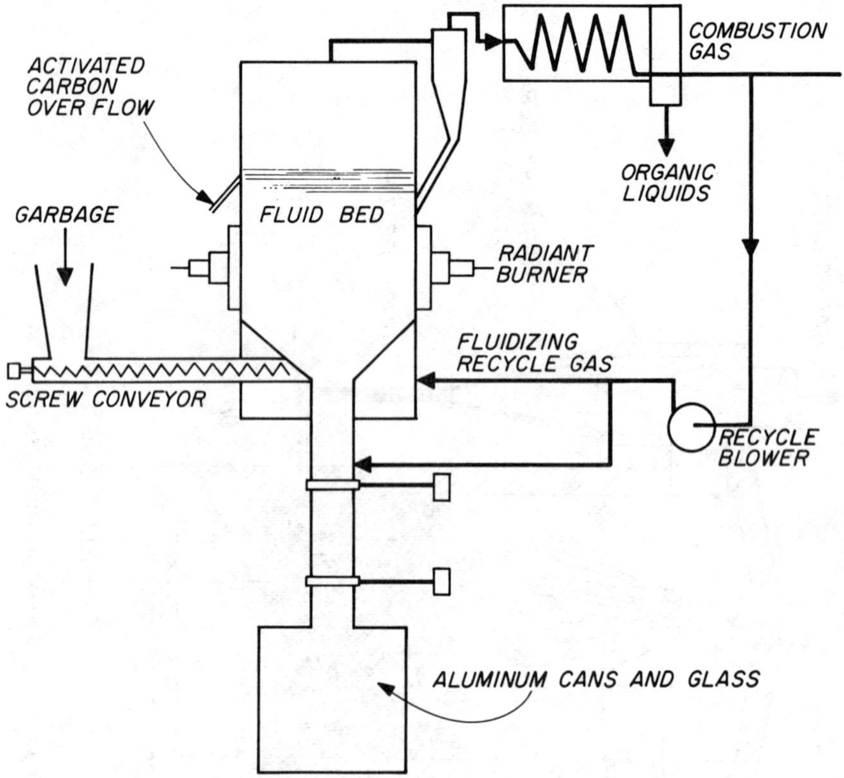

Figure 15-5. Schematic diagram of pyrolysis unit for municipal garbage.

CHAPTER 16

EFFECTIVENESS OF HEAT WHEELS

Stanley E. Gifford

Regenerative Heat Corp.
Middletown, Conn. 06457

Where applicable from an engineering standpoint, rotary air-to-air heat exchangers are an effective method of reducing thermal pollution, and in addition are economically advantageous.

Figure 16-1 shows a cylindrical rotor, incorporating heat transfer material, all mounted in a steel frame. The rotor usually revolves at about 14 rpm. The heat transfer media can be of various materials, such as knitted wire mesh of aluminum, stainless steel, or other substance that can be drawn into wire, knitted, and hold their set under operating conditions. These types of heat wheels (regenerators) have individually formed segments which are removable.

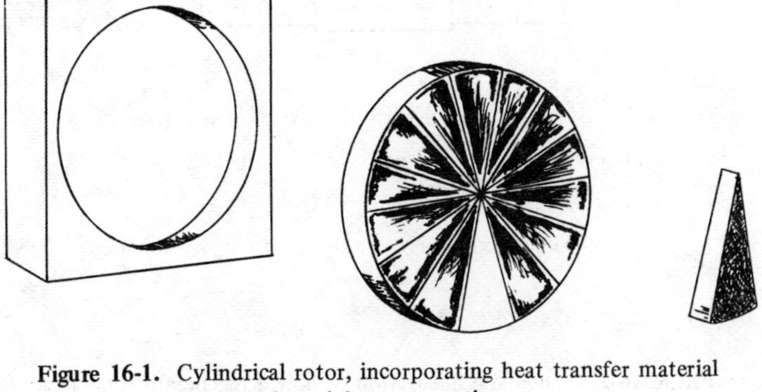

Figure 16-1. Cylindrical rotor, incorporating heat transfer material and steel frame mounting.

Other types of heat wheels are made of light-gauge corrugated metal or a honeycomb impregnated with desiccant. One type of heat transfer material is not suitable for all applications. For example, the desiccant honeycomb cannot be washed down with water, steam, or solvent. Regenerators are available in single units with capacities up to 50,000 cfm and can be used in pairs to handle larger air volumes.

Figure 16-2 shows how a regenerator operates. Hot contaminated air is passed through one side of the regenerator, giving up its heat to the transfer material. Note that the contaminants can be eliminated upstream or downstream of the heat wheel. A counterflow outside fresh air stream is passed through the other side of the regenerator. As the hot heat transfer material rotates into this cold air, it gives up its stored heat, creating a continuous, significant preheating process. It should be pointed out that similar heat transfer takes place with cold contaminated effluent, used to precool hot, wet outside air.

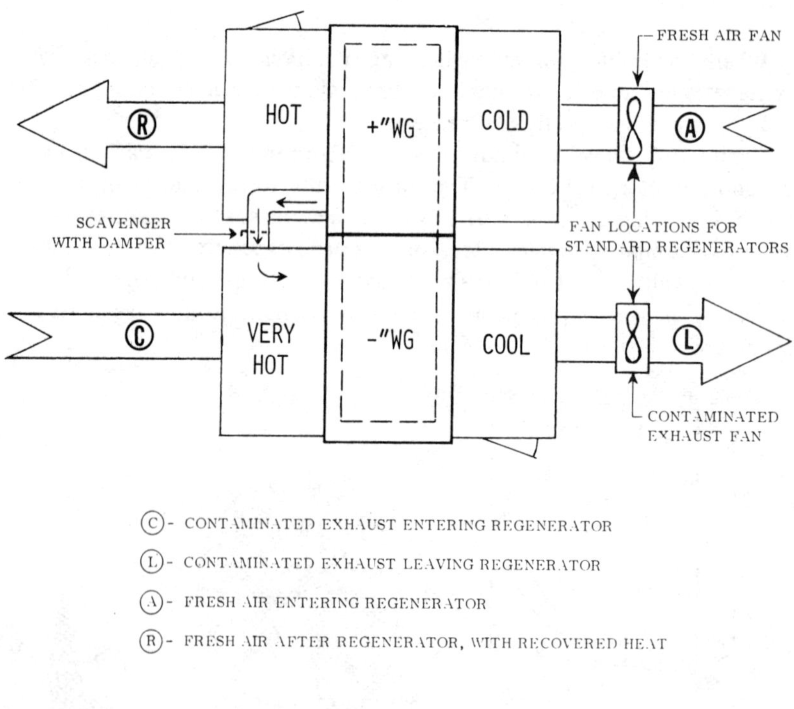

Figure 16-2. Schematic illustration of how a regenerator operates.

Since the equipment is of the regenerative type, as compared to recuperative, there will be cross-leakage between the air streams. The direction of this cross-leakage is controlled by maintaining appropriate air pressures in each air stream. Figure 16-2 also shows the preferred fan arrangement, with the exhaust fan inducing the effluent from the process, through the regenerator, and on out, thereby keeping the contaminated air stream at a negative pressure. The outside air fan should blow through the regenerator, thereby assuring a positive pressure for the fresh air. Now any cross-leakage is from fresh to contaminated air.

The other possible source of cross-contamination is from effluent trapped in the voids of the rotor passing from the dirty to the clean side as the rotor revolves. Figure 16-2 shows a sheet metal collecting box or scavenger located in the regenerated air stream. A small amount of outside air, equal to the rotor volume times its rpm, is used to flush the voids as they enter the clean side. It is relieved by the system air pressure back to the dirty side for heat recovery gain. The scavenger volume is about 3% of the regenerator capacity. Cross-leakage volume depends on the pressure difference between the streams, temperature, construction of unit seals, etc., and thus will vary widely.

The preferred fan arrangement prevents cross-contamination in case of mechanical failure of either fan. If the exhaust fan fails, the outside air stream is at a higher pressure than the dirty stream, so the leakage is inward. Should the fresh air fan fail, the exhaust side is at a negative pressure, and the clean side is still at a higher pressure. Other fan arrangements are of course possible. During operation, however, even with appropriate dampers and/or controls they can maintain appropriate air pressures but cannot stand the test of mechanical failure.

Heat wheels are generally of benefit only when there is exhaust—either because it is contaminated or because the codes require it. Occasionally, regenerators can be used to exchange heat within a single process, but this is not common practice.

BASIC DATA

The economic feasibility of a regenerator installation depends on four basic items:

1. Quantity of exhaust—acfm at a given temperature, scfm, and pounds per minute
2. Temperature of exhaust—°F dry bulb, °F wet bulb (if moisture is present), or grains or lb water/lb dry air
3. Hours of operation—hours, days, weeks
4. Cost of energy delivered to point of use—\$/million Btu or some other convenient parameter.

A fifth item which must be considered is whether you can get the energy you need. If you have been cut back on energy or are adding new processes, you may need a regenerator to allow you to operate within your energy allotment. This factor alone may make all others insignificant.

A sixth item is also important for any final determination, but general knowledge is usually sufficient for preliminary investigations–volumetric analysis of the effluent. This information is important for determining condensation temperatures of the contaminants. The results of this analysis will aid in determining the material from which the heat transfer media should be made.

EFFECTIVE USE OF HEAT

Coincident with gathering basic data, you should determine the most effective use for the regenerated heat. There are four general areas of use that cover most situations.

First, to save the most operating dollars, try to recycle the regenerated heat back to a process. Note that the regenerated heat does not have to be brought back to the same process; rather, it can be delivered anywhere heat is needed.

Second, if the heat cannot be recycled to a process, the regenerated heat can be used for plant heating, usually by providing makeup air to overcome negative plant pressure. The savings can be gained during the heating season.

A third consideration is the saving of capacity on basic equipment for new facilities and processes, or to reduce the demand on existing equipment or energy supplies. Aside from the obvious reduction in heating capacity, regenerators can make a significant impact on cooling equipment. On cold contaminated exhausts from a laboratory, closely controlled manufacturing area, or refrigeration process, for example, a regenerator will probably save about 60% of the total heat requirement of the outside air component of the total cooling load. This will generally represent enough dollar savings in high side equipment–compressors, cooling towers, pumps, piping, etc.–to cover the cost of the regenerator, plus a good portion, if not all, of its installation cost.

A fourth area of use has come to the forefront recently. This application involves the cooling of contaminated effluent (regardless of whether the regenerated heat can be used) to assist the operation of solvent recovery systems or low-energy pollution control systems. The unique heat transfer characteristics of regenerators using knitted wire mesh make this application increasingly important. The ability of knitted wire mesh to transfer or transport heat is not directly proportional to the specific heat or thermal conductivity of the material used, nor do dirt loading or films of condensed liquids significantly influence the rate of heat transfer. Thus where oils condense and/or dirt builds up—from a heat treating exhaust, for example—the typical reduction in heat transfer rate is not apparent. Regenerators become the most effective method of obtaining constant, high-performance characteristics.

TEMPERATURE

Having established the most effective use of the recovered heat, it is important to determine at what outside temperature level the performance of the proposed heat recovery system is to be analyzed. In all cases, establish performance at outside winter-design conditions. When dealing with processes, I favor the 99% value as shown in the American Society of Heating, Refrigeration and Air Conditioning Engineers (ASHRAE) guide. This indicates the temperature which will be exceeded only 1% of the hours in December, January, and February—about 22 hours in the average season. If the process operation is critical, the "median of extremes" can be used: the middle value of the coldest temperatures recorded each year for periods up to 25 or 30 years. This particular level is important because it indicates the reduction in demand on the energy source at the time when it is probably subject to its greatest draw. It could mean that recovery from several effluents would provide enough energy to run an additional or new process. Further, it can indicate the degree by which the heat-producing equipment within the process can be reduced. It can also serve as a check on the performance of the operating controls at substantially reduced capacity levels.

If you have a system where you can recycle the regenerated heat back to the process, the analysis should be made at the average year-round temperature. This serves to indicate the 12-month operating cost savings and, again, provides a check on the controls under normal operation.

If recycling is not possible and you plan to use the regenerated heat for plant heating, your analysis should use the winter average temperature. This information is available from the ASHRAE guide, along with the number of "degree days." This analysis indicates the dollar savings in operating cost during the heating season. This temperature level is perhaps the most difficult at which to calculate savings.

The percentages must be converted to Btu and the Btu to dollars to help management understand what is proposed. When it comes to converting Btu to dollars, a very detailed analysis can be made. However, adequate values can be obtained by using an equivalent heating season based on "average winter temperature" and "degree days." A *degree day* is based on temperature difference and time and is used in estimating fuel consumption.

If we assume a heating season of 6,000 degree days with an average winter temperature of 40°F, then

$$\frac{6{,}000 \text{ deg days}}{(65 - 40)} = 240 \text{ days or } 34 \text{ weeks}$$

To get appropriate operating parameters if the regenerator is being considered for effluent cooling only, analyses should be made at winter and summer outside-design conditions as well as at the year-round average temperature.

There are many processes which involve significant evaporation of moisture, and these can be excellent candidates for heat regeneration. Let us limit our discussions and examples to hot, dry exhausts. By dry, assume that the effluent is under 130 grains/lb of dry air. The moisture conversions have to be taken into account, and the calculations are somewhat more complex.

Let us refer to Figure 16-3, which illustrates the performance of regenerators on hot, dry exhausts. The first basic concept is that

$$\text{Mass A} \cdot \text{R} \times \text{Temp Rise} = \text{Mass C} \cdot \text{L} \times \text{Temp Drop}$$

Convert all flows to pounds of dry air. Tables of specific volumes of dry air at various temperatures are readily available in most handbooks. They can also be calculated by using the ratio of absolute temperatures from a base of 70°F or 530#/R (70°F + 460 = 530#/R), having a specific volume of 13.34 cu ft/lb of dry air.

Note that if the masses are equal in the above equation, they cancel out and

$$\text{Temp Rise} = \text{Temp Drop}$$

In most processes the mass exhausted must be replaced by an equal weight; so let us assume that we will have equal masses of air in each air stream.

Since we are dealing only with hot, dry exhausts, we need only define the sensible heat efficiency with balanced weights of air. A mid-range is 80% when velocities through the media are less than 900°Fpm. That is, the recovery from A to R will be 80% of the potential difference between A and C. In Figure 16-3, we will recover 80% of the potential difference of 400 - 50, or 350°F (0.80), for a recovery of 280°F. Adding the base outside temperature of 50°F, we find that the regenerated air temperature will be 330°F.

What has been saved? Assume a mass flow of 1,000 lb/min. Therefore

$$1{,}000 \text{ lb/min} \times 60 \text{ min/hr} \times 0.24 \text{ Btu/lb F}^{\circ} \times (330 - 50) =$$

$$4{,}032{,}000 \text{ Btu/hr}$$

Equivalent heating season for: 6,000 degree days
40°F average winter temp.

$$\frac{6{,}000}{(65\text{-}40)} = 240 \text{ days or } 34 \text{ weeks}$$

$\therefore$ 240 days at a constant 40°F $\rightarrow$ 6,000 degree days

Mass A · R x temp rise = mass C · L x temp drop
and if Mass A · R = Mass C · L
then temp rise = temp drop (at equal weights)

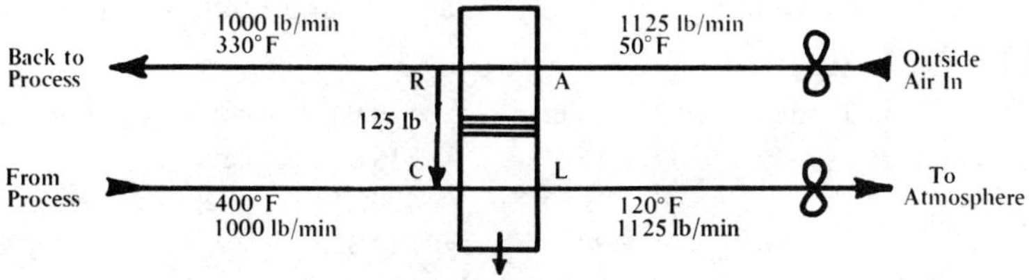

$$\eta_s = 80\% = \frac{A - R}{A - C} \quad \text{or } A - R = 0.80\ (A - C) = .8\ (400 - 50) = 280^\circ\text{F temp rise}$$

$$R = A + 280^\circ = 330^\circ\text{F}$$

$$\text{and} \quad L = 400^\circ - 280^\circ = 120^\circ\text{F}$$

Savings for 1000 lb/min mass flow:

1,000 lb/min x 60 min/hr x 0.24 Btu/lb $^\circ$F x (330 - 50) = 4,032,000 Btu/hr

Economics at \$2.50/million Btu energy cost and
2-shift, 5-day, 50-week operation or 4000 hours

4.032 million Btu x \$2.50/million Btu = \$10.08/hr
x 16 hrs = \$161.28/day
x 5 days = \$806.40/week
process $\rightarrow$ x 50 weeks = \$40,320/yr
plant heat $\rightarrow$ x 34 weeks = \$24,418/season

Regenerator Cost

In stainless at 20,000 acfm avg hot side $\rightarrow$ 70¢/acfm
$\therefore$ 20,000 x \$0.70 = \$14,000 budget cost

Figure 16-3. Performance of regenerators on hot, dry exhausts.

Using the specific heat of the effluent at 0.24 Btu/lb F° will not materially influence our results for a preliminary evaluation.

If we assume a work schedule of two shifts, five days per week, and fifty weeks per year, we have

2 shifts x 8 hrs	=	16 hr/day
		x 5 days
		80 hr/wk
		x50 weeks
		4,000 hr/yr of operation

Using a cost for energy delivered to the point at $2.50/million Btu, the dollar savings will be

4.032 million x $2.50/million	=	$10.08 per hour
x 16	=	$161.28 per day
x 5	=	$806.40 per week
x 50	=	$40,320 per year

Note that if this were used for plant heat, we would have only 34 weeks in the equivalent heating season. The savings would be $806.40 x 34 weeks, or $24,418. Actually, they would be somewhat higher. We would have calculated the Btu savings with 40°F average winter temperature. However, this example is an easy way to show the marked advantage of recycling the regenerated heat back to a process.

How much will the regenerator cost? Looking at a possible dollar savings of $24,000, and considering a one-year payback for the regenerator alone, we find that $24,000 looks to be a good, justifiable figure. Figure 16-4 shows approximate prices per average hot side at about 90% capacity. Our average hot side temperature is (400°F + 120°F) / 2, or 260°F, the specific volume is 18.13 cu ft/lb. Using a net process mass flow of 1,000 lb/min, plus 125 lb/min for scavenge and cross-leakage, we have a volume of 20,396 at the average hot side condition. According to the graph, at 20,000 units the price per unit is 70¢ for stainless steel, giving a unit cost of about $14,000.

Installation costs for regenerators vary widely throughout the country. However, a rough estimate for a stainless steel regenerator in this case would be approximately another $14,000. This should cover the fans, plenums, supporting base (but no heavy structural reinforcement of a roof), setting the equipment with a small crane (but no long hauls over a roof), and wiring from adjacent power and ducts into and out of the process (but nothing too sophisticated or extensive). Therefore the total installed cost might be $28,000, versus savings of $24,000 for plant heat, or $40,000 for the process.

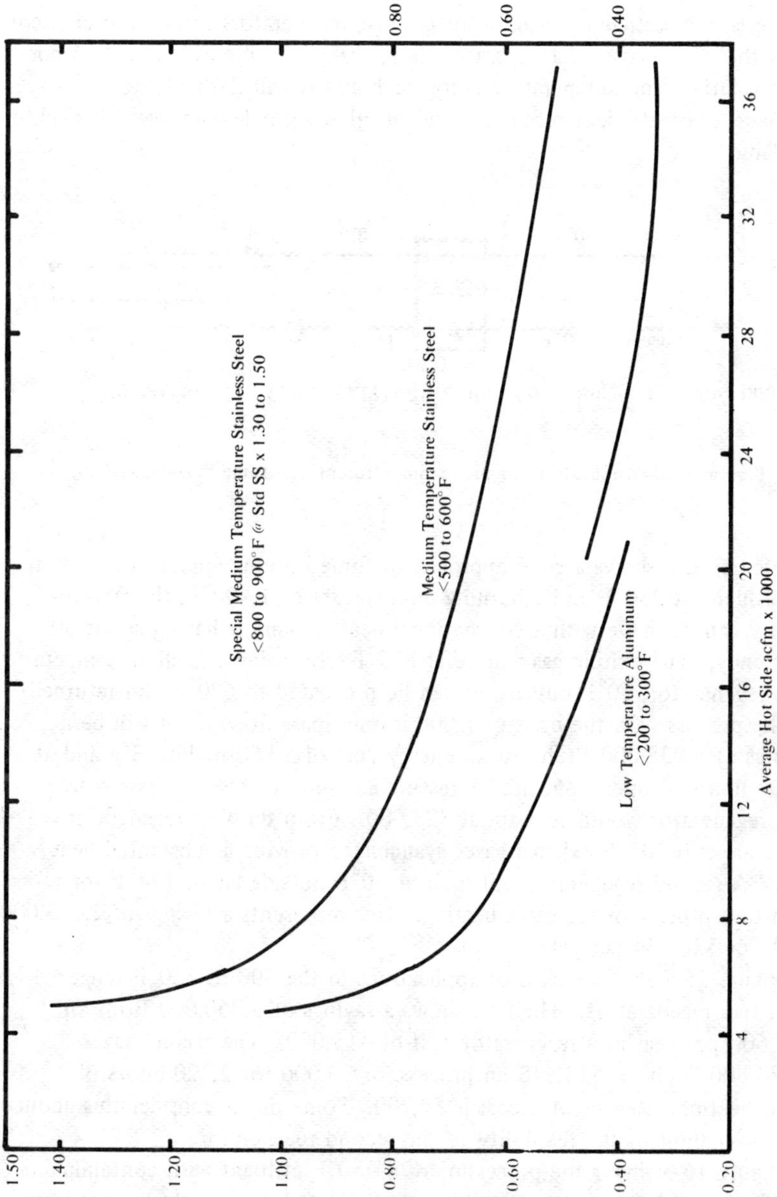

Figure 16-4. Approximate prices per average hot side at about 90% capacity.

Figure 16-5 shows a possible application for preheating the effluent in a fume burning system. Even though there is a high Btu recovery, this kind of use is not particularly significant because regenerators can be too efficient. Note that the recovered temperature is 1,170°F. With an effluent that contains volatiles, this temperature is too high and it will flash. It doesn't pay to make a less efficient regenerator when other suitable equipment is readily available.

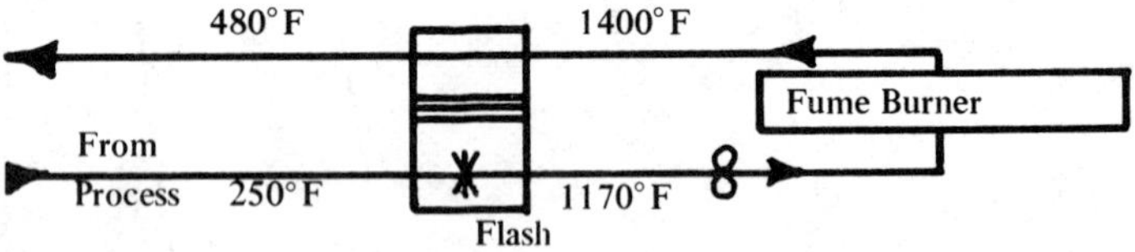

1000 lb/min x 60 min/hr x 0.24 Btu/lb of x (1170°-250°) = 13,248,000 Btu/hr

Figure 16-5. Application for preheating effluent in a fume burning system.

Figure 16-6 shows a good approach to fume burning applications. With an effluent at 250°F and a burning temperature of 1,400°F, the first recovery can be made with a conventional heat exchanger having about 50% efficiency. The exiting gas will be at 825°F. By using a medium temperature regenerator, 50°F outside air can be preheated to 670°F and returned to the process. On the basis of 1,000 lb/min mass flow, there will be a savings of 8,928,000 Btuh. At an energy cost of $2.50/million Btu and at 4,000 hours of operation, this represents a savings of $89,280 per year. The regenerator would cost about $23,000. From the first regenerator there are still 205°F exhaust gases available to provide a substantial heat sink. A second regenerator will preheat 50°F outside air to 174°F for return to a process or for plant heating. This represents a savings of 2,053,000 Btuh, or $20,534 per year.

Figure 16-7 shows a class of applications in the 500 to 600°F range with two regenerators. The first shows a savings of 5,760,000 Btuh, or $57,600 per year at a regenerator cost of $15,000. The second saves 1,334,800 Btuh, or $13,248 on process, or $9,000 for 2,720 hours of plant heating. Regenerator cost is $7,700. Some might consider this about the lower limit of the feasibility of the second regenerator.

Figure 16-8 shows the processing of a 250°F effluent with contaminants suitable for solvent recovery or low-energy air pollution control equipment. Note that at an average year-around temperature of 50°F the regenerator will reduce the effluent temperature from 250°F to 90°F. This represents

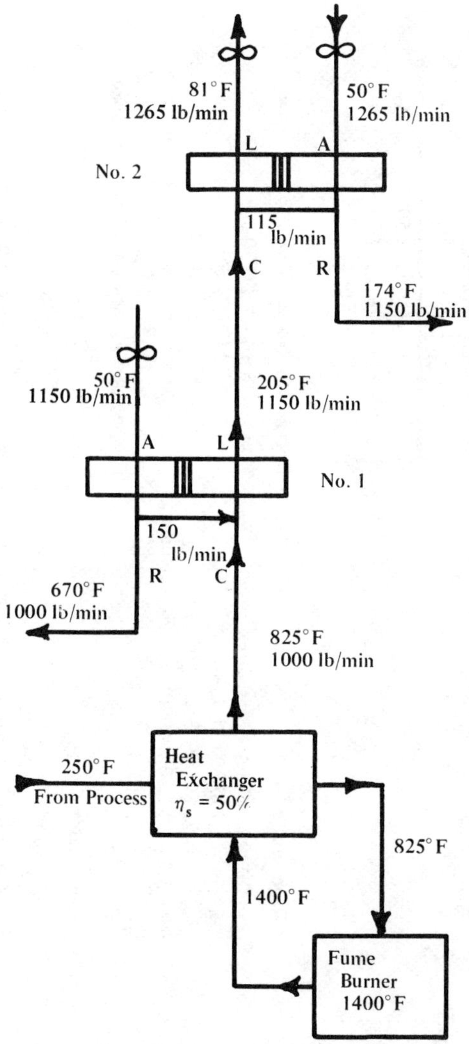

Regenerator No. 2

1150 lb/min x 14.4 x (174-50)
= 2,053,440 Btuh
at $2.50/$10^6$ Btu and
4,000 hr
2.053 x $2.50 x 4,000 =
$20,534 savings

Regenerator Cost

≈ $7,800 in aluminum
≈ $14,000 in stainless

Regenerator No. 1

1000 lb x 14.4 x (670-50)
= 8,928,000 Btuh
at $2.50/$10^6$ Btu and
4,000 hr
8.928 x $2.50 x 4,000 =
$89,280 savings

Regenerator Cost

≈ $23,000 special stainless

Figure 16-6. Approach to fume burning applications.

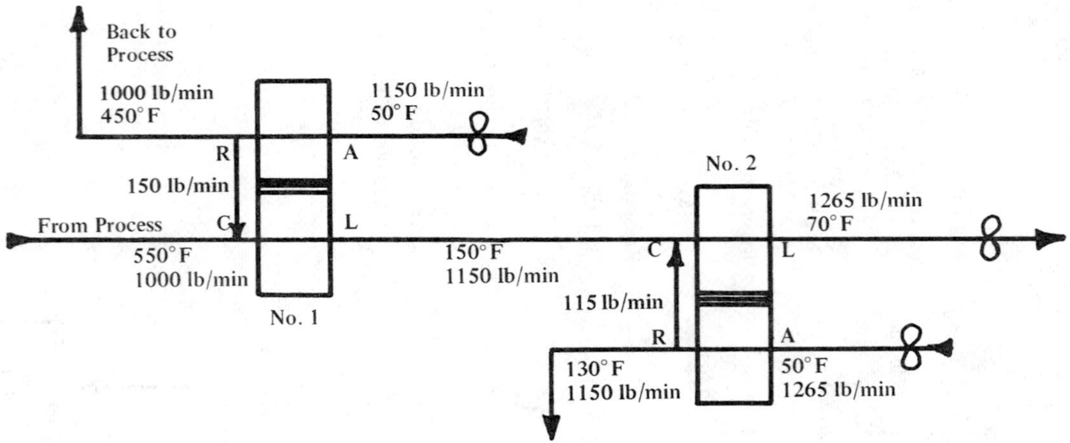

Regenerator No. 1

1000 lb x 14.4 x (450-50)
= 5,760,000 Btuh
at $2.50/10^6$ Btu x 4000 hr
5.76- x $2.50 x 4000 = $57,600 savings

Regenerator Cost

≈ $15,000 in stainless

Regenerator No. 2

1150 lb x 14.4 x (130-50)
= 1,324,800 Btuh
at $2.50/10^6$ and 4000 hr for process
1.3248 x $2.50 x 4000 = $13,248 savings
at 2720 hr for plant heat
1.3248 x $2.50 x 2720 = $9,009 savings

Regenerator Cost

≈ $7,700 in aluminum

Figure 16-7. Class of applications in the 500-600°F range with two regenerators.

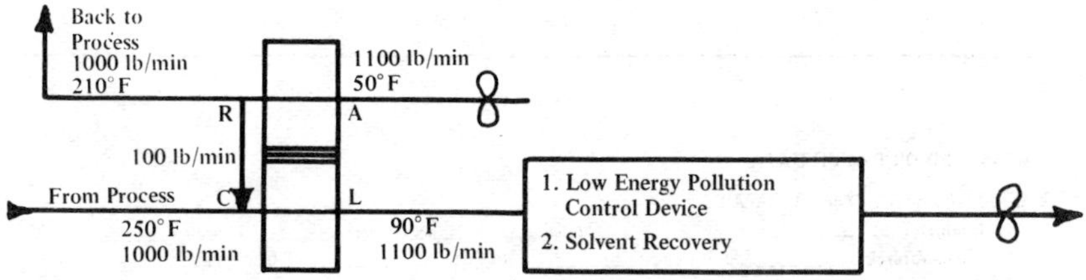

Outside Air Heat Regeneration

1000 lb x 14.4 x (210-50) = 2,304,000 Btuh
at $\$2.50/10^6$ Btu and 4000 hr
2.304 x \$2.50 x 4000 = \$23,040 savings

Regenerator Cost

≈ \$7,700 in aluminum
≈ \$13,100 in stainless

Effluent Cooling with Regenerator

1100 lb x 14.4 x (250-90) =
2,534,400 Btu/hr

Equivalent Water Cooling

$$\text{GPM at } 10^\circ \text{ rise} = \frac{2{,}534{,}400 \text{ Btu/hr}}{60 \text{ min/hr} \times 1.0 \text{ Btu/}^\circ\text{F} \times 10^\circ\text{F} \times 8.34 \text{ gal}}$$

GPM = 506 at 10° rise

Figure 16-8. Processing of a 250°F effluent with contaminants suitable for solvent recovery or low-energy air pollution control equipment.

a savings of 2,534,400 Btuh, or the equivalent of 506 gpm of water over a 10 degree range—perhaps the equivalent of a 150-ton cooling tower. In addition, the efficiency of the regenerator will remain constant, allowing predictable performance. Other types of heat transfer equipment may be subject to wide variations in performance, due to a build-up of excessive films.

On top of all this, 2,304,000 Btuh can be recovered and recycled back to the process. This represents some \$23,000 per year, which can first be used to amortize the pollution control system or, in conjunction with the value of salvaged solvents, the solvent recovery system. After this, the dollars can reflect lower operating costs for the system in question. Regenerator cost is \$7,700 in aluminum and \$13,100 in stainless steel.

Table 16-1 is a Survey and Selection Data Form which, used in conjunction with the principles stated in this chapter, should permit the environmental engineer to determine the feasibility of using heat wheels in exhaust gas streams.

Table 16-1. Survey and Selection Data Form

MEASURED OR DESIGN DATA:

A. CONTAMINATED EXHAUST STREAM:

1. **Quantity:** ________ cfm @ ________ °F; or ________ Std. cfm; or ________ #/min; or ________ #/hr
2. **Temperatures:** ________ °F Dry Bulb; ________ °F Wet Bulb; ________ °F Dew Point
3. **Contaminants:** (Use separate sheet if more space required) Describe individual constituents including weights or volumes of each:

B. FRESH AIR STREAM:

1. **Quantity:** (Usually equal to weight of contaminated exhaust stream)
   a. ☐ Check here if balanced weights.
   or b. ________ cfm @ ________ °F; or ________ std. cfm; or ________ #/min; or ________ #/hr
   c. Fresh air will be induced from ☐ Outside, ☐ Inside
2. **Temperatures:**
   a. Design (for mechanical equipment selections):
      Summer: ________ °F Dry Bulb; ________ °F Wet Bulb
      Winter: ________ °F Dry Bulb.
   b. Average (for economic analysis)
      Summer: ________ °F Dry Bulb; ________ °F Wet Bulb, for ________ day cooling season
      Winter: ________ °F Dry Bulb, for ________ day heating season

C. FUEL:

1. Type and Heating Value ________
2. Present consumption for this process ________
3. Cost ________ per ________
4. Approximate present operating efficiency ________ %

D. GENERAL:

1. Characteristics of Power Available: ____volts ____ phase ____cycles
2. Hours of operation: ____ hrs/day; ____ days/wk; ____ wks/yr = ________ hrs/yr
3. Is condensation of exhaust gases permissable: ☐ Yes; ☐ No
4. End use of transferred heat:
   a. From Process Exhaust:
      ☐ Space Heating ☐ Recycle to Process
      ☐ Space heating in winter and recycle in summer
   b. From Ventilation Exhaust:
      ☐ Return to space in winter ☐ Dispose of heat in summer
      ☐ Return to air conditioned space in summer

REGENERATIVE HEAT CORPORATION

P. O. BOX 78 MIDDLETOWN, CONN. 06458

Project Identification: ________

____

____

CHAPTER 17

A CRITIQUE OF INCINERATION, APPLICABLE CODES, AND TESTING

John R. Dervay II

Air Preheater Company
Wellsville, N.Y. 14895

Generally, one conducts a source test for the following reasons: (1) to show compliance with air pollution codes and/or to prove equipment performance; (2) to evaluate the effect of operating and process changes on equipment performance; (3) to identify and solve a problem with equipment or with a process.

Source testing at relatively low temperature and steady flow conditions is straightforward. The testing of solid waste incinerators is, on the other hand, fairly complicated. Incinerators do not meet these two conditions. Additionally, there is inconsistency of incinerator source testing standards.

SOLID WASTE INCINERATION TODAY

Incineration has always been an economic alternative for the disposal of solid wastes. Unfortunately, to the general public the word incineration often brings to mind a picture of black smoke and obnoxious odors emanating from a stack. Basically, this is a result of a history of public apathy concerning pollution. Due to lax or nonexistent air quality standards, emissions were not considered in equipment design.

The enactment of stringent air pollution codes has given impetus to the development and improvement of solid waste incinerators. This newer generation of incinerators is called by a number of names–fixed air, controlled air, or starved air. Generally, these types of incinerators are characterized by a large main chamber, which is tightly sealed to eliminate ambient air leakage, and a secondary chamber, which is generally the stack. All the combustion air entering the system is introduced by a main fan, with the air controlled to each discrete section by fixed orifices or other flow control devices. The primary emphasis is on controlled combustion.

Conceptually, the controlled air incinerator carries out the combustion processes in two steps, the first of which initiates the burning process under less than stoichiometric air conditions in the main chamber. The heat released in this limited combustion is partially used in the volatilization (gasification) of the remaining solid waste. The second step carries the combustion products, together with the gaseous hydrocarbons, into a second chamber (the stack) where the controlled injection of additional air permits the gases to go to complete combustion either spontaneously or by ignition with an afterburner.

This two-step process results in low stack emissions since the only source of particulate emission is the main chamber. Since the main chamber is operating under starved air conditions, one does not have the highly turbulent combustion which encourages the entrainment of noncombustible particles into the gas stream.

Figure 17-1 is a schematic diagram of a controlled air incinerator—a typical installation of the C-E Combustall controlled air incinerator.[1]

INCINERATOR AIR POLLUTION CRITERIA

Before an incinerator manufacturer can accept an order, he must assure himself and the local air pollution control officer that the incinerator can meet the applicable emission criteria. This can be difficult for two main reasons:

1. At least six different incinerator emission standards are used across the nation
2. Proving that an incinerator can meet the codes is expensive, and sometimes arbitrary.

Table 17-1 shows a number of source tests conducted on various Combustall models. The measured particulate loadings have been calculated against the most commonly used emission standards.

It should be pointed out that there are no conversion factors that can relate one of these emission bases to another. For example, the Incinerator Institute of America (IIA) states that by multiplying an emission loading expressed as "pounds particulate per 1000 pounds of flue gas by 0.53 results in the emission loading expressed as grains particulate per standard cubic foot." Factors such as this proliferate in the literature and are used by enforcement agents. Only through proper analysis of the applicable

Figure 17-1

Table 17-1. Results of Combustall Source Tests

Model	Waste Type	Grain Loading Corrected to 12% CO_2 Gr/SCF		Grain Loading at 50% X's Air		#Part per 100# Waste Charged		#Part* per 1000# of Dry Flue Gas		#Part per 1000# Dry Flue Gas Corr. to 50% X's Air		Lb/Hr*	
		A**	B**	A	B	A	B	A	B	A	B	A	B
200	IIA Type 0	0.077	0.072	0.084	0.079	0.13	0.122	0.145	0.137	0.142	0.134	0.204	0.191
200	HEW Type 1	0.062	0.056	0.084	0.075	0.091	0.078	0.170	0.151	0.170	0.151	0.180	0.162
800	HEW Type 1	0.076	0.063	0.0745	0.0671	0.10	0.083	0.128	0.106	0.133	0.110	0.174	0.144
800	HEW Type 1	0.056	0.0508	0.0612	0.055	0.021	0.0191	0.089	0.0808	0.101	0.092	0.514	0.466
800	HEW Type 1	0.065	0.059	0.0692	0.062	0.027	0.0245	0.104	0.094	0.111	0.100	0.593	0.538
800	IIA Type 0	0.047				0.061						0.379	
2000-B	IIA Type 1	0.027	0.0157	0.0281	0.0163	0.046	0.0267	0.044	0.0255	0.045	0.026	0.662	0.384
2000-B	IIA Type 1	0.052	0.032	0.0595	0.036	0.057	0.035	0.0595	0.037	0.059	0.036	1.647	1.013
1200-H	Oil Filters	0.078		0.0614		0.102		0.069		0.088		0.608	

* Not corrected to either 12% CO_2 or 50% X'S AIR

**"A" includes condensibles, "B" without condensibles per EPA guidelines 12-23-71.

variables, which must be experimentally obtained, can one relate the various emission standards.[2]

The best way to prove that a particular incinerator can meet the applicable emission code is to test that incinerator on-site. However, the expense can be prohibitive. Depending on code requirements, a source test can run from $2,000 to more than $15,000. For a small incinerator selling at $5,000, the minimum cost for a test represents an economic deterrent to the sale of that incinerator.

Recognizing this, the Environmental Protection Agency (EPA) and some of the states accept "representative incinerator testing." This means that for a line of "geometrically" similar incinerators one needs to source test only one, two, or three of the models in that line to obtain approval for all the models. Theoretically, this provides a viable economic approach for showing compliance. However, on an individual application basis a number of problems can be encountered.

If the measured emission from a representative test is not substantially below the applicable emission code, an air pollution control official may not approve the incinerator line. This difference between actual emissions and maximum emissions (allowed by the code) varies from locality to locality and from enforcement official to enforcement official even in the same locality.

In addition, the requirements for representative testing vary considerably between the states. For example, New York state requires that the IIA standard waste test charges (cardboard and water) be burned while conducting representative source tests. The state of Illinois, on the other hand, requires the use of the Public Health Service test charges (shredded newspaper and cardboard plus potatoes, plastics, and lard) and will not accept the IIA test charges. Thus representative testing, even where allowed, can be expensive.

Last but not least is the attitude of many enforcement officers, who are just simply against incineration. A case in point is the following, an excerpt from a rejection letter written by a local director of environmental services, who is a Registered Professional Engineer, to an incinerator manufacturer:

> It is our contention that regardless of the efficiency of the unit with respect to its particulate emission reduction, it is always possible to hypothesize a set of local conditions that would make the installation unacceptable. An example was offered of a hypothetical unit having a zero particulate emission rate with only hot air discharged from the stack. Such a unit would be objectionable to a human receptor in a house six feet away from the stack's mouth at a window at the same elevation as the emission point.

This letter ends with

> We will consider with complete objectivity all submissions for proposals such as yours presented to this office.

INCINERATOR TESTING METHODOLOGY

At least four particulate sampling trains are specified for use in testing an incinerator:

1. Environmental Protection Agency, Method 5
2. American Society of Mechanical Engineers (ASME), PTC-27
3. American Society for Testing Materials (ASTM), D-2928-71
4. Incinerator Institute of America, T-6

Table 17-2 compares the major features of each of these trains. Of these, the most widely accepted is that promulgated by the EPA. Nevertheless, there has been, and still is, much disagreement as to which of the above sampling trains yields the most accurate information. The discussion centers on two main areas: (1) particulate collection efficiency and (2) consideration of condensible hydrocarbons as particulates.

Both of these considerations center on the definition of particulate matter. EPA and most states have defined particulate matter as a solid and/or liquid at 70°F and atmospheric pressure, with the exception of water. By this definition, one must theoretically change the particulate-bearing flue gas sample to the standard temperature and pressure and then filter the gas stream for "particulates." This, however, is not practical. Water vapor would condense on the filter and blind the filter, so that filtering could not take place. EPA's train maintains a sufficiently high temperature at the filter so that water vapor will not condense on the filter and blind it. The ice-cooled impingers are used to condense the remaining particulate.

If one were to define particulate matter as a solid and/or liquid at the *stack conditions*, the filter would have to be maintained at the stack temperature and pressure. The simplest way to do this is to place the filter directly in the stack while sampling. This definition of particulate matter and sampling procedure is being promulgated by ASTM. The ASME's PTC-27 defines particulate as "*solid* matter greater than 1 micron." As such, the test would require an in-stack filter, although PTC-27 does not speak directly to condensible matter considerations.

EPA's position is that the total emission of an incinerator should be measured, not just a portion of it. Those materials that form solids and/or liquid droplets as the stack gas cools in the atmosphere should be measured. It would be illogical to do otherwise.

There is evidence, however, that some of the matter in the impingers of an EPA train is formed by chemical reaction that would not occur as the stack gas cools in the atmosphere. For this reason, EPA eliminated the condensible analysis of its Method 5, as published in the *Federal Register*, dated 23 December 1971. However, Method 5 applies only to incinerators

Table 17-2. Major Features of Sampling Trains

Element	Environmental Protection Agency (Method 5)	American Society of Mechanical Engineers (PTC-27)	American Society for Testing & Materials (D2928-71)	Incinerator Institute of America (T-6)
Nozzle	Gooseneck	Gooseneck or Elbow	Gooseneck or Elbow	Elbow
Probe	Glass-lined and heated	Stainless steel	Stainless steel	Stainless steel
Filter	Cyclone, glass fiber filter	Any of 99% efficiency for particles	Filter or thimble in stack	Cyclone and bag filter
Cooling Section	Smith-Greenberg Wet Impingers	Condenser	Condenser	Condenser
Flow Measurement and Control	Pitot tube and nomograph	Totalizing gas meter	Totalizing gas meter or flowmeter	Null nozzle
Prime mover	Vane pump	Vane pump	Vane pump	Vacuum pump

with capacities of 50 tons/day or greater. For less than this capacity, which includes all of the controlled air incinerators, the condensible portion must still be included.

A number of experimental studies have been reported in the literature trying to compare the EPA train with the other sampling trains (both in and out of stack filters). Some of these studies show the EPA train collects more particulate matter than the other trains; some show the opposite. In actuality, none of these studies employed experimentally valid methodology and analysis.

For two sampling trains with their filters outside the stack, one must collect the same quantity of particulate matter if the filter media of the two trains had the same collection efficiency, saw the same particulate concentration, and the particulate at the points of collection had the same size distribution. When comparing an in-stack against an out-of-stack sampling train, one must analyze the resultant particulate for chemical composition. This would shed light on any differences in particulate collection between the two trains. These considerations have not been evaluated in published comparison tests.

A final note must be made on the use of a glass-lined sampling probe, as promulgated by EPA. The stack temperatures on the controlled air incinerators (such as the Combustall) are generally around 1,700°F. As such, stainless steel sampling probes are not acceptable. This is due to oxidation of the probe and to the scale's being collected as particulate. As an example, a comparative source test conducted by San Bernardino County, California, on a Model 1200 Combustall showed the following.

Table 17-3. Total Particulate Loading

Probe	Grains/SDCF corrected to 12% CO_2
316 Probe	0.81
Glass, air cooled	0.06

The filters, condensers, etc., were the same in both trains. Analysis of the particulate collected by the No. 316 stainless probe train showed it to be primarily metallic oxides. The glass-lined air-cooled probe is shown in Figure 17-2.

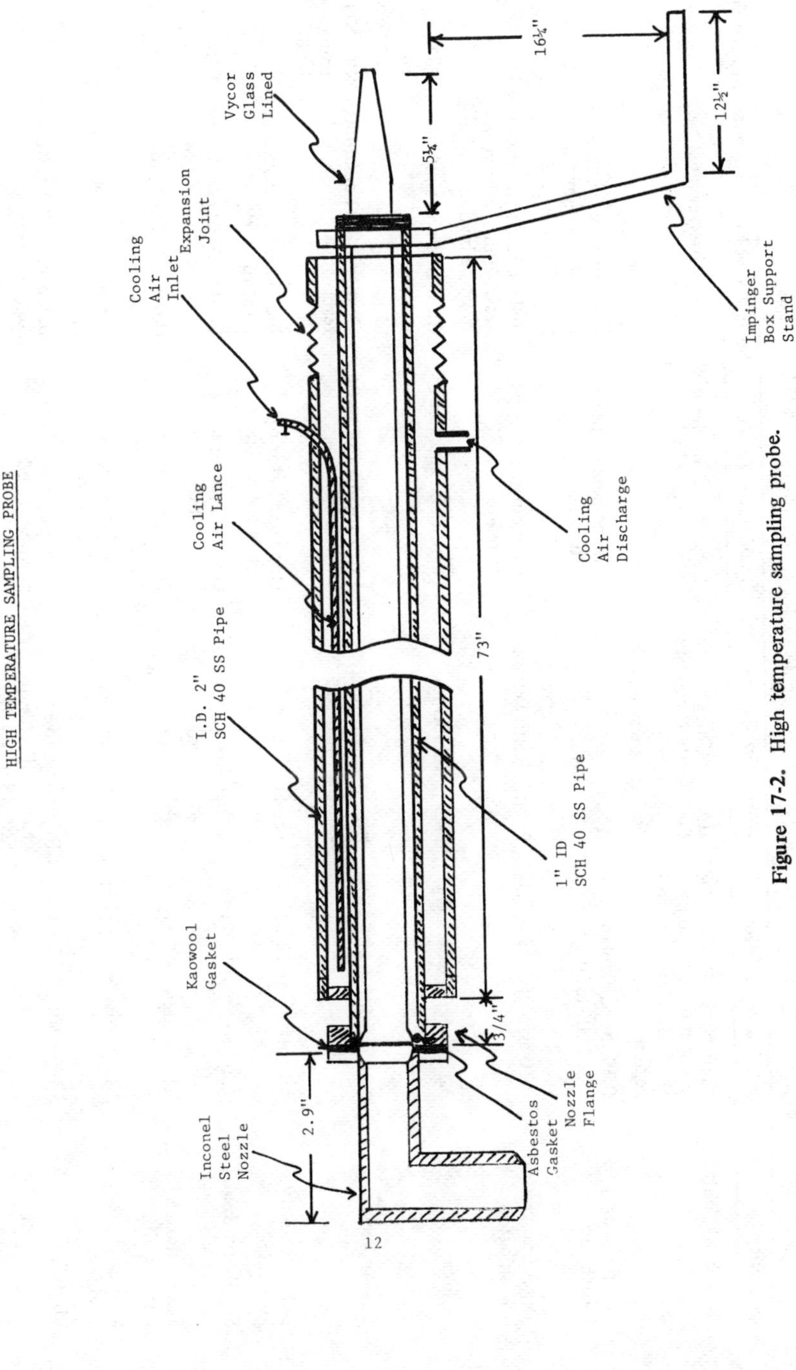

Figure 17-2. High temperature sampling probe.

Table 17-4

Locality	Allowable Emission	Waste Test Charge	Incinerator Tested	Special Test Requirements	Special Test Measurements
New York State	0.20 gr/sdcf @ 12% CO_2 but can be reduced by the 9 air quality control regions.	IIA Test Charges Each waste type tested separately Two tests/waste type	Largest and Smallest to certify entire line	State observer must be invited	
Wayne County, Michigan	0.30 $\frac{\text{lb particulate}}{\text{1000 lb flue gas}}$	Actual waste	One incinerator to certify entire line	County observer must be invited–test must take place in county	
Puget Sound, Washington	0.10 gr/sdcf @ 12% CO_2	Public Health Service	One incinerator to certify entire line	Independent lab must run test. County observer must be invited	
Illinois	0.1 gr/sdcf @ 12% CO_2	PHS or Illinois test charges	Largest and smallest to certify entire line	State observer must be invited	CO–maximum emission 500 ppm @50% x's air
Los Angeles, California	0.10 gr/sdcf @ 12% CO_2	Actual waste–6 days of tests	Every incinerator must be tested	County APCO conducts tests	CO, NO_X, SO_X, halogens, hydrocarbons
Dekalb County, Georgia	0.20 gr/sdcf @ 12% CO_2	Actual waste	One incinerator to certify entire line	Independent consultant must run test	
San Bernardino, California	0.30 gr/sdcf @ 12% CO_2	Actual waste	One incinerator to certify entire line	County APCO conducts tests.	SO_X
San Francisco Bay Area, California	0.15 gr/sdcf @ 6% O_2	Actual waste	One incinerator to certify entire line	County APCO conducts tests	CO, hydrocarbons
Monterrey, California	0.10 gr/sdcf @ 12% CO_2	Actual waste	One incinerator to certify entire line	Independent laboratory must run test	CO
Virginia	0.14 gr/sdcf @ 12% CO_2	Representative waste charges. Each waste type tested separately	Each incinerator size must be tested	Independent laboratory conducts test	

CURRENT EFFORTS IN INCINERATOR TESTING

At present, at least three major organizations are preparing new incinerator testing standards: (1) American Society of Mechanical Engineers, (2) American Society for Testing Materials, and (3) American National Standards Institute. Each of these committees is wrestling with such considerations as the definition of particulate, the type of sampling train to be used, the type of waste to be burned during tests and whether prototype testing is valid, or should each incinerator be tested? Should waste volume and/or weight reduction be defined? Should regulations on combustible content in the ash be imposed? Should gases such as SO_X, NO_X, CO, etc., be measured?

None of these considerations can be easily answered due to a lack of technical data. Since these committees are working independently of each other, one can expect their respective standards to be different.

This poses a future problem for the states, municipalities, and especially the incinerator manufacturers who must use such standards. Table 17-4 is indicative of the myriad standards currently in use. The new standards will, in all probability, just add to the confusion.

REFERENCES

1. Theoclitus, G., *et al.* "Concepts and Behavior of the Controlled Air Incinerator," *Proceedings of 1972 National Incinerator Conference* (1972), p. 211.
2. Dervay, J. R., *et al.* "An Evaluation of Current Incinerator Emission Standards," *Proceedings of 1972 National Incinerator Conference* (1972), p. 111.

CHAPTER 18

WET SCRUBBING FOR ODOR CONTROL

Michael Y. Yang

W. R. Grace & Co.
Fords, N.J. 08863

Paul N. Cheremisinoff

New Jersey Institute of Technology
Newark, N.J. 07102

WET SCRUBBING

Wet scrubbing is often ideally suited to the treatment of large flows of air contaminated with low odor threshold compounds such as mercaptans or amines. For gas scrubbing, large flow rates are considered to range from several thousand to over 100,000 cfm, and odor levels range from several hundred to tens of thousands odor units per cubic foot.

Wet scrubbing may effectively remove odorants by any of the following mechanisms:

1. Odorous vapor may condense from air into a cold liquid contact tower, which may be either a spray or packed tower, with cooling or mechanical refrigeration.
2. Odorous molecules may be attached to particulate matter, and the odor and the particulate can be removed from the air by wet scrubbing.
3. Odorous vapor may be highly soluble in the scrubbing liquid; therefore it could be removed by absorption. The odorous solution would then be deodorized chemically or biologically.
4. Odorous gases may be emulsifiable in a chemical reagent or emulsifying agents, and removed from solution by emulsion breaking and separation.
5. Odorous vapor may react with a strong oxidizing or chlorination agent in the gas phase to form water-soluble, odorless, or less offensive products. These secondary products can be removed by wet scrubbing.

6. Odorous vapor may react with a chemical scrubbing liquid in the scrubber by gas-liquid contact; the reaction products could be odorless or less offensive.
7. Odors can be removed by a combination of more than one of the mechanisms mentioned above, such as condensation, followed by chemical wet scrubbing or gas phase chlorination, followed by wet scrubbing.

Success in removing odors by wet scrubbing depends upon such factors as the identification of the odorous substances and their sources, including the determination of the odor threshold number and its chemical nature. The identification of odorous substances is important—because the next step is finding a suitable scrubbing liquid. The scrubbing liquid absorbs or reacts rapidly with the odorants according to one of the mechanisms mentioned above.

After the proper selection of a scrubbing liquid, along with some kind of laboratory or pilot scale tests, the next step is to design an odor control system. The design may include a packed tower scrubber or spray tower, some type of particulate removal devices such as dry or wet cyclones, a venturi scrubber, or probably a gas phase reactor to provide enough contact time for gas phase oxidation or chlorination.

SCRUBBING SOLUTIONS

Scrubbing Liquid Selection

The selection of a scrubbing liquid depends largely on the properties of the odorants to be removed. Useful information for the selection of a scrubbing liquid for a specific odorant is scarce in the published literature.

The scrubbing liquid in the wet scrubber system may be water, an aqueous solution or suspension of a reactive compound, or an organic solvent. Water-soluble gases such as hydrogen sulfide, organic sulfur gases, ammonia, organic nitrogen compounds such as amines, organic acids, chlorine, and other chlorine-containing compounds may be removed by water scrubbing (Table 18-1). For odorous compounds that are insoluble in water, chemical reagents may be used. Many organic substances and some specific types of odor—for example, the characteristic odor of aldehydes, sulfides, and phenols, their functional groups, or the bond by which it is attached to the molecule—are usually unsaturated, which makes these functional groups readily susceptible to chemical oxidation.

Reaction products are usually odorless or less offensive than the original pollutants. The chemical reagents can be simply an alkali or acidic solution to neutralize the odorous compounds, or a strong oxidizing reagent either in gas or liquid phase, to completely oxidize the odorous compounds to odorless matter. Oxidizing agents used successfully in commercial applications are potassium permanganate, ozone, chlorine, chlorine dioxide, and hypochlorites. Application of emulsifying agents, organic solvents, and homogeneous catalysts in wet scrubbing also offers possibilities in special cases.

Table 18-1. Solubilities of Odorants in Water at Different Temperatures

Temperature (°C)	Ammonia	Hydrogen Sulfide	Sulfur Dioxide
0	85.6	0.7066	22.83
4	79.6	0.6201	19.98
8	72.0	0.5466	17.40
12	65.1	0.4814	15.09
16	58.7	0.4287	13.05
20	53.1	0.3846	11.28
24	48.2	0.3463	9.76
26			9.06
28	44.0	0.3130	8.43
30			7.80

Solubilities are in grams of gas dissolved in 100 grams of water at a total pressure of 1 atmosphere.

Neutralization

Some odorants are either acid or base (Bronsted) in nature and can be deodorized simply by neutralization. Ammonia and amines are Bronsted bases. They have the ability to combine with protons, an aqueous solution of hydrochloric acid, and sulfuric or sulfamic acid, and have been successfully applied in amine and ammonia neutralization. Amine salts, owing to their ionic character, contrast with amines in physical properties. They are all odorless, nonvolatile solids, even though the amines from which they are derived are odoriferous gases or liquids. They are also insoluble in ether or hydrocarbon solvents, which dissolve the typically organic amines.

With the exception of substances of very high molecular weight, the salts are readily soluble in water and exist in the solution in ionized condition. The chlorides are soluble in water. The sulfates are less fusible and less soluble in water. Solubility in water is decreased by addition of an excess of the appropriate mineral acid, and amine salts can be crystallized by this common ion effect. Amine salts can also be recovered by addition of sodium hydroxide to the aqueous solution. The alkali is a much stronger base than the amine. Liberated amine can be collected by adsorption or extraction with ether.

Phenols have a distinctive, weak acidic character, attributable to a combination of a hydroxyl group with an unsaturated ring, or to the presence of an enolic group [–CH = C(OH)–]. Phenols form salts with sodium or calcium hydroxide, but not with the carbonate. Either sodium or calcium hydroxide is suitable as a scrubbing liquid for phenols, but calcium hydroxide is less expensive. Its final product, calcium phenolate, can be converted into liquid phenol and calcium carbonate by the addition of carbon dioxide. This process is quantitative when twice the stoichiometrically needed carbon dioxide is applied. The solution with potassium chloride makes it possible to skim off the phenol floating on the surface.

Calcium hydroxide scrubbing is economically advantageous, compared with the straightforward application of sodium hydroxide as a scrubbing liquid. Moreover, the carbon dioxide content of the flue gas (which is by no means negligible) would convert the more expensive sodium hydroxide into sodium carbonate, instead of the rather inexpensive calcium hydroxide. The sodium phenolate output would be additionally contaminated by the sodium carbonate. The three lower carboxylic acids have sharp, acrid odors; acids from butyric through caprylic have rank, disagreeable odors. The lower carboxylic acids usually have a higher water solubility, as shown in Table 18-2. Sodium or calcium hydroxide or sodium carbonate can be used as the scrubbing liquid for carboxylic acid removal.

Oxidation and Chlorination

An important part of chemical wet scrubbing is taken up by those methods which cause odorants to be deodorized either by oxidation or chlorination. Since oxidation deodorization depends on the presence of nascent oxygen or chlorine, the scrubbing liquid selected should be able to generate one of them. There is a variety of substances whose molecules contain surplus oxygen or chlorine atoms which split off easily, producing nascent oxygen or chlorine. The most commonly used oxidation and chlorination agents are ozone, potassium permanganate, chlorine, sodium hypochlorite, and chlorine dioxide.

Ozone

Using ozone for industrial odor control is a classic application. It has been used in sewage treatment plants for years. The chemical oxidation of odorants by ozone can reduce or eliminate the odor of a large variety of materials because ozone is a powerful oxidation agent. When applied properly, by releasing its nascent oxygen, ozone can completely oxidize

Table 18-2. Aqueous Solubility of Some Osmogenic Substances

Compound	Aqueous Solubility
Acetone	Infinite
Acetyl methyl carbinol	Slightly soluble
Allyl mercaptan	Insoluble
Amyl acetate (iso)	0.25(15°C)
Butyric acid (n)	Infinite
Butyric acid (iso)	20(20°C)
Caproic acid (n) and (iso)	Very slightly soluble
Caprylic acid	Very slightly soluble
Carbon disulfide	Insoluble
Catechol	45(20°C)
Diethylamine	Very soluble
Dimethylamine	Very soluble
Ethyl acrylate	Slightly soluble
Ethylamine	Infinite
Ethyl mercaptan	1.5(15°C), soluble in alkali
Formic acid	Infinite
Methyl acrylate	Insoluble
Methyl mercaptan	Soluble in alkalli
Phenol	8.2(15°C)
Phenol	Infinite (65.3°C)
Propionic acid	Infinite
Resorcinol	123(20°C)
Trimethylamine	41(19°C)
Valeric acid (n)	3.3(16°C)
Vanillin	1(14°C)
Vanillin	5(75°C)

Solubilities are in grams of compound per 100 grams of water at temperature (°C).

many odorants. The instability of ozone dictates the need for an on-site generation facility when ozone is used.

An ozonation-wet scrubbing system is suitable for odor abatement of heavily loaded odorous air. This type of system can be used to oxidize water-insoluble odorants into water-soluble products. In turn, these products can be scrubbed with water to remove most of the soluble odor load prior to a final polishing ozone treatment. The dosage of ozone in the system has to be carefully controlled because ozone is toxic, corrosive, and has a distinctively pungent odor. Even at very low concentration—usually in the ppm range, with a retention time of 1 to 5 seconds—ozone is satisfactory for most industrial odor control applications.

Potassium Permanganate

Given sufficiently vigorous reaction conditions, potassium permanganate ($KMnO_4$) is capable of oxidizing a great many organic compounds to carbon dioxide, water, sulfates, and nitrates. In most industrial odor control systems, permanganate oxidation reactions are carried out in relatively mild conditions. For example, most permanganate scrubbing systems use permanganate in mild alkaline solutions, containing 1 to 4% potassium permanganate. The solution is recirculated at ambient temperatures.

Under these conditions, potassium permanganate does not, however, react with all organic compounds. Some odorous compounds which are susceptible to oxidative degradation under these conditions are aldehydes, reduced sulfur compounds, unsaturated ketones, phenols, amines, hydrogen disulfide, and sulfur dioxide. These odorants are related mainly to the functional group of the compounds–the unsaturated carbonyl group of the aldehydes or the divalent sulfur of a mercaptan. Compounds with these functional groups usually can be oxidized rapidly to form different kinds of compounds, which are generally odorless or have less offensive odors. The permanganate oxidation of simple sulfur compounds, such as hydrogen sulfide and sulfur dioxide, is quantitative and virtually instantaneous. Other odorants can also oxidize to odorless compounds under a wide variety of reaction conditions; however, pH, concentration, contact time, and reaction temperature are important variables.

When potassium permanganate is used in air scrubbing (most of the time in alkaline, neutral, or slightly acid media), permanganate is reduced to insoluble manganese dioxide:

$$2KMnO_4 + H_2O \rightarrow 2\ KOH + 2\ MNO_2 + 3\ (O)$$

In a strong acid media, permanganate is reduced to bivalent manganese:

$$2\ KMnO_4 + 3\ H_2SO_4 \rightarrow K_2SO_4 + 2\ MnSO_4 + 3\ H_2O + 3(O)$$

In both cases, the nascent oxygen formed will immediately react with the oxidizable odorants to form odorless or less offensive compounds. Oxidation with permanganate is greatly influenced by the pH value of the aqueous solution. Depending on substrate nature, the rates of many oxidation reactions increase with increasing alkalinity; others, being strongly catalyzed by hydrogen ions, proceed faster at a lower pH.

Case Study

In a foundry it was found that odors were being emitted to the atmosphere.[2] These odorous compounds were identified as phenol, formaldehyde, and amine. A 2% $KMnO_4$ solution, at a pH of 9, was initially tried as the scrubbing liquid. This proved to be successful for about half of the desired process volume. Then the pH was increased to 11 to 12. The results were improved, removing most of the phenol and formaldehyde on 75% of the designed flow volume. However, the contaminant decomposed in the form of ammonia and was still present. Therefore a pH of 5 to 6.5 with a 3% $KMnO_4$ solution was used, and an acceptable effluent concentration on all the contaminants in the full stream was obtained.

Generally speaking, in a packed tower recirculating scrubber a $KMnO_4$ solution, unless continuously monitored and adjusted, will reach a pH of ~8.5, corresponding to a bicarbonate buffer system due to the formation of CO_2 during the reaction and its subsequent absorption by and reaction with the scrubbing liquid. For many applications the pH range of 8 to 9 is quite useful. Reactions with phenols and alkenes, for example, occur rapidly in this pH range. However, for alcohols and ketones a pH of 11 to 12 is more desirable. For deodorizing amines a lower pH value of ~5 is desirable.

The effectiveness of $KMnO_4$ in wet scrubbing deodorization was determined by a series of quantitative experiments.[3] The experiments were carried out in a laboratory scale-simulated scrubbing unit. A 1% $KMnO_4$ aqueous solution (pH = 8.5) was used as the scrubbing liquid, with an aqueous solution (pH = 8.5) used as a control. The contact time was close to one second. Each experiment was run for one hour after the water scrubber became saturated with contaminant. The experimental results are shown in Tables 18-3 and 18-4.

Since these experiments were carried out with no $KMnO_4$ makeup, it is reasonable to expect better results under actual scrubbing conditions with fresh $KMnO_4$ fed in to make up for what is consumed by the chemical reaction. From Table 18-4 and other references[4] it seems that $KMnO_4$ is most effective for the deodorizing of mercaptans, aldehydes, and phenols.

The MnO_2 formed during the oxidation reaction is, however, a potential problem. Without a properly designed particulate removal system, MnO_2 will clog the flow distributors, especially when spray nozzles are used. This will cause decreases in efficiency or will completely break down the system.

Chlorine

Chlorine has been successfully used for many years to disinfect and control odors of water supply. Chlorine has also been used to deodorize gases from rendering plants, fish meal processing, and pharmaceutical manufacturing. Chlorine will react with many hydrocarbons, replacing one or more hydrogen atoms and forming hydrogen chloride as a by-product. With unsaturated hydrocarbons, chlorine produces saturated chloride.

Aqueous solutions of chlorine (maximum solubility 1% at 9.6°C) are strongly acidic and contain some elemental chlorine, which may chlorinate

Table 18-3. Results of Laboratory Odor Reduction Tests Using Gas Washing Bottle[3] Scrubbing Solutions Containing $KMnO_4$ (0.5%)

Compounds Tested	Odor Units Before Scrubbing	Odor Units After Scrubbing at pH 9	pH 11	pH 12
Acrolein	3,200	200	200	400
Acrylonitrile	1,600	12	4	1
1,3 butanediol	16	2	2	2
1-butanol	200	50	40	30
Butyraldehyde	25,600	32	40	20
Isobutyraldehyde	64	10	20	20
Butyric acid	409,600	100	100	8
Diethylamine	512	32	40	20
Ethyl sulfide	25,600	80	64	2
Ethyl disulfide	25,600	2	50	2
Indole	3,200	512	2	2
Ethanol	16	8	2	2
Methyl ethyl ketone	160	8	72	18
Methyl isobutyl ketone	240	80	52	6
Phenol	1,280	2	0	0
h-propyl sulfide	160,000	800	3,000	800
Skatole	32	2	2	2
Valeraldehyde	25,600	3	4	3

organic compounds. In weakly acidic solutions, chlorine is converted to chloride ion and hypochlorous acid. In neutral to alkaline solutions, chlorine is converted to chloride ion and hypochlorite ion.

$$Cl_2 + H_2O \rightarrow HOCl + HCl$$

$$Cl_2 + 2NaOH \rightarrow NaOCl + NaCl + H_2O$$

Because of its toxic and odorous nature, chlorine, when it is used in conjunction with wet scrubbing units, is usually added to the odorous gases before entering the scrubber; a sufficient retention should be given for maximum effect. It is reported that chlorine concentrations of ~20 ppm for a rendering plant[4] and 20 to 50 ppm for pharmaceutical manufacturing[5] have given satisfactory results.

Hypochlorite

The hypochlorites are strong oxidizing agents. The application of hypochlorites in odor control usually achieves the same result as chlorine gas.

Table 18-4. Threshold Odor Number Reduction in Polluted Air by $KMnO_4$ [3]

Pollutant	Threshold Odor Number: H_2O, pH 8.5	Threshold Odor Number: $KMnO_4$, 1% conc., pH 8.5
	Mercaptans	
Butanethiol	200,000	33*
Pentanethiol	100,000	16*
Hexanethiol	85,000	10.5*
Heptanethiol	3,200	20*
Octanethiol	3,500	6.5*
	Other Sulfur Compounds	
Mercaptoacetic acid	65	1**
2-Mercaptoethanol	30	1**
Allyl isothiocyanate	2,500	1**
Thiophenol	1,300	13*
Thiophene	4,000	13*
	Amines	
Dimethylamine	1,300	20
Trimethylamine	2,700	20
Triethylamine	60-70	50-65*
Cadaverine	20	1**
Indole	5	1**
Skatole	60-100	1**
	Phenols	
Phenol	11	1**
o-Cresol	20	1**
o-Chlorophenol	200	1**
m-Chlorophenol	45	25*
p-Chlorophenol	5	1**
	Miscellaneous Organic Compounds	
Styrene	2,000	10*
Allyl acetate	1,700	25*
Acrolein	140,000	1**
Benzaldehyde	80	1**
Acetaldehyde	1,700	200
1-Butanol	150	40*
Off-gas of bone elevators (Rendering Plant)	100-140	20-25*
Cooker condensate (Rendering Plant)	4,000	250*
Off-gas of asphalt plant	15-25	1**

*Residual odor characteristics much improved. **Odorless.

However, hypochlorites are relatively odorless and harmless. The active ingredient is hypochlorous acid (HOCl), which is generated on hydrolysis.

$$NaOCl + H_2O \rightarrow HOCl + Na^+ (OH)^-$$

The available chlorine in hypochlorite solution at pH 7.5 will be present as 50% HOCl and as hypochlorite ion (OCl^-). At pH 10, only 0.3% of the available chlorine will be present as HOCl. At pH 11 or 12, the HOCl is almost completely dissociated to the ineffective hypochlorite ion. The inefficiency of HOCl at higher pH's is due to the equilibrium of HOCl dissociation and shifts forward at a higher pH.

$$HOCl \rightarrow H^+ + OCl^-$$

In a scrubbing liquid application for odor control, hypochlorites are available in a variety of forms, such as sodium hypochlorite, calcium hypochlorite, lithium hypochlorite, etc. Factors that govern the choice of the hypochlorite to be used are cost, type of metering equipment to be used, the system maintenance required, and availability. Other factors being equal, the choice would be sodium hypochlorite. NaOCl is commercially available in approximately 5 to 15% aqueous solutions. Most producers limit the strength to 15% because stronger solutions form sodium chloride, presenting purification problems.

The NaOCl solution is bulky, costly, and inconvenient in terms of transportation and storage. It has a short half-life in concentrated solution, especially at high temperatures. An NaOCl solution with 10% available chlorine at 77°F has a half-life of 220 days, and only 3.5 days and 0.079 days at 140°F and 212°F respectively.[6] The rate of deterioration of NaOCl solutions is also catalyzed by the presence of ions, such as copper, nickel, and cobalt, which makes long-term storage impractical. The ideal solution is on-site generation of sodium hypochlorite for direct application to the wet scrubbing system.

Modular sodium hypochlorite generation systems are commercially available, as shown in Figures 18-1 and 18-2. These systems produce sodium hypochlorite by electrolytic decomposition of salt or sea water, with low electrical consumption and capacities as required for any application. Modular on-site NaOCl generation units have the advantages of safety, instantaneous chlorine generation, flexible output, high efficiency, and moderate costs.

Figure 18-2 shows an electrolytic chlorite generating cell. The anode portion of the electrode is coated with platinum, which increases the life of the cell to over five years. Normally, power requirements for chlorite generation are very low, in the range of 1.7 to 2.8 kW/lb of equivalent chlorine generated, depending on the resistance of the water. As salt water

Figure 18-1. Electrolytic cells assembly of a 6 lb/hr equivalent chlorine-sodium hypochlorite generating cell. (Engelhard Industries, Inc.)

flows through the electrolytic cells, a current is passed through the water, breaking it down into its basic components, which then react to produce the dilute NaOCl solution. The over reaction is

$$NaCl + H_2O \rightarrow NaOCl + H_2$$

Concentration of NaOCl is proportional to the current. It can be closely controlled, through the constant-current power supply, to provide a constant NaOCl output with changing water salinity, temperature, and flow rate. For even higher efficiency and flexibility, and cost saving, a closed-loop system can be used.

Sodium hypochlorite wet scrubbing has been successfully applied in industrial odor control of emissions from feather meal dryers, blood dryers, feather meal cookers, cooker gases, and perk pans in rendering plants.[7] The results of a sodium hypochlorite wet scrubbing pilot plant are shown in Table 18-5. In Table 18-6, four industrial odor control systems, using an NaOCl solution as the scrubbing liquid, are tabulated in order of flow rates, odor sources, NaOCl consumption rates, and operating costs. The results indicate the advantages of low-cost and high odor abatement efficiency of this scrubbing liquid.

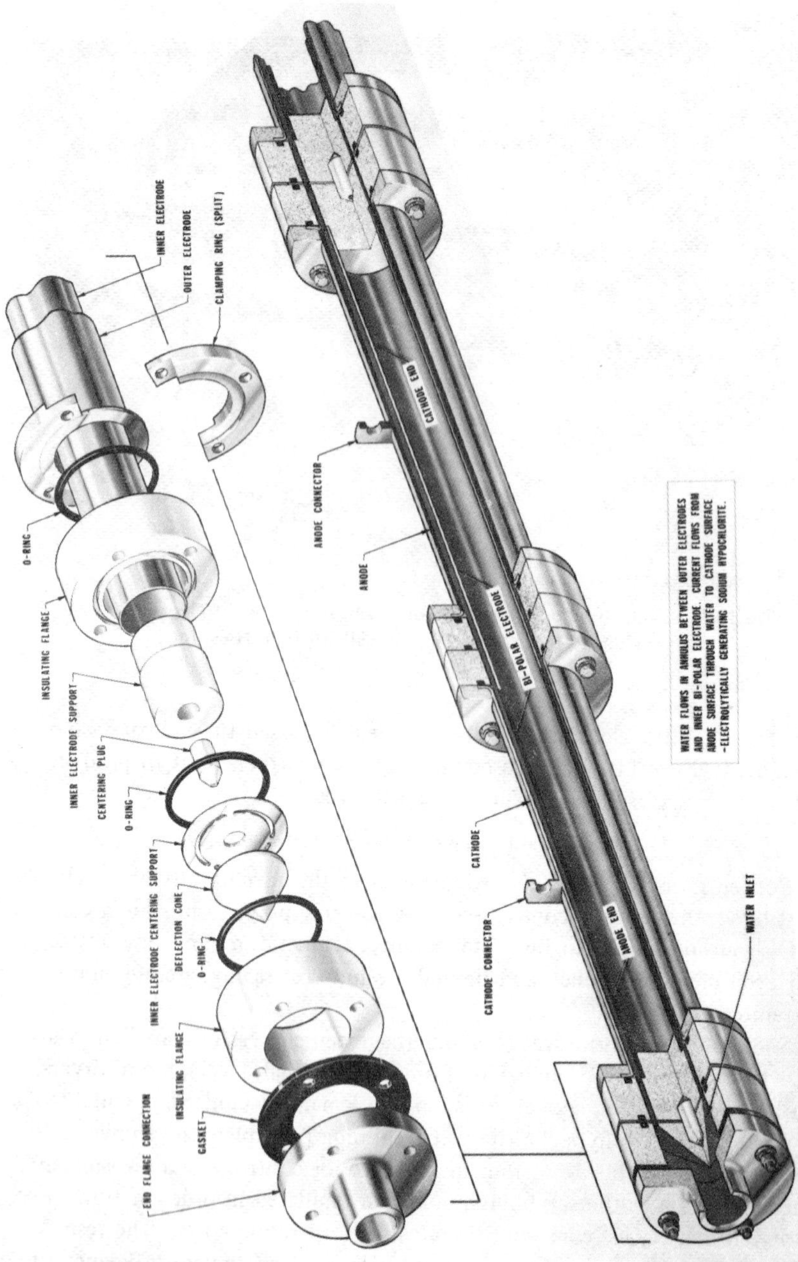

Figure 18-2. Cutaway view of an electrolytic sodium hypochlorite generating cell. (Engelhard Industries)

Table 18-5. NaOCl Wet Scrubbing–Pilot Plant Test Results[7]

Odor Source	Odor Concentration*		Gas Residence Time in Packing
	Inlet	Outlet	
Tallow Cooker	1,200	23	1 sec
Tallow Cooker	1,200	5	3 sec
Fish Meal Dryer	6,000	35	1 sec
Fish Meal Dryer	6,000	2	3 sec
Feather Cooker	20,000	10	1 sec
Feather Cooker	20,000	14	3 sec
Wet Rendering	2,000	32	1 sec
Wet Rendering	2,000	10	3 sec
Paint Pigment Processing	100	14	1 sec
Paint Pigment Processing	100	5	3 sec
Varnish Cooker	50,000	2,000	1 sec
Varnish Cooker	24,000	175	3 sec

* Odor units/CF
**All gas samples were analyzed for odor content by use of an odor panel per ASTM Designation D 1391-57 (reapproved 1967), Atmospheres (Dilution Method).

Sodium hypochlorite has also been used as a scrubbing liquid in a Canadian rendering plant.[8] The wet scrubbing system consisted of a once-through venturi scrubber using river water and followed by a packed tower. The gas flow rate was 32,000 cfm, with an inlet odor loading ranging from 50,000 to 77,000 O.U./cf. Originally this scrubber used potassium permanganate as the oxidizing agent in the packed tower. The subsequent substitution of the NaOCl solution reduced chemical costs and achieved a slightly greater odor abatement efficiency. In addition, the manganese oxide sludge removal problem was eliminated.

Chlorine Dioxide

Chlorine dioxide (ClO_2) is entirely harmless in aqueous solution. Although it is extremely soluble in water and five times more soluble than chlorine, it does not react with water as does chlorine. Aqueous solutions containing up to 0.8% of ClO_2 will retain their strength for several months if properly stored in the dark. In light, the solution decomposes slowly to HClO and $HClO_3$, and is easily expelled from solution in water by contact with air.

ClO_2 reacts rapidly with organics, which makes it a very active oxidizing agent. Applied in a wet scrubbing system for odor control, satisfactory results can be obtained with a ClO_2 solution having a strength between 100 and 500 ppm.[9] The exact strength and amount recirculated will depend upon

Table 18-6. Odor Test Results of Scrubber* Installations in Rendering Plants[3]

Odor Sources	Flow Rate cfm	Chemical Consumption Rate Glns per Hour	Cost	Odor Concentration/cf Inlet	Outlet
I Two feather meal dryers One Duke processors cooker gases from six feather meal cookers and feather storage area	32,000	15 gph	$0.34/hr	5000 to 25000	50 to 200
II Gas from: Raw material receiving area. Three open type perk pans, and two expellers	21,000	8 gph	0.2/hr	Up to 1000	10 to 20
III Cooker gases Raw material receiving room, cooker room, perk pans and expellers	21,000	8 gph	.20/hr	Up to 25000	<10
IV Raw material receiving areas. Three animal cookers. Three perk pans. Two expellers. One feather cooker. Feather meal dryer. Grinding Room	55,000	15 gph	.34/hr	5000	10 to 15

*Scrubber installations include venturi scrubber followed by packed tower except application number 2 which has packed scrubbing tower only. Chemical scrubbing liquid is 1% NaOCl aqueous solution. Cost based on 1972 dollar and cost of NaOCl.

the physical dimensions of the scrubbing tower, the amount of air being scrubbed, and the nature and concentration of odorants.

Due to the explosive nature of ClO_2 gas (with even a greater tendency to explode in compressed liquid state), ClO_2 gas or liquid cannot be transported and is always manufactured on site. There are several ways to manufacture ClO_2 on site, the most popular method being the chlorine-chlorite process:

$$2NaClO_2 + Cl_2 \rightarrow 2ClO_2 + 2NaCl$$

Both chemical solutions are fed to the base of a glass or PVC reaction tower packed with porcelain rings. As the mixture flows upward (with a

Table 18-7. Chemical Scrubbing Liquid Selection

Compounds	Odor Source	Scrubbing Liquid
Aldehydes	Rendering and related industries	$NaHSO_3$, $KMnO_4$
Amines	Rendering, food processing, pharmaceutical industries	ClO_2, Cl_2, NaOCl, HCl, H_2SO_4, $KMnO_4$
Ammonia	Rendering, fish processing, fertilizer	HCl
Hydrogen Sulfide	Pulp and paper manufacturing, water treatment facilities, rendering	NaOCl, Na_2CO_3
Mercaptans	Rendering, pharmaceutical, pulp and paper, waste treatment	NaOCl, ClO_2, Cl_2, NaOH, $Ca(OH)_2$, $KMnO_4$
Phenols	Plastics, pharmaceutical	NaOH, $Ca(OH)_2$
Carboxylic acids		NaOH, $Ca(OH)_2$, Na_2CO_3,,$KMnO_4$

contact time of approximately one minute), chlorine dioxide forms and is discharged at a concentration of 500 to 5,000 ppm of ClO_2, which can be directly used for wet scrubbing odor control.

Chlorine dioxide is also available in a stabilized 5% aqueous solution, with the manufacturer claiming a shelf life of over one year.[10] However, since this is a stabilized complex, it is necessary to reactivate the ClO_2 when it is applied to the scrubbing system. This can be done by either adjusting the pH of the scrubbing liquid to 7 or lower (the lower the pH, the faster the ClO_2 releases) or introducing the solution into a media containing free available chlorine. Certain precautions must be taken in a ClO_2 wet scrubbing system. The ClO_2 must not be used in the presence of ammonia, hydrogen sulfide, phosphide, or methane because the combination forms an explosive mixture. However, the ClO_2 solution is an effective scrubbing liquid for amines and mercaptans.

Sodium Bisulfite

A reaction characteristic of aldehydes and some ketones in the addition of sodium bisulfite is

$$RCHO + NaHSO_3 \rightarrow RCH(OH)SO_3Na$$

With excess $NaHSO_3$, the carbonyl component can be converted almost entirely into the addition product. The addition product is a crystalline salt and has the usual characteristics of an ionic metal compound. It is very soluble in water, but is subject to being precipitated out by the common ion effect. It is also infusible and nonvolatile. Because of these properties, bisulfite addition can be employed for the collection of aldehydes, as a scrubbing liquid in odor control.[11] Aldehydic compounds that are water insoluble can often be converted into an additional product by dissolving them in alcohol prior to treatment with aqueous bisulfite solutions.

Emulsifying Agents

Emulsifying agents have been tried in wet scrubbing liquid formulation to emulsify water-insoluble odorants and to increase the efficiency of absorption by increasing interfacial mass transfer. The foam generated by emulsifying under controlled conditions is also helpful in controlling the odorants' emission from scrubbing liquid and interception of particulate matter in the gas stream. Care must be taken in selecting emulsifying agents. They must be biodegradable if they are discharged into the local sewer system.

Emulsifying agents may be used with other chemical reagents, such as neutralizing or oxidizing agents. A combination of emulsifying agent with either an acidic or alkaline neutralization agent is ideal for scrubbing odorants of alkaline or acidic nature. However, when used with strong oxidizing agents, the emulsifying agent is also subject to oxidation. This causes high chemical consumption, which may not be economical or practical.

Homogeneous Catalysts

Homogeneous catalysts may be used in a chemical wet scrubbing liquid formulation to increase the scrubbing efficiency by increasing the reaction rate and reducing chemical consumption. Aqueous solutions with 0.019 M sodium metaperiodate ($NaIO_4$) and 0.0034 M potassium permanganate at pH 7 to 8 cleave an olefinic double bond rapidly at 25°C.[12] The permanganate is reduced only to the manganate stage, from which it is regenerated by the periodate, which by itself does not attack the olefin. This homogeneous catalyst system can also be applied to water-insoluble compounds, which can be oxidized in aqueous dioxane or aqueous iso-butyl alcohol.

An example of using homogeneous catalysts is the removal of hydrogen sulfide from the exhaust air of a nylon manufacturing plant.[13] In this case the concentration of hydrogen sulfide was small and the volume of air large. The process removed the odorous pollutant in a single stage, using ferric hydroxide as a catalyst.

$$6\ H_2S + 4\ Fe(OH)_3 \rightarrow 2\ Fe_2S_3 + 12\ H_2O$$

$$2Fe_2S_3 + 3\ O_2 + 6\ H_2O \rightarrow 4\ Fe(OH)_3 + 3\ S_2$$

Elementary sulfur is the final product.

Other homogeneous catalysts systems are also available. The application of homogeneous catalysts has its potential in chemical wet scrubbing.

WET SCRUBBING SYSTEM DESIGN

Once a scrubbing liquid has been selected, a wet scrubbing system has to be designed. The scrubber is obviously the heart of the system since it is the unit that removes the odorants from the gas stream. Usually a packed tower is used in odor control because it has the design potential for providing almost any degree of odor reduction (limited by equilibrium conditions), provided the odorous gases can be effectively absorbed and chemically deodorized by the scrubbing liquid. Other kinds of absorption devices such as spray chambers and venturi scrubbers find less application in odor control. Venturi scrubbers are used extensively for particulate removal. Detailed design of scrubbers can be found in many selected references.[14,15,16]

Absorption efficiency is limited by the equilibrium conditions. Usually the scrubber is operated at countercurrent gas-liquid flow. Fresh scrubbing liquid or recirculated scrubbing liquid with makeup is introduced into the top of the packing. The scrubbing liquid is distributed either by spray nozzles or other types of flow distribution, such as a weir.

The scrubbing liquid flows in countercurrent to the gas stream, leaving the packing with a maximum vapor content. The overall height of the packed bed is determined by calculating the height of the transfer unit and the number of transfer units needed for absorption of the odorous gas. The final results must be modified by experience to allow the ineffectiveness to reach equilibrium in absorption.

The ratio of liquid to gas flow rate is important since increasing the liquid-gas ratio will reduce the number of transfer units needed. The diameter of the tower depends upon the volume of gas to be cleaned. The limiting velocity of the gas stream is the flooding velocity, which depends on the physical properties of the gas and tower packing. Usually the optimum gas velocity is about 50 to 70% of the flooding velocity.

One of the most important factors in scrubber design is the interfacial area between gas and liquid. Total absorption is directly proportional to this area. Tower size can be appreciably reduced if the interfacial area per unit of tower volume can be increased. An obvious method of increasing interfacial area is to distribute the liquid absorbent into the gas stream in fine droplets. This can be done by high-pressure head spray nozzles. Another way to increase interfacial area is by changing the tower packing. Packing materials are continually being invented and range from dense packings such as Pall rings, spheres, and saddles to low-density packings. Glass fibers are often used for operations related to fine mist. The selection of the tower packing will depend upon economics, pressure drop limitations, and the surface area requirement. To prevent carryover of the absorbent by the gas stream, an entrainment separator should be installed. Usually mist eliminators are made of knitted mesh of plastic or steel.

Pretreatment of Gas

Odor control associated with particulate matter requires special consideration, especially in cases where the particulate has absorbed the odor. In poultry rendering, for example, emissions are predominantly aldehydes and small amounts of dialkyl disulfide. The odors associated with the dry rendering of dead stock and slaughterhouse trimmings contain a variety of carbonyl compounds. Odors from fish meal processing are mainly amino compounds, from trimethyl amine to putrescine. In nearly all cases mentioned, the odors are accompanied by particulate matter and fatty mists.

A control system for handling odor-laden particulates may consist of a combination of particulate removal devices such as a venturi scrubber, or dry or wet cyclones, followed by the packed tower. Even if the particles have no adsorbed odorants, the pretreatment of the particulates and odor-laden gas before they enter the wet scrubbing system will reduce the chemical consumption rate and minimize clogging of the packing. A mist eliminator should be used to remove any particle-laden scrubbing liquid from the venturi scrubber on wet cyclone before it passes to the packed tower section.

Corrosion

Because of strong oxidation and chlorination reagents, and acidic or caustic neutralization agents used in the wet chemical scrubbing system, corrosion is a major problem. It is necessary to fabricate scrubbing systems of materials which resist chemical attack. Materials of construction for wet scrubbing systems range from mild steel to thermoplastics, or fiberglass-reinforced plastics to stainless steel and even special alloys. Final selection will depend upon the corrosiveness of the air stream and the type of scrubbing liquid to be used. Usually, thermoplastics and fiber-reinforced thermoplastics deserve first choice for small or medium-size units and for most of the piping and duct systems. For units handling large volumes of gas, fiberglass-reinforced plastics (FRP), stainless steel, or chemical resistant resin-coated mild steel is required. For example, sodium hypochlorite solution is very corrosive, but PVC and CPVC are resistant to hypochlorite, and are also resistant to acidic or caustic solutions, and up to 10% $KMnO_4$ solutions. Selection of fabrication materials can be based on information in the published literature or from steel and resin manufacturers.

Water Treatment

In odor control wet scrubbing systems, whenever the odorant is transferred into the scrubbing liquid from air it becomes a water pollution problem. The absorbed odorants could be completely oxidized by chemical reagents with the wet scrubbing system, or could form other reaction products which may or may not be directly discharged to the local sewer system or receiving water body. This problem should be reviewed with local authorities.

Most wet scrubbing systems recycle the scrubbing liquid for economic reasons. Particles, carried by the gas stream generated by the chemical reaction, will build in concentration in the system, which may plug flow distributors and deposit on packing. The particle accumulation problem can be solved by pretreatment of gas before it enters the chemical wet scrubbing system. Filters, centrifugal separators, or clarifiers can be installed in the recycle loop to remove particles. The selection of particle removal equipment depends upon the particle size distribution and loading, flow rate, and available pressure drop through the equipment.

In some cases particles may be generated during the deodorization. For example, in potassium permanganate oxidation the permanganate is reduced to manganese dioxide, which is an insoluble precipitate. A 5% sodium bisulfite ($NaHSO_3$) solution can be used to flush the MnO_2 deposit from the system, along with any MnO_2 sludge removal device. If there is a waste treatment system in the plant, the MnO_2 will probably provide a benefit to the treatment system. Manganese dioxide will oxidize slightly and act as a coagulant aid. In some cases MnO_2 may also act as a collector, although in most cases it remains completely inert.

Miscellaneous System Design Criteria

Because of the odorous nature of gas emissions from odor sources, the wet scrubbing system should be designed and fabricated air-tight and operated under a slightly negative pressure. This will ensure that odorous emissions are captured and discharged through the scrubbing system. When gas phase oxidation reagents, such as chlorine and ozone, are applied to the system, the gas feed system should also be under negative pressure.

Gaseous oxidizing reagents can be fed into a scrubbing system, as shown in Figure 18-3. The gas can be fed at a predetermined concentration. Dry chemicals can also be metered into the system in concentrated form. Potassium permanganate feed rate can be controlled by its characteristic color, while a sodium hypochlorite concentration can be controlled by changing the flow rate from a recycle loop.

Many odor emissions are mixtures of odorants and require at least a two-stage scrubbing system, using acidic reagents to neutralize amines.[17] Other types of reagents in the second stage, such as NaOCl, deodorize the remaining odorants. Chlorine gas can also be injected upstream of the scrubber, to oxidize the odorants so they become soluble and/or less odorous.

Biological oxidation processes have been used successfully in wastewater treatment. Microorganisms are used to convert organic waste into innocuous inorganic forms. Biological oxidation processes, such as activated sludge and trickling filter processes, can also be combined effectively with wet scrubbing odor control systems. Water-soluble odorants can be collected by absorption with a scrubber, then pass through the bio-oxidation facilities to remove the odorants and permit the treated water to be recycled to the scrubbing tower.

Case History

An odor control system was developed for a 2,4-D (2,4-dichlorophenoxyacetic acid) production plant using a wet chemical scrubbing system to abate odors.[18] The odor source was the reaction kettles. Since phenolic compounds are acidic, aqueous sodium hydroxide was selected as the scrubbing liquid to be used in a countercurrent packed tower scrubber. A small bench-scale pilot unit was constructed of a 2-in.-diameter glass tube packed to a height of 4 in. with 1/4-in. ceramic saddles. Test runs were then carried out. Based upon the pilot plant results, a plant scrubber was designed to handle 10,000 cfm of air. The packed tower was scaled up to 7.5 ft diameter with a packed height of 13.5 ft. The process was also changed to reduce the odorant emission by installing a reflux condenser at the reaction kettle, which condensed a large portion of the organic vapor before the gas entered the scrubbing system.

The system reduced the chlorophenols below 1 ppm. Although the system's performance was satisfactory, improvements have been made. The ceramic packing was replaced by polypropylene saddles because the ceramic packing was observed to be softening and disintegrating. The original rubber-lined steel ducting lasted only six months because of the phenolic organic content in the gas. Fiberglass-reinforced furan was finally used, and it solved the corrosion problem.

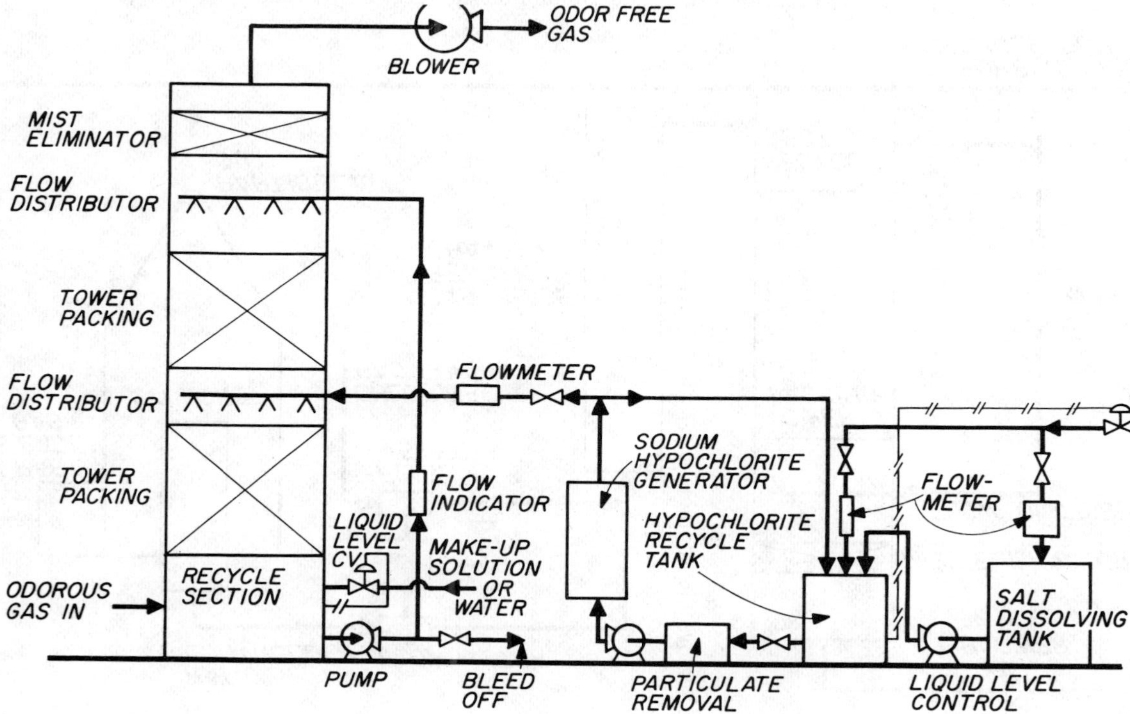

Figure 18-3. Sodium hypochlorite odor control system with on-site NaOCl generation wet scrubbing.

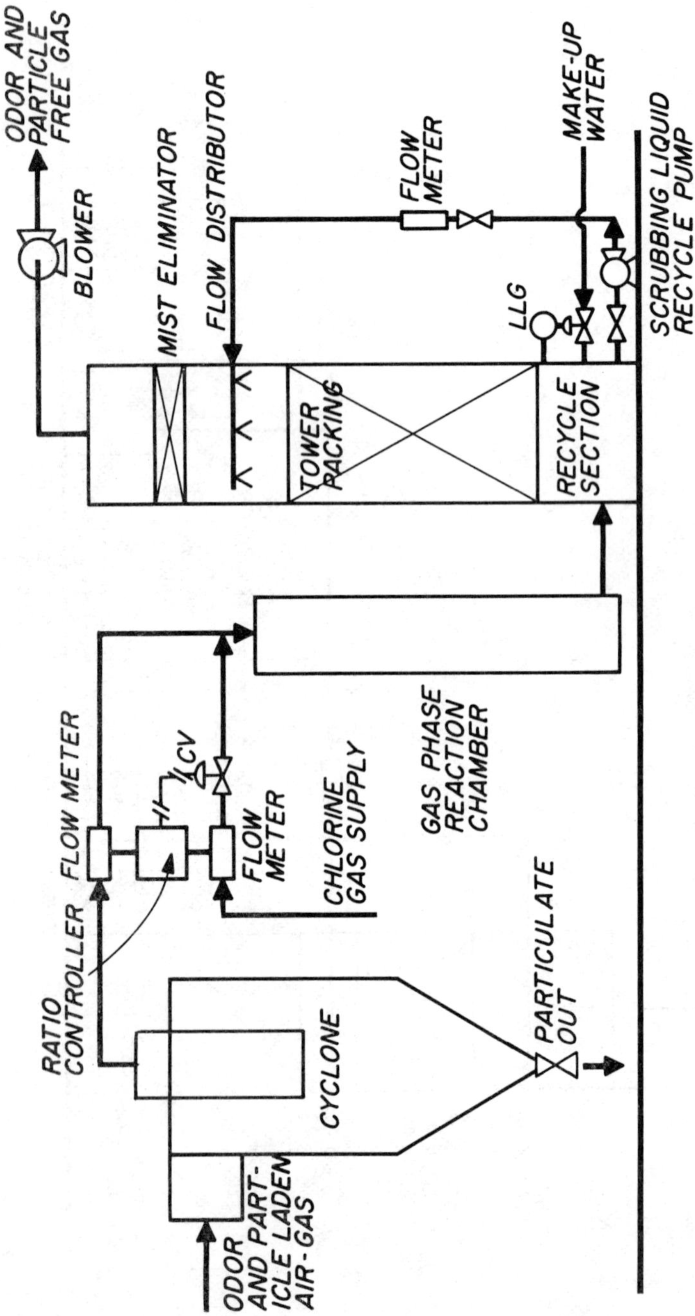

Figure 18-4. Schematic of particulate control–chlorination wet scrubbing odor control system.

This example demonstrates the approach to solving a plant odor control problem by wet scrubbing. This approach can be summarized as (1) Define the odor source and odorous component; (2) Select a suitable scrubbing liquid, based on the chemical nature of the odorants; (3) Do a pilot test; (4) Scale up to a full-size unit; (5) Improve process equipment to reduce odor emission from its source; and (6) Improve the system by adapting the latest developments, *i.e.*, new packing or new corrosion-resistant material.

REFERENCES

1. Anderson, C. E. and H. R. Adolf. Paper No. 71-22, presented at the 64th APCA Annual Meeting (1971).
2. Sohr, R. T. *Modern Casting* (August 1973).
3. Posselt, H. S. and A. H. Reidies. *I&EC, Prod. Res. & Dev.*, **4**, 48 (1965).
4. Mills, J. L., *et al.* Paper No. 67-10, presented at the 60th APCA Annual Meeting (1967).
5. Johnson, W. C. and L. L. Kempe. *J. Air Poll. Control Assoc.*, **9**, 32 (1959).
6. Allied Chemical Co. "Chlorine Bleach Solution," Solvay Div., Bull. No. 14, 2nd ed. (1960).
7. Lundgren, D. A., *et al.* Paper No. 72-116, presented at the 65th APCA Annual Meeting (1972).
8. Bethea, R. M., *et al. Env. Sci. & Technol.*, **7**, 504 (1973).
9. *Chem. Eng.*, **59**, 174 (1952).
10. International Dioxide Inc. Bulletin No. 501 (1966).
11. Murthy, B. N. Paper No. 73-275, presented at the 66th APCA Annual Meeting (1973).
12. Fieser, L. and M. Fieser. *Advanced Organic Chemistry*. (New York: Reinhold Publishing Co., 1961).
13. British patent No. 80381 (1958).
14. Treybal, R. E. *Mass Transfer Operations*, 2nd ed. (New York: McGraw-Hill, 1968).
15. Sherwood, T. K. and R. L. Pigford. *Absorption and Extraction*, 2nd ed. (New York: McGraw-Hill, 1952).
16. Perry, J. H., ed. *Chemical Engineers Handbook*, 5th ed. (New York: McGraw-Hill, 1973).
17. Burgwald, T. A. Final report on I.I.T. Research Institute Project No. C 8122 (1971).
18. Henshaw, T. B. *J. Air Poll. Control Assoc.*, **15**, 516 (1965).

CHAPTER 19

ACTIVATED CARBON AND THE CONTROL OF ODOROUS AIR POLLUTANTS

W. D. Lovett, R. L. Poltorak

Calgon Corporation
Pittsburgh, Pa. 15230

Increased emphasis on improved air quality and the associated control of odors has focused attention on activated carbon adsorption as a viable abatement method in many odor control problems. This has led to the increased use of activated carbon in diverse industries and has fostered the development of new carbon systems and technology for many odor/air pollution problems.

In many applications, activated carbon adsorption is only one unit process in a control system. It is often used to complete odor removal after condensation, water scrubbing, chemical scrubbing, or other techniques have been applied. In other instances activated carbon is used to *concentrate* odorous emissions, producing substantial reductions in the energy required for subsequent destruction of the odors.

Emphasis is given here to carbon systems for controlling odors from typical low-concentration/large-volume emissions and small-volume sources, such as storage tanks and reactor vessels.

ADSORPTION WITH ACTIVATED CARBON

Adsorption can be defined as the phenomenon whereby molecules adhere to a surface they contact. It is usually explained in terms of surface tension or surface energy per unit area of given solid. This tension or energy is caused by molecules in the surface layer which are subjected to unbalanced forces. Molecules within the solid are completely surrounded with identical molecules and thus are subjected to equal forces.

When the surface energy of the carbon is sufficient to overcome the kinetic energy of a molecule in proximity to the carbon surface, the molecule is adsorbed by the carbon. This is known as physical adsorption and the forces responsible are called Van der Waals forces.

The phenomenon of carbon adsorption has been applied in gas mask filters, air conditioning systems, and sewage treatment ventilation. For example, toxic vapors, odorous compounds, or otherwise undesirable contaminants adhere to the large surface area provided by the intricate pore structure of carbon granules. Hassler[1] presents a good description of activated carbon and adsorption principles. Faulkner *et al.*[2] report on odor control using granular activated carbon.

To maximize adsorption efficiency, the greatest possible surface area should be contained in the smallest practical volume. This principle is first applied in the manufacture of granular carbon. The objective is to assure that each granule is provided with maximum surface area and controlled pore size while maintaining a desired density and hardness. This surface is created by permeating a carbonaceous material, such as coal or coconut char, with minute holes or pores.

Figure 19-1 illustrates the minute pores in a carbon granule. The number and diameter of these pores determine the total available surface area. Surface areas for commercial activated carbons vary from 500 to 1,400 sq m per gram, or about 125 acres in as little as one pound.

Due to the nonpolar nature of its surface, activated carbon has the ability to adsorb organic and some inorganic materials in preference to water

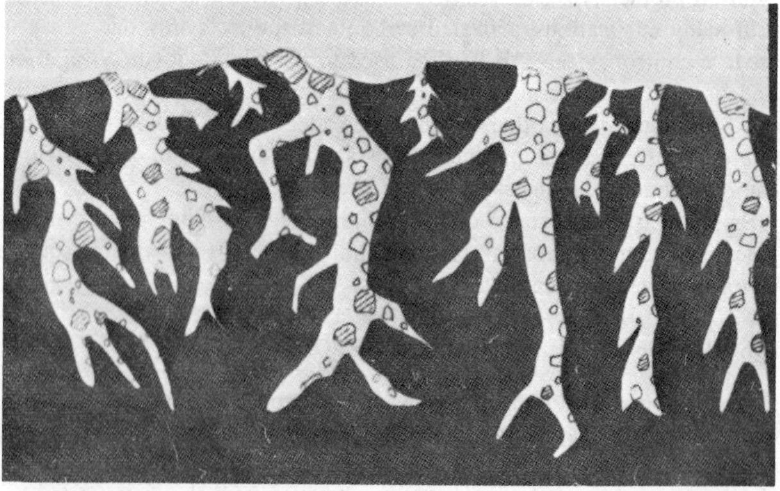

Figure 19-1. Minute pores in a single carbon granule are an important factor in determining maximum surface area and adsorption efficiency.

vapor. The amount of materials adsorbed is partially dependent upon the physical and chemical characteristics of the specific compound or compounds. In general, organics having molecular weights over 45 and boiling points over 0°C will be readily adsorbed. Adsorption is general among organics; that is, it is not strongly influenced by such characteristics as chemical class, solubility, oxidizability, etc. The latter is an important advantage in odor control, where complex mixtures of dissimilar materials are frequently encountered.

Figure 19-2 shows the ability of activated carbon to adsorb toluene at various temperatures and pressures. The adsorption capacity of activated carbon varies directly with concentration or adsorbate partial pressure, and inversely with temperature. Under normal circumstances, adsorptive capacities can reach 5 to 40% of the weight of activated carbon. When plotted graphically, adsorption isotherms are usually flat, showing retention of adsorptive capacity even at *low temperatures*—another important characteristic in many odor control applications.

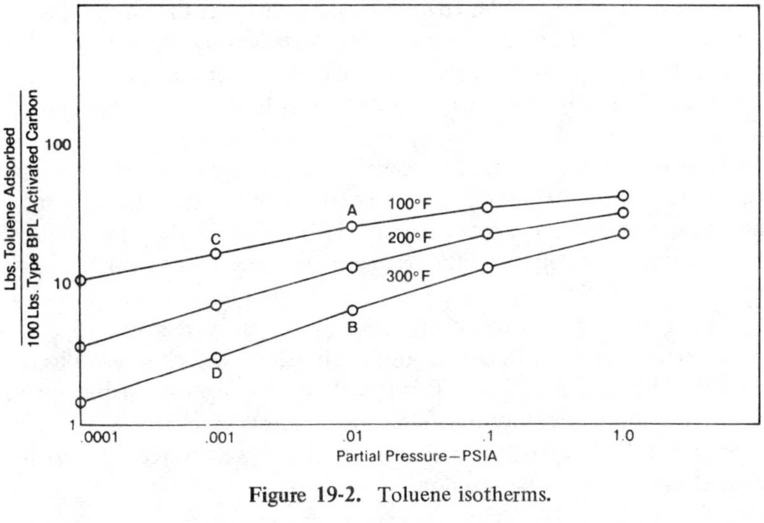

Figure 19-2. Toluene isotherms.

Activated carbon will adsorb at a given set of conditions until the capacity indicated on the isotherm is reached. Once saturated, the carbon can be reused after a portion of the adsorbed material has been removed by regeneration. This regeneration can be accomplished thermally in a furnace or by passing hot gas or steam through the carbon bed until sufficient material has been desorbed. The amount desorbed will be the design working capacity of the bed. It is usually preferable, for economic or safety considerations, to allow some adsorbed material to remain in the pores of the carbon.

Economic considerations are related to the increased regeneration costs resulting from the effort to strip the carbon clean of all adsorbed organics. Experience has shown a disproportionate rise in costs of trying to remove final traces of organics from the carbon.

Safety considerations are necessary when easily oxidized organics at high concentration are being adsorbed. Allowing a residue of the volatile material to remain in the pores of the carbon minimizes the temperature rise associated with heat of adsorption (similar to heat of condensation) and reduces the possibility of ignition in the carbon bed.

Regeneration can sometimes be accomplished with a pressure swing, such as a vacuum, or by a combination of heat and vacuum, solvent washing, or chemical treatment.

Refer again to Figure 19-2. If incoming air at 100°F has a toluene partial pressure of 0.01 psia (680 ppm toluene by volume at atmospheric pressure), the carbon saturation capacity would be 26 weight percent (A). If desorption occurred at 300°F (B) for a sufficient time, the working capacity of the carbon would be 20% (A minus B). Discounting vaporization cooling effects, a 10:1 pressure reduction (C) would give a capacity of 10 weight percent (A minus C), whereas a combination of heat plus vacuum (D) could increase the working capacity to 23% (A minus D).

Hot gas or steam regeneration generally has an advantage in lower capital cost, but vacuum regeneration has the potential of producing a greater concentrating effect. A vacuum also allows for much higher regeneration temperatures because most of the oxygen is removed, thus minimizing high-temperature oxidation. Another advantage of high temperature plus vacuum is that very strongly adsorbed compounds can often be removed.

In several air pollution/odor control applications, activated carbon has the ability to act as a catalyst. This capability is utilized primarily for the removal of odorous hydrogen sulfide. In the presence of free oxygen, hydrogen sulfide is catalyzed to elemental sulfur by activated carbon, as indicated by the following reaction:[3]

$$H_2S + \frac{1}{2} O_2 \xrightarrow{\text{active carbon}} H_2O + S$$

The elemental sulfur remains on the carbon surface. The reaction will take place with varying proportions of hydrogen sulfide and air.

As an illustration, activated carbon has been used to remove low concentrations (2 to 20 ppm) of hydrogen sulfide from ventilation gases of sewage treatment plants.[2] In this case, odorous organic compounds are simultaneously removed by physical adsorption while hydrogen sulfide is oxidized to sulfur on the carbon surface. Tests have indicated that activated carbon has removed more than 20% of its weight of H_2S.

ACTIVATED CARBON AS A CONCENTRATOR

The ability of activated carbon to concentrate the dilute impurities typical of air pollution problems is of great economic significance to odor control considerations. After the carbon has reached the design breakthrough point, it can either be regenerated in situ or transferred to a separate regeneration facility.

The primary methods utilized for separate regeneration are a high-temperature kiln at the carbon use site or a manufacturer's reactivation services. With either approach, attrition losses are quite low (usually less than 10%), and the carbon can be returned to essentially virgin capacity. A large foundry is presently concentrating a 100,000 cfm stream, containing 1 to 10 ppm of odorous compounds, and utilizing manufacturer reactivation.

For many years, steam has been used as the primary heat source for in situ carbon regeneration. This method is used primarily in solvent recovery systems where the solvent has sufficient reclamation value to justify such an installation. However, in odor control, any value from recovered solvents or other organics can help to offset the cost of control equipment. If the organics have no recoverable value, the regeneration stream at least contains impurities which have been concentrated substantially, so that they might be used as a fuel to offset the high energy consumption of an incinerator. The ability of activated carbon to increase the concentration of a combustible material, so as to decrease the fuel requirement for an incinerator, is particularly significant.

Hot gases–either heated air, heated "inert" gases, or stack gases–can also be utilized for regeneration. The choice of regenerating gas is dependent on a number of factors, including availability of high-temperature stack gas, heat content of the gas, heat exchange opportunities, and the ignition temperature of the saturated carbon bed.

After breakthrough of the carbon bed, high-temperature gas, at a rate much lower than that of the air stream being treated, is passed through the carbon bed, causing desorption and producing a stream with higher concentrations of organics, usually in excess of 15 times the initial concentration in the stack.

It should be noted that the system designer does not always have a choice of the regeneration methods mentioned. For example, in most of the above cases an implicit assumption is that fuel is available for regeneration heating. Also, in some cases organic compounds could undergo auto-oxidation in a saturated carbon bed, thus liberating sufficient heat to ignite the bed at temperatures below those necessary for air regeneration. In circumstances of fuel shortage or potential hazard, a possible alternative would be regenerating with a pressure swing, *i.e.*, vacuum cycle.

In a system using vacuum desorption, the pressure of the system is reduced to a point below the final partial pressure of the adsorbate, and desorption then occurs very quickly at extremely high concentrations.[4] In some cases these concentrations exceed 90% vol, thus making condensation more attractive. If incineration is chosen, zero fuel cost is a realistic objective.

A system utilizing the vacuum regeneration concept has been technically proved for the control of gasoline vapors emitted during refueling operations (Figure 19-3). Hydrocarbon vapors at 10 to 40% vol are controlled

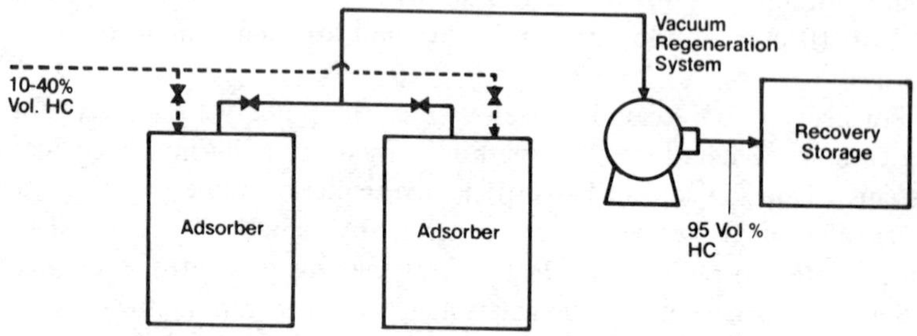

Figure 19-3. Gasoline vapor control system.

in canisters of activated carbon until the beds are completely saturated. Vacuum regeneration produces a stream containing 95% vol–well above the upper explosive limit. It should be noted that while vacuum techniques are proved with high vapor concentrations and highly saturated carbon, only limited data on the dilute streams typical of odor emissions are currently available. In those cases where additional working capacity is desired, a combination of heat and vacuum regeneration has proved successful. In a large lithographing company, this system, followed by condensation and recovery, was compared with a thermal incineration system. Capital costs were comparable. Thermal incineration had an operating cost five to six times greater–even with the utilization of heat recovery.

ODOR CONTROL APPLICATIONS

Sewage Treatment

As stated by Stone,

> Increasing urbanization and its encroachment upon sewage treatment plant sites, combined with public demand for improved environmental quality in general, have produced a need for reduced air pollution from sewers and treatment plants. The major air pollution problem associated with sewage systems is odor.[5]

The seriousness of this problem was confirmed in a nationwide survey of representative sewage treatment plants.[6] Of the plant superintendents questioned, 40% indicated they had received complaints about odors. Odor complaints were not restricted geographically but were found to be relatively uniform throughout the United States. In addition, 37% of the plants have had to, or are about to, take action to deal with their respective odor problems, either through process modification or the installation of odor abatement equipment.

The principal odors are H_2S and organic compounds, which are generated by the anaerobic decomposition of compounds containing nitrogen and sulfur. The odorous emissions include mercaptans, indole, skatole, organic acids, and other decomposition products. These numerous olfactory offenders can be found in many locations in a sewage system, including collection, transport, and treatment phases. The production of odors in sewage treatment is affected by several factors, including

1. Temperature
2. Sewage strength
3. Velocity of sewage flow
4. pH
5. Sulfate concentration
6. Aerobic conditions

The effects of each of these are discussed in detail by Dague.[7]

Nearly all stages of sewage treatment are potential sources of odor, but settling areas are more likely to be the source. Those plants employing secondary treatment and those using polymers in conjunction with vacuum sludge dewatering are more apt to experience odor problems. In many instances an adverse environment is created by sewage flows in excess of plant capacity.

The characteristics of activated carbon (described previously) make it a useful tool for sewage odor abatement. The diversity of odor sources within a treatment plant and their control with activated carbon is illustrated in a large installation where odors from the headworks, primary clarifiers, trickling filters, secondary clarifiers, and other areas are controlled with scrubbing techniques and subsequent carbon adsorption, or with carbon adsorption alone. With the objective of complete odor control, the headworks at this plant are covered as shown in Figure 19-4 and the 32,000 cfm odorous emissions are vented directly to the carbon beds in the foreground.

Likewise, the primary clarifiers have been covered with fiberglass lids (Figure 19-5), and the blowers pull 4,000 cfm from each clarifier directly into the beds of activated carbon.

As shown in Figure 19-6, the odorous trickling filters have been covered with domes. The resultant vapors (30,000 cfm from each filter) are collected and ducted, first through a series of hypochlorite and water scrubbers and subsequently to the segmented carbon beds in the foreground.

In addition to control of these odor sources, large carbon canisters have been installed on secondary clarifiers at this plant to control odor emissions from a turbulent water exit area. In other sewage treatment systems, activated carbon has been used to control odors emitted by pumping stations in the sewage transport system.

Rendering

Objectionable odors are emitted in all phases of the rendering process, but particularly during the cooking operation. The chief components of rendering odors include aldehydes, alcohols, ketones, acids, alkanes, acrylonitrile, and dipropylsulfide. All can be adsorbed by activated carbon.

The use of activated carbon to control rendering odors is discussed by Teller,[8] who discusses the use of an activated carbon bed to follow the spray section of his Teller Tower. Activated carbon was chosen because of its high effectiveness for removal of organic materials at low concentrations and its low cost. In the range of concentration treated and with the class of compounds to be adsorbed, the capacity of the carbon was estimated at 10 lb of noxious materials per 100 lb of carbon. In this system, activated carbon was used on a throwaway basis.

In his paper on the use of condensers for odor control in rendering, Strauss also describes the use of activated carbon following condensation of the cooker effluent.[9] No regeneration was undertaken because of the small quantities involved, but he concluded that a system utilizing a condenser, followed by an activated carbon filter, "has proven wholly satisfactory and virtually eliminated odors from the rendering plant."

In addition to treating the noncondensible portion of cooker effluent, activated carbon has been utilized in air filters to remove odors from rendering plant ventilation air. One plant utilized 36 small filters to treat 18,000 cfm. The activated carbon removed all odors and was reactivated every three months.

Figure 19-4. Headworks at this plant are covered and the 32,000 cfm odorous emissions are vented directly to the carbon beds in the foreground.

Figure 19-5. Primary clarifiers have been covered with fiberglass lids, and the blowers pull 4,000 cfm from each clarifier directly into the beds of activated carbon.

Figure 19-6. Odorous trickling filters covered with domes. Resultant vapors from each filter are collected and ducted, first through hypochlorite and water scrubbers and subsequently to segmented carbon beds in foreground.

Other Applications

There is a variety of industries and applications where adsorption technology has been used to control odors. For example, in the food industry activated carbon is used to alleviate the odor associated with a 100,000 cfm stream in an onion processing operation. In this instance, mist eliminators have been used as the adsorbers; activated carbon replaces the felt pads.

A carpet manufacturer received neighbors' complaints about biphenyl odors (biphenyl is used as a dye carrier). The odorous emissions are now scrubbed, passed through a particulate filter, and subsequently treated with activated carbon. The carbon system employs four beds, each 10 ft long by 2 ft wide with about 4 in. of carbon in each bed.

Activated carbon is also used to control odors associated with pesticide production, as well as acetic acid fumes associated with vinegar production.

New Systems

With the advent of air pollution control, activated carbon systems were developed which could handle

1. Large volumes of air
2. Low concentrations of materials
3. Operations within reasonable pressure drop limitations.

These systems will be discussed in detail later, but first consider the alternatives that are available with activated carbon and how these considerations are applied to systems development.

Figure 19-7 is a representation of the many different carbon alternatives that can be considered for new odor control applications. The circumstances surrounding the proper selection of a system are complex and should be fully considered for each individual problem.

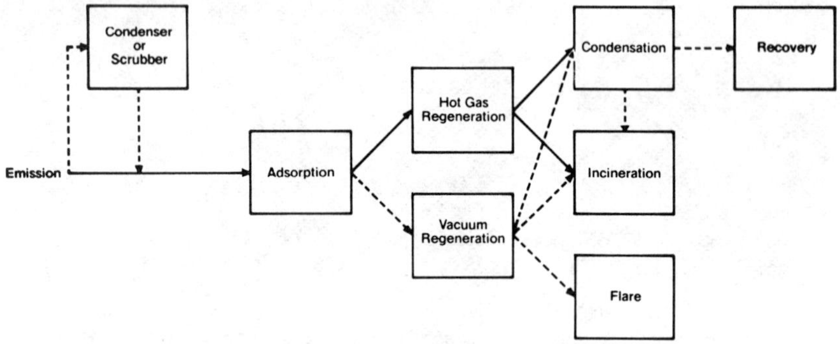

Figure 19-7. Activated carbon air pollution control systems.

Since activated carbon is most efficient in removing impurities at lower temperatures, cooling inlet air may be justified. Generally, if inlet air temperature exceeds 150°F, cooling should be considered to enhance adsorption. After the bed has been saturated, it can be regenerated by either hot gas or perhaps a vacuum, as described earlier. An additional adsorption bed may be necessary if adsorption must occur on a continuous 24-hour basis.

During hot air regeneration, concentrations up to 25% of the LEL are permissible, and thus regeneration gas could be sent directly to an incinerator for odor destruction. Condensation prior to incineration has the advantage of eliminating surges on the incinerator. In those cases in which desorbed material can be reused or reprocessed, condensation after regeneration should always be considered.

In vacuum regeneration systems, concentrations exceeding 90% vol are seen. In this case, flaring or condensation are the preferred downstream routes. If incineration is used, air or stack gas can be used for dilution.

Foundry Odors

The need for new systems to handle the low concentration, large-volume odor problems currently being encountered is illustrated by the odor control system installed at the Chrysler Huber Avenue Foundry.[10] This system utilizes activated carbon in a moving bed.

Figure 19-8 depicts the panel bed adsorber "house" where adsorption occurs. This novel adsorption system is employed to treat odorous emissions originating from three shakeout stations where pyrolysis products of the hotbox resin were identified[11] as the sources of odor. The resin is used to bind sand into cores, which serve as molds over which molten metal is poured and shaped as desired. After collection, the odorous emissions are sent through a wet multicyclone dust collector and then directed through the odor adsorber.

The adsorber house has several compartments, each containing 2,400 lb of activated carbon and each capable of handling 10,500 cfm. As the carbon approaches saturation, a portion is dropped to the spent carbon loadout bin and fresh carbon is added at the top. In addition to adsorbing odors, the carbon beds serve as a final dust collector. As reported by Engel *et al.*[11] laboratory studies showed odor removal efficiencies over 99%. Tests on the stacks themselves agreed closely.

Each compartment in the odor unit is approximately 7.5 ft high by 5 ft square and contains five vertical panels, each 7 ft high. A cross-section of one adsorber panel is shown in Figure 19-9. Each is 2 in. thick and is filled with activated carbon. Weirs 5 in. wide protrude from the sides to support the carbon, effecting a working carbon thickness of more than 4 in. Panels are arranged in a series of "V" configurations so that the odorous shakeout emissions coming from one side must go through the carbon before exiting through a "V" on the other side of the compartment. As the carbon flows down through the bed, it tumbles gently and exposes new surfaces for maximum adsorption and dust filtration.

Paint Bake Ovens

Paint bake ovens have been cited as large emitters of photochemically reactive organics which ultimately produce ozone, a known cause of damage to both animals and vegetation. The reactive materials are basically derived from organics that are used as paint solvents. However, in addition to solvent vapors emitted from a bake oven, aerosols–which are also offensive

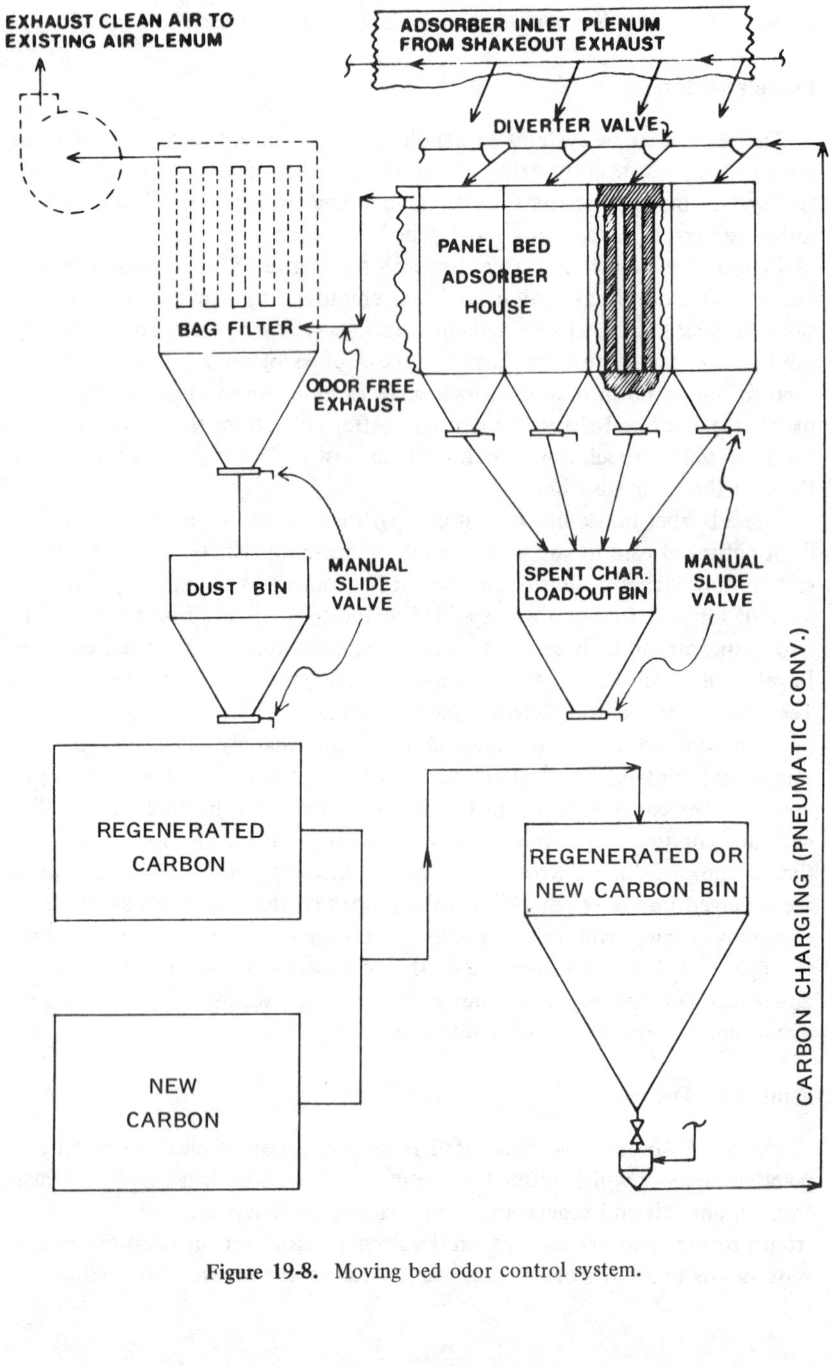

Figure 19-8. Moving bed odor control system.

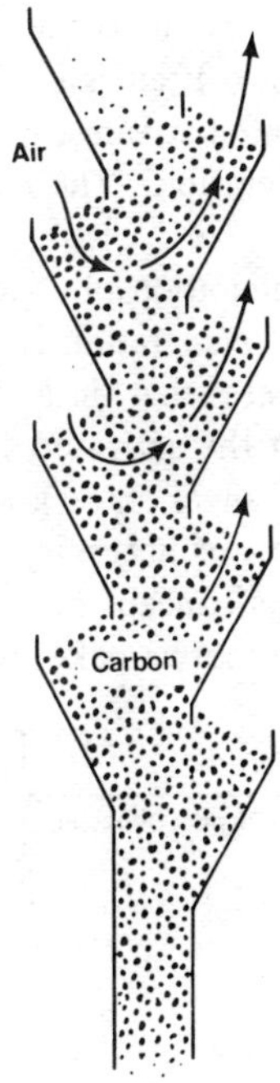

Figure 19-9. Moving bed adsorber panel.

from an odor standpoint–are formed from the partial oxidation and/or polymerization of organic solvents and resins.

In some localities the primary method of compliance with hydrocarbon regulations has been to reformulate by using exempt solvents. Recently, attention has also been directed toward water-based paints. However, while the photochemical oxidant problems may be alleviated by these methods, odor problems persist. Thus, in all probability, some method of control will be necessary on most paint bake ovens.

Since passage of Rule 66 in California, the recommended control methods for removal of photochemically reactive hydrocarbons were incineration and adsorption. However, the capital cost of a steam-regenerated activated carbon system is sometimes higher than incineration. In addition, at low concentrations the amount of steam required, relative to the amount of pollutant desorbed, is so large that separation and recovery of pollutants from the stream becomes less practical than in traditional solvent recovery.[12] Thus thermal incineration became the preferred control technique. However, even with heat recovery equipment which adds significantly to the capital cost, incinerators need fuel to operate–a large amount of fuel. And today fuel is in extremely short supply, and expensive.

A promising alternative is to reduce the volume of air to the incinerator by concentrating the pollutant on activated carbon. If air is used to regenerate the activated carbon bed, a highly concentrated stream can be produced in a substantially reduced volume for incineration. The resultant savings in supplemental fuel are quite significant.

Hardison[13] reports a successful adsorption/incineration pilot-scale test on an acrylic spray booth, but he concluded that incineration is the best control technique for bake ovens. Mattia[15] indicates that, in a similar system, inert gas was needed to prevent oxidation of the activated carbon. Since no references could be found on the use of air or stack gas to regenerate activated carbon beds in large systems, it was necessary to test this concept in a pilot unit. An illustration of this concept can be seen in recent work performed by Calgon[16] on an automotive bake oven (Figure 19-10).

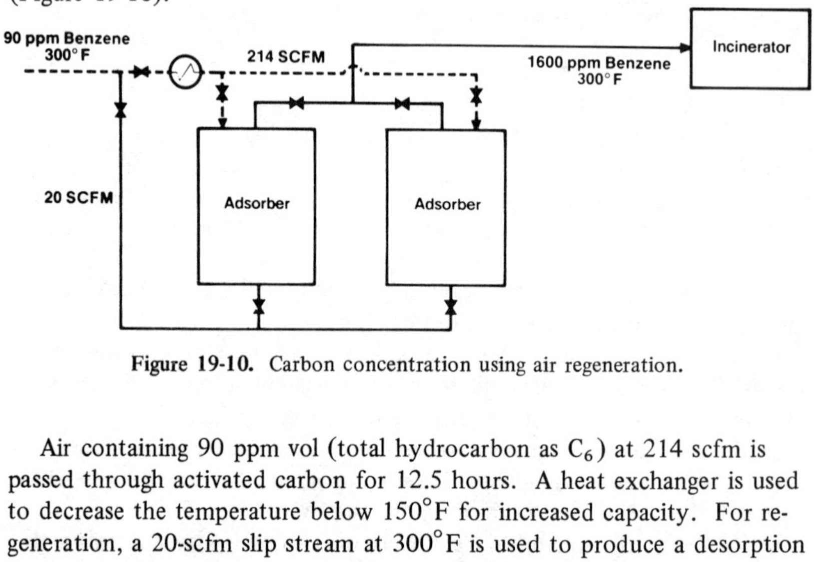

Figure 19-10. Carbon concentration using air regeneration.

Air containing 90 ppm vol (total hydrocarbon as C_6) at 214 scfm is passed through activated carbon for 12.5 hours. A heat exchanger is used to decrease the temperature below 150°F for increased capacity. For regeneration, a 20-scfm slip stream at 300°F is used to produce a desorption stream containing an average 1,600 ppm for 5 hours.

Figure 19-11 shows the results of analysis using a gas chromatograph to identify the organic components in the adsorber inlet and outlet streams. As can be seen, essentially all components having higher molecular weights (retention time) than C_1 have been removed by adsorption on the activated carbon. Actual measurements indicate that more than 90% of all organics above C_1 were removed by the activated carbon.

Table 19-1 illustrates the effect of carbon concentration for this system. The total volume of air to be treated has been reduced by a factor of 27 (10.7 x 2.5), while the concentration of emissions has been increased 17 times. If the initial stream were incinerated directly at 1,400°F, 2,500,000

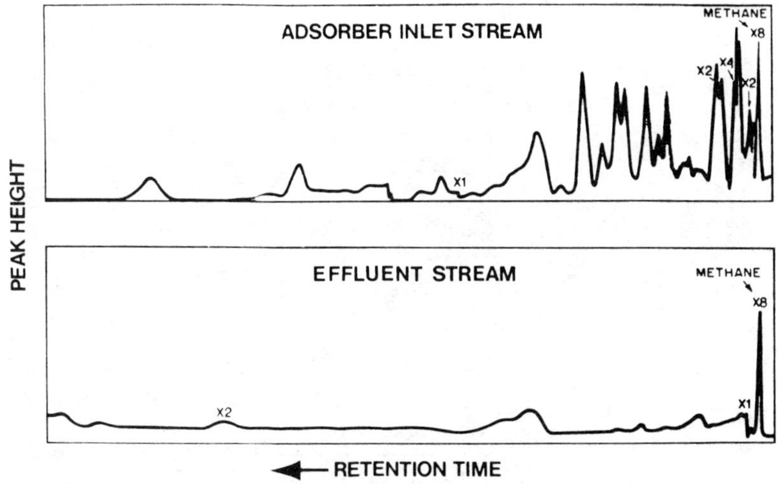

Figure 19-11.

Btu's of additional fuel are required per cycle. The same stream that has been first concentrated on activated carbon requires only an additional 62,000 Btu's to incinerate, due to its smaller volume and higher fuel value. Therefore the effect of carbon concentration is a 40:1 reduction in energy requirements. Even if the odor had negligible fuel value, an energy saving of 27:1 could still be effected. If catalytic incineration at 900°F is chosen instead of direct incineration, an energy savings of 59:1 is possible because the fuel value of the richer regeneration stream is a greater proportion of the energy needed to reach incineration conditions. A comparison of the following two systems illustrates the ramifications of this new technology.

Table 19-1.

	Flow Rate	Cycle Time	Odor Concentration
Initial stream	214 scfm	12.5 hr	90 ppm
Regeneration stream	20 scfm	5.0 hr	1,600 ppm
Concentration ratio	10.7:1	2.5:2	17:1

Incineration with Heat Recovery

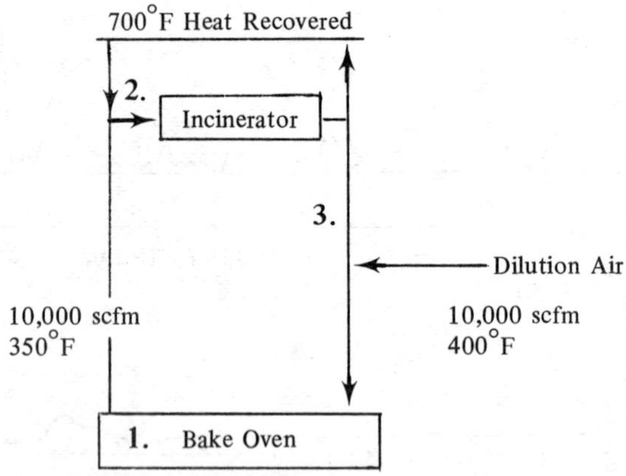

1. Heat required to operate oven at 400°F:
   10,000 scfm x 60 min x 6 Btu/ft^3* = 3.6 mm Btu/hr
2. Heat required to operate incinerator at 1,400°F:
   10,000 scfm x 60 min x 7 Btu/ft^3* = 4.2 mm Btu/hr
3. Heat recovered to oven:
   10,000 scfm x 60 min x 6 Btu/ft^3 = 3.6 mm Btu/hr
4. Net heat needed for system (1 + 2 - 3 above)
   3.6 $ 4.2 - 3.6 = 4.2 mm Btu/hr
5. Net heat for system

$$\frac{4.2 - 3.6}{3.6} = 17\% \text{ more}$$

Adsorption/Incineration

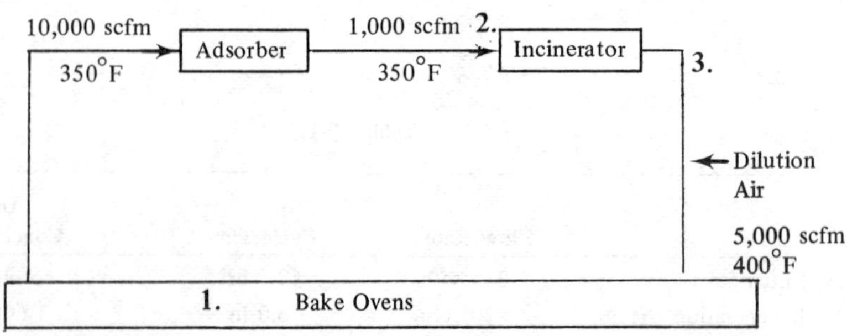

*Net enthalpy.

1. Heat required to operate oven at 400°F:

   10,000 scfm x 60 min x 6 Btu/ft^3 = 3.6 mm Btu/hr

2. Heat required to operate incinerator at 1,400°F

   1,000 scfm x 60 min x (20.7 - 6.3*) Btu/ft^3 = 0.9 mm Btu/hr

3. Heat recovered

   5,000 scfm x 60 min x 6 Btu/ft^3 = 1.8 mm Btu/hr

4. Net heat needed for system (1 + 2 - 3) above)

   3.6 + 0.9 - 1.8 = 2.7 mm Btu/hr

5. Net heat for system

   $$\frac{2.7 - 3.6}{3.6} = 25\% \text{ less}$$

Consequently, a system utilizing incineration with maximum heat recovery needs 17% additional heat to operate the pollution control system. The adsorption/incineration system shows a 25% reduction in heat needed. Or, in simpler terms, incineration with heat recovery uses over 50% more fuel than an adsorption/incineration system.

Styrene Removal

Similar styrene odor studies have been conducted by Calgon. Styrene has a threshold odor level of approximately 4 ppm vol and is considered photochemically reactive. Since styrene vapors are also toxic, large volumes of ventilation air must be used to reduce concentrations in plant air, producing an odorous exhaust gas. Manufacturers of glass-reinforced polyester products made by lay-up and spray techniques are particularly concerned with styrene odor emissions.

In a typical emission stream the concentration of styrene vapor is 100-200 ppm vol at ambient temperature. Studies conducted by Calgon under similar conditions indicate that activated carbon can adsorb styrene vapors to a capacity in excess of 25% weight before odor breakthrough occurs. Subsequent regeneration with hot air produced a working capacity in excess of 10% weight. It was also found that styrene does not polymerize on activated carbon during air regeneration at temperatures up to 450°F. Since more than 95% of the recoverable styrene is desorbed as pure styrene, recovery of 95% purity styrene by condensation is possible.

In a manner similar to the control of bake oven odors, styrene vapors in the highly concentrated regeneration stream could also be destroyed in a thermal or catalytic incinerator.

---

*Net enthalpy - fuel value of concentrated organics.

Styrene has also been recovered by conventional solvent recovery techniques using activated carbon. Air at ambient temperature containing styrene in a concentration of 100 ppm vol was adsorbed on activated carbon. The saturation capacity of the carbon reached 25% weight of styrene. Steam at 210°F was used as the regenerant, and styrene was recovered by condensation. Normal steam requirements were 9 to 12 lb per pound of styrene recovered. The cyclic working capacity of the activated carbon bed was 15 to 20% weight.

Adsorption-Oxidation System

A combined adsorption-oxidation system (ADSOX) for odor control has been reported by Wagener and Winter.[17] In this process, large-volume odor emissions, characterized by a low concentration of organic vapors, are passed through activated carbon, allowing odor-free air to be exhausted. A small volume of inert gas, produced in the upper portion of the adsorber, is used to regenerate the adsorber. This produces a stream with a high concentration of organics in a greatly reduced flow for subsequent destruction in an incinerator.

Before passing through the carbon adsorbers, the inert gas is cooled to the required desorption temperature by mixing with a circulating gas current. To effect the fastest desorption, the desorption temperature should be 100 to 500°F above the boiling point of the adsorbed organics. The dilution gas not only lowers the temperature of the inert gas to the required desorption temperature but increases the flow rate of inert gas through the carbon bed. When desorption is completed, inert gas generation is turned off and a cooling blower is brought into service. When the temperature has fallen to under 250°F, adsorption can be restarted.

Prototype systems include one on a rendering plant where a 600-cfm odorous emission from a condensate tank at 86°F had a concentration of 17 ppm weight. Also, odorous air from a quick-frozen food operation is cleaned with an Adsox unit. The concentration of organics in this instance is 9 ppm. Adsorbers are regenerated biweekly.

A 10-adsorber ADSOX unit is planned for treating a 236,000-cfm odorous emission from an aluminum cold-rolling mill. Organics in the stream total 4,250 ppm. Compared with an inlet volume of 236,000 cfm, the inert gas flow will generate only 3,900 cfm for subsequent incineration—a concentration factor of 60.

Prefab Carbon Filter

In addition to the recent emphasis on odor abatement systems for large-volume, low-concentration emissions, there are many sources of odor where small amounts of gases are vented. Activated carbon's capacilities have been applied to many such situations by using a unique concept in air pollution control: Vent Sorb (Figure 19-12).

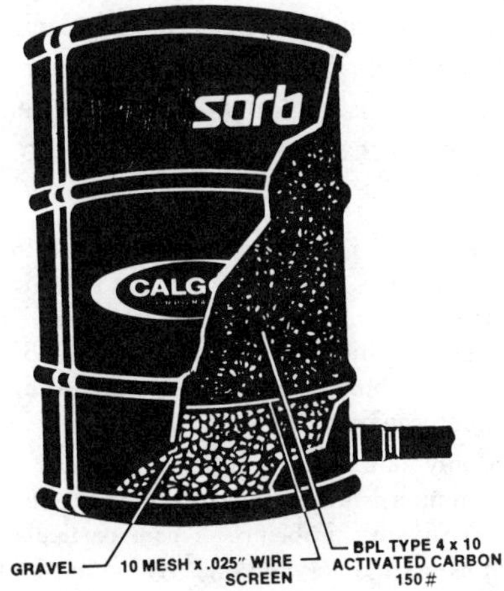

Figure 19-12. Vent sorb.

Vent Sorb is a prefabricated, miniature granular activated carbon filter containing all the elements of a full-scale carbon system, *i.e.*, vessel, support medium, activated carbon (approximately 150 lb), and inlet distribution nozzle. The prefab unit uses the very high adsorptive capacity of granular activated carbon to remove unwanted gases from air and other carrier gases vented to the atmosphere by storage tanks, reactor vessels, and other sources in which small amounts of gases are involved. Contaminants emitted from these vents include hydrogen sulfide, mercaptans, SO_2, amines, ammonia, oil vapors, halogens, hydrocarbons, solvents, aldehydes, monomers, organic decomposition products, chlorinated hydrocarbons, and other uncharacterized odorous gases.

Applications

The storage of volatile solvents presents problems due to tank filling and emptying and tank breathing. Odorous vapors normally emitted to the atmosphere have been controlled by the adsorption capabilities of activated carbon. Vent Sorb has found wide acceptance, for example, on crude sulfate turpentine storage tanks. The carbon adsorbs turpentine vapors along with sulfur compounds such as H_2S, mercaptans, and trace amounts of other odor-producing substances which have extremely low odor threshold levels. Other areas where the unit has been employed on storage tanks include asphalt plants, fuel oil storage facilities, and rendering plants.

The control of reactor vent gases also utilizes the carbon approach. After a reaction is completed in a closed system, reactors are often vented to the atmosphere prior to discharging their contents. The vent gases in most cases are hot and contain high concentrations of unreacted starting materials, by-product, and product. It was sufficient in the past to merely remove as much as possible by condensation or scrubbing techniques, venting the remainder to the atmosphere. With more stringent air pollution codes going into effect, this practice is no longer applicable generally, and further vapor control is required. Areas employing Vent Sorb for this purpose include:

1. Removal of methylamine and diethylamine following a caustic scrubbing operation. The amines are a by-product of a pesticide operation and are extremely odorous at low concentrations (threshold level 0.02 ppm). In this case the unit handles a 30-cfm air flow from the caustic scrubbing system, which contains approximately 15 ppm amine vapors.
2. Vent systems on batch polymerization reactors. During the venting operation, unreacted monomer is discharged to the atmosphere, resulting in serious odor or toxicity problems. This is particularly true of monomers such as methyl acrylate, methyl methacrylate, acrylonitrile, or vinyl chloride. Vent Sorb has been applied to these odorous situations to remove the last traces of odor after condensation.

Another application is in controlling odors emanating from an API oil-water separator. The activated carbon primarily removes hydrocarbon odors and sulfurous odors formed in the oil-water separation sump.

Miscellaneous applications include

1. Sludge thickener tanks at sewage treatment plants
2. Wastewater holding tanks at chemical plants
3. Vents from hood exhausts of chemical laboratories
4. Rendering plant storage tanks
5. Sewer gas vents at chemical or municipal waste treatment plants.

SUMMARY

Activated carbon has proven ability to remove odors. It is effective in treating low- and high-concentration emissions in small- and large-volume streams—all of which are found among the diverse odor problems we now encounter. Some of the processes described are still in the pilot plant stage, but most are fully developed and are currently in use.

In some of the applications discussed, activated carbon's ability to concentration emissions makes it a very attractive abatement technique. By reducing the total volume of air to be treated and increasing the concentration of odorous materials, activated carbon can, for example, produce an enriched stream for subsequent incineration. Not only does this substantially decrease the operating cost of an odor control system, it also offers a greatly reduced usage of scarce energy. As a result, odor control and energy conservation no longer need be mutually exclusive.

Activated carbon adsorption has been shown to increase or complete the removal of odors as part of a larger environmental improvement system. If odor control codes become more stringent, activated carbon systems will not become obsolete. Simple changes in operational parameters will increase system efficiency. For example, if increased odor removal efficiency is needed, additional system capacity can be obtained by decreasing cycle times and/or adsorption temperatures. This can normally be accomplished without additional equipment.

REFERENCES

1. Hassler, J. W. *Activated Carbon,* rev. ed. (New York: Chemical Publishing Co., 1963).
2. Faulkner, W. D., W. G. Schuliger, and J. E. Urbanic. "Odor Control Methods Using Granular Activated Carbon," presented at 74th National Meeting of the American Institute of Chemical Engineers, New Orleans (March 12, 1973).
3. Egorov, N. N., *et al.* "Purification of Gases with Activated Carbon," *Metallurgizdat* (1950).
4. Fusco, R. A. and R. L. Poltorak. "The Use of Activated Carbon to Control Gasoline Marketing Loss Emissions," presented at National Petroleum Refiners Association Meeting (September 12-13, 1973).
5. Stone, R. "Sewage Treatment System Odors and Air Pollutants," *J. San. Eng. Div., Proceedings of the American Society of Civil Engineers* (August 1970), pp. 904-909.
6. Survey conducted by Calgon Corporation, Activated Carbon Division (1973).
7. Dague, R. A. "Fundamentals of Odor Control," *J. Water Poll. Control Fed.,* **44** (4), 583-594 (April 1972).
8. Teller, A. J. "Odor Abatement to the Rendering and Allied Industries," presented at APCA meeting (May 1962).
9. Strauss, W. "The Development of a Condensor for Odor Control from Dry Rendering Plants," *J. Air Poll. Control Assoc.,* **14** (10) (October 1964).
10. Schaum, J. "Chrysler Initiates Odor Control at Huber Avenue Foundry," *Modern Casting*, no. 8 (August 1973).
11. Engel, J. H., Jr., R. Kanczuzewski, and J. K. Samuels. "Odor Control Program," presented at 75th National Meeting, American Institute of Chemical Engineers, Detroit (June 5, 1973).
12. Environmental Protection Agency, "Package Sorption Device System Study," EPA-R2-73-202.
13. Hardison, L. C. "Controlling Combustible Emissions," *Paint and Varnish Prod.* (July 1967).
14. Hardison, L. C. U.S. Patent 3,395,972.

15. Mattia, M. M. "Process for Solvent Pollution Control," *Chem. Eng. Prog.* (December 1970).
16. Urbanic, J. E. and W. D. Lovett. "The Use of Activated Carbon for Control of Paint Bake Oven Emissions," presented at the 67th APCA Annual Meeting (June 1974).
17. Wagener, K. and K. Winter. "ADSOX–A Combined Adsorption and Oxidation Process for Eliminating Emissions of Organic Vapors," presented at ACHEMA, Frankfurt/Mainz (June 6, 1973).
18. Poltorak, R. L. "Foundry Odor Control with Activated Carbon," *Modern Casting* (August 1973).
19. Faulkner, W. D., W. G. Schuliger, and J. E. Urbanic. "Odor Control Methods Using Granular Activated Carbon," presented at 74th National Meeting, American Institute of Chemical Engineers, New Orleans (March 12, 1973).
20. Lovett, W. D. and F. T. Cunniff. "Activated Carbon Methods for Controlling Stationary Source Air Pollutants," presented at the 76th Annual Meeting, American Institute of Chemical Engineers, Philadelphia (November 1973).
21. Joyce, R. S., J. R. Lutchko, R. K. Sinha, and J. E. Urbanic. "New Developments in Ambient Odor Control Adsorption," presented at the Conference on Odors: Evaluation, Utilization and Control, New York (October 1-3, 1973).

CHAPTER 20

ACTIVATED CARBON SOLVENT RECOVERY

Carleton H. Boll

RaySolv, Inc.
Highland Park, N.J. 08904

Activated carbon adsorption can, in general, be used in situations where vaporized hydrocarbons must be removed from a gas stream. Some typical operations where these gas streams are present are coating, degreasing, gravure printing, laminating, and the desolventizing of molded rubber goods. There are, however, situations where activated carbon can be used only with great difficulty. The first of these is where there are extremely low concentrations (below 500 ppm) of solvent in the air stream. It is possible to adsorb down to 50 ppm, but the size of the plant and the cost of regeneration go up dramatically with these low concentrations, usually making it an impractical approach.

A second situation is where there are gross amounts of contaminants in the solvent-laden air that would, over a period of time, poison or blind the pores of the carbon. As a matter of protection for the carbon, every adsorption system should have a filter ahead of the adsorber beds. However, if there is, for example, paint overspray, entrained plasticizer, entrained ink, or finely divided particles such as titanium dioxide or silicones, the cost of eliminating these particles from the air stream can become significantly high—perhaps enough to eliminate carbon adsorption as a desirable process.

Third, there may be chemical reactions on the activated carbon bed. Under certain conditions, activated carbon can act as a catalyst for decomposition. The breakdown compounds produced can be many times worse than the original compound. Methyl ethyl ketone (MEK), for example, was one of these types of compounds a few years ago. Now technology has been developed to prevent the production of diacetyl and other compounds of the MEK, which are detectable by nose at parts per billion.

A fourth situation might exist when a number of solvents and different solvent blends are used, with many changes over the course of a normal working period. It is possible to recover the solvent, but the separation of these solvents into any sort of usable form becomes prohibitively expensive.

Fifth, in surge operations—for example, a 25,000 scfm gas stream for a period of two minutes, with nothing for two hours—carbon adsorption would be an expensive air pollution control method from a capital investment standpoint.

In summary, activated carbon solvent vapor recovery should be considered for almost every operation where the solvent is used as a carrier in producing a product and is eventually evaporated off the product and into an air stream.

WHEN TO USE ADSORPTION

An activated carbon system in the 25,000 scfm range (and up) will more than likely have to be specially engineered. The cost of a solvent vapor recovery plant is, in general, based on the air volume handled. It is an air handling plant that just happens to pick up a solvent as it goes through the plant. The materials of construction dictated by the kind of solvents being handled also affect the cost. Therefore the engineer should look at reducing the air volume to a minimum and still produce the product.

Air is not free. It costs money to handle the air leaving the plant. In some cases the air volume can be reduced simply by experimenting with closing dampers on fans to the system. Second, increasing the temperature at which the desolventizing is taking place, thereby getting a higher concentration at lower air rates, should be investigated. Three, the possibility of recycling air within the system should also be considered. Four, leaks permitting extraneous amounts of clean air to get into the exhaust should be reduced.

In many cases reductions in the order of 50% or more are possible, which directly translates into a reduction of the plant costs.

The solvent composition should also be examined, particularly if it is possible to reuse the solvent after it has been recovered. Approach the makeup of solvent blends from a historical basis and with a very jaundiced eye. Historically, solvent has been thrown away. The ideal solvent system is a single solvent that is stable in the presence of carbon and is not soluble in water.

HOW TO USE ADSORPTION

The "how" of using activated carbon is selecting the final design parameters for the system, finding a supplier, and having the system installed. The air flow must be measured, and measured at the maximum design capacity of the system. If various products are running on different machines, it is necessary to calculate which of these products uses the maximum air on each particular machine. Data should be gathered (when possible) during actual production runs.

While measurement of air flows is being made, it is also necessary to take the temperature of the air. This is important in case an air cooler may be required. Carbon adsorbs most efficiently below 95°F. Therefore a hotter air stream must be cooled to that temperature prior to going to the adsorbers.

Next, it is necessary to know the pounds of solvent per minute contained in the air stream. There are several different ways to measure this. It can be done by taking air samples and having them analyzed on a gas chromatograph by an air pollution testing firm. An alternative method is to back-calculate from the pounds of solvent being applied and make the assumption that 100% of it leaves in the air stream. This number can be checked in reviewing the annual machine hours and the annual purchases of solvent.

The next thing that has to be determined is the availability of utilities. As a rule of thumb, 4 lb of low-pressure steam are required for every pound of solvent to be desorbed. Power must be available for the fans. Instrument air is typically required and decisions must be reached as to the location of the control panel (normally in an explosion-proof area) and whether to enclose the condenser, subcooler, and decanter—those parts of the system typically susceptible to freezing—in a building or to provide winterizing.

The presence of carryover must be checked for solvent-laden air as well as plasticizers and other compounds that might tend to foul the carbon bed. A filter may be necessary. As a rule of thumb, if the present outlet ducts look clean, there will probably not be a problem. If they are heavily coated, however, carryover must definitely be taken into account in the design of the system.

The next task is to pick a supplier. Talk to as many as possible on an informal basis, without asking for a formal quotation. Out of those you have interviewed, pick the top three or four and reverse the interview process. Travel to their engineering facility. Then you can meet not only their sales people but their engineering staff, to enable you to get a feel for their capabilities and experience. Then give the firms you pick any information they require to bid on your specification.

Pick the supplier on the basis of your estimate of the quality of its engineering capabilities, the lowest utility usage guaranteed, and delivery time—not just first cost. The system will still be operating in 10 to 20 years, so even a 10% additional steam usage with forecast fuel increases will eat up a 10% savings in initial cost very quickly. First-cost savings are important, but annual maintenance and operating costs are even more important.

In Europe, where solvents have been considerably more expensive, their recovery has been the standard mode of operation, not originally from an air pollution control standpoint but primarily for strictly economic benefits. The same can be true in the United States. Table 20-1 points out some of the economic benefits.

If solvents are universally captured and returned for reuse, a very big step will be taken toward reducing the present shortage of solvents, petrochemicals, and energy.

Table 20-1. Coating Operation
Using 285 gal/hr Toluene in 30,000 scfm of Air

1. Replacement cost of toluene 1,600,000 gal/yr @ $0.40/gal			$560,000
2. Utilities			
Steam 34 x 10^6 lb/yr @ $1.25/1000 2200 lb/100 gal		2.75¢/gal	$ 42,500
Cooling Water 220 x 10^6 gal/yr @ $0.20/1000 gal 14,000 gal/100 gal		2.9¢/gal	44,000
Electricity 140 KW connected @ $0.01 KwHr $50 KwHr/100 gal		0.5¢/gal	8,000
			$ 94,500
3. Operation and maintenance No direct operation is required.	Assume	0.6¢/gal	10,000
4. Total operating cost		6.75¢/gal	$104,500
5. Gross return per year (1) $560,000 minus (4) $104,500			$455,500
6. Investment in plant 30,000 scfm @ $10/scfm			$300,000
7. Return on investment after taxes $455,500 x 50% income tax - investment of $300,000			76%

CHAPTER 21

PROCESS SMOKE CONTROL WITH LOW-VOLTAGE ELECTROSTATIC PRECIPITATORS

William A. Cheney

United Air Specialists, Inc.
Cincinnati, Ohio 45200

New developments in low-voltage precipitators offer exciting possibilities for the control of a wide range of emissions from process control ovens. These devices are lower in capital cost than most other equipment used competitively and offer advantages in operating costs, plus the ability to recover, in a recyclable form, the discharge that normally goes to the atmosphere. As an energy-saving bonus, the system employed for air cleaning can often recover a very high proportion of the heat that is coming from the oven.

The original Cottrell precipitator industrial application dates back prior to World War I. The original design was a high-voltage wire suspended in the center of a hollow cylinder with dirty air passing from end to end. Particles were charged and precipitated on the walls of the cylinder as they passed through. Later modifications employed several wires between parallel plates normally 4 to 6 inches apart. Typical voltages on the ionizing wires or emitters ranged from 50,000 to 100,000 volts. This is the basic design that is employed on power plant emissions today. It is not well suited to stack volumes below 50,000 cfm because of cost, and it has some limitations on efficiency in the submicron particle range.

In 1932 Westinghouse assigned Gaylord Penney the task of devising a precipitator that would be satisfactory for use in air conditioning ducts. He responded with the invention that is now known as the two-stage, low-voltage, or Penney-type precipitator. It is normally used with the ionizing wires at 12 kv before the collecting plates and with the collecting plates at

6 kv spaced about 1/4 in. apart. Penney overcame ozone generation by using the lower voltages and employing positive ionization instead of negative. This basic design has been widely used in air conditioning and heating ducts in commercial and residential applications.

Some early work was done to apply the low-voltage precipitator to industrial stack applications in southern California. Essentially, the basic unit in air conditioning ducts was lifted onto a rooftop and put in the process stack. In order to get acceptable efficiency, the face velocity was slowed from the normal 450 fpm to as low as 100 fpm. The applications were not satisfactory, largely because of the high maintenance required.

Among the problems that existed:

1. Most service for the electrostatic components had to be performed from inside the duct, or a crane had to be used to remove the cell for service.
2. A short anywhere in the unit ionizer or collecting cell would shut down the entire system. The ionizing wires were up to 17 ft long and subject to frequent breakage.
3. There was a potential fire hazard because the collector was powered from one large transformer that had to be designed for a relatively large current.

With passage of the Clean Air Act of 1970 it became apparent, however, that the low-voltage precipitator had a significant place in controlling certain types of stack emissions. With the help of Professor Penney, the parameters of voltage, plate spacing, and wire spacing were maximized for the most efficient precipitator design. The mechanical design defects that had plagued the application of air conditioning-type precipitators to industrial processes were also corrected. These modifications included the following:

1. Modular design of components to allow ease of handling and low power requirements. The total energy for each module is so low that it permits use of a power pack that meets UL requirements. All high-voltage lead wires were removed from the air stream to eliminate the high maintenance associated with these wires.
2. Easy access during operation of the process so that maintenance could be performed without stepping into the dirty air stream or shutting down the process.
3. Developing cleaning techniques for the various contaminants.
4. Removing insulators from the air stream so their tendency to short out under load would be greatly reduced.
5. Developing remote panels to indicate the condition of the rooftop unit.

Just where does the low-voltage precipitator fit in the pollution control picture? First of all, the emission must be visible—vapors from highly volatile materials such as trichlorethylene are not suitable for precipitators.

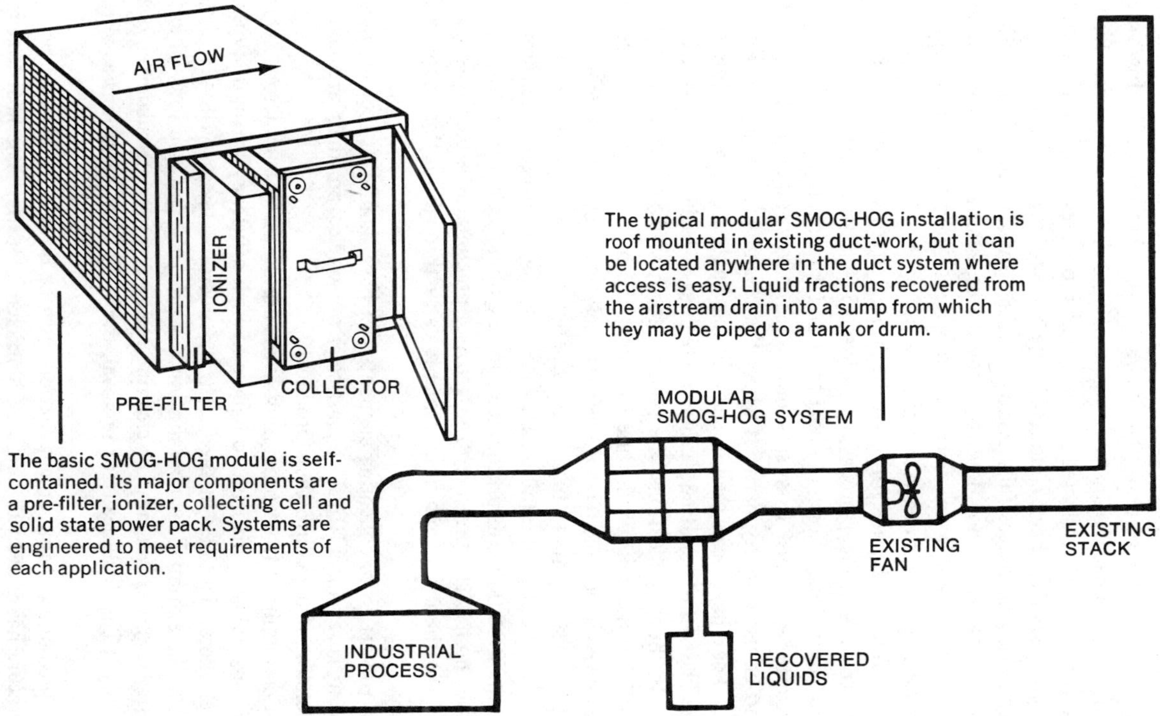

Figure 21-1. Low-voltage electrostatic precipitator. Recently developed small capacity units of this type are currently finding a wide range of applications such as vinyl curing ovens; rubber process smoke; welding and glass fumes; heat-treat and quench operations; and range hood exhausts from steak houses.

A general guideline would be "wherever oily material is vaporized with the application of heat during the process." The oil vapors condense and become visible as they are cooled. They are precipitated as they are drawn through the low-voltage unit. The characteristic that makes the low-voltage precipitator practical for this wide range of applications is that the oil runs off the plates as it is collected, and the collectors are therefore largely self-cleaning. Tests have been conducted where up to one gallon per hour per thousand cfm was successfully removed with no visible emission. If the collected material did not flow from the plates, such a unit would require cleaning about every hour. Because the material does flow, the elements require cleaning about every two months.

Successful applications for low-voltage precipitators include (1) vinyl curing ovens, (2) rubber processing smoke, (3) smoke from textile texturizing and dying operations, (4) heat-treat and quenching operations, and (5) range hood exhaust from steakhouses. The low-voltage precipitator unit can also be used on dry particles where the cleaning frequency can be tolerated. They have been used in such diverse applications as welding smoke and the glass fumes emitted during the sealing of high-intensity quartz lamps. In these applications, the precipitator has advantages over other methods that make it worthwhile to accept the higher cleaning frequency.

The low-voltage precipitator has extremely low power requirements to strip the contaminants from the air. Energy must be applied only to the contaminant–not to the full air stream. The electrical power consumption to operate a 10,000 cfm precipitator, for example, would be in the range of 1¢ per hour. The pressure drop across a precipitator is usually about 1/2 in.; so the low energy requirements carry through on this characteristic also.

The run-off from collecting elements can have significant reclaim value. One plant is reusing 110 gal of plasticizer per week from their precipitator and blending it back into their product.

Particles emitted from processing ovens are normally submicron. This particle size range is difficult to handle at high efficiency for most types of air cleaners. Particles in the 0.2 to 0.3 micron range may flow with the air stream or exhibit Brownian motion. A properly designed low-voltage precipitator can exceed any air cleaning regulation for particulate removal or opacity. This includes odor removal, even for such operations as vinyl curing.

The precipitator requirement that oil must be brought to its condensation temperature has proved to be a desirable characteristic. The heat that must be removed from the air stream can be reused in the process or building with a fast payoff on the heat exchanger.

In summary, the low-voltage precipitator (1) performs at high efficiency, (2) has exceedingly low energy requirements, and (3) permits energy and material recovery.

CHAPTER 22

SAFETY REQUIREMENTS FOR ELECTROSTATIC PRECIPITATORS ON ORGANIC EMISSIONS

William B. Rossnagel

Rossnagel & Associates
Cherry Hill, N.J. 08003

Electrostatic precipitators are being used more and more to remove air pollutants from textile, organic, printing, floor tile, machine shop, and other similar emissions. These precipitators include both the "dry" and "wet" type. Wet precipitators are those which operate with either continuous internal water sprays or with continuous sprays in the inlet ducting to the precipitator. A precipitator which is spray washed with water after shutdown is not considered a "wet" precipitator.

Precipitators, both wet and dry, have also been combined in series with several types of scrubbers, with direct and indirect cooling coils,[1] with ozone generators, and with other devices to eliminate various pollutants. Both dry and wet units have been used with increasing success on various organic emissions. The organic emissions range from oil mists and floor covering oven exhausts to carpet underlay ovens and oil quench oven exhausts.

While these precipitator installations have proved very successful when applied to specific organic emissions (those with a high ratio of condensible hydrocarbons to noncondensible hydrocarbons), they must be installed in a system designed to prevent overheat, fire, or explosions. Fires can start on the product within the oven, on the accumulated deposits of combustibles (solid, liquid, and gaseous) in the ducts, or in the heat exchanger/precipitator/blower system.

A safety standard for precipitator systems has been generally established by Factor Mutual's Document D.S. 7-2. Between 1971 and 1972 Factory Mutual had nine losses from fires and explosions, totaling over $2 million on precipitator systems on cement plants alone. As a result, their safety requirements have been tightened.

The basic types of fires which can occur in a typical precipitator air pollution control system for a textile oven are shown in Figure 22-1. The degree of control will vary with the type of oven (directly or indirectly heated) and the flash point ignition temperature of the solvents, oils, plasticizers, etc.

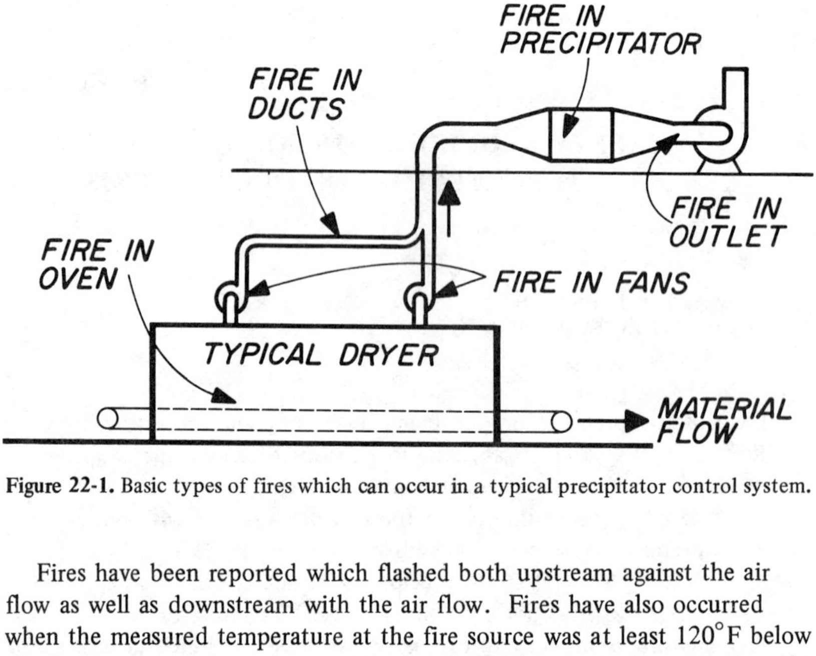

Figure 22-1. Basic types of fires which can occur in a typical precipitator control system.

Fires have been reported which flashed both upstream against the air flow as well as downstream with the air flow. Fires have also occurred when the measured temperature at the fire source was at least 120°F below the flash point of the organic materials. To explain how these fires in pollution control systems can occur, we will analyze a specific case history.

CASE HISTORY

An Eastern plant had an effective pollution control system utilizing a heat wheel and electrostatic precipitators, as shown in Figure 22-2. The sprinkler head and bypass gate system successfully handled a dryer-generated fire. The next month, an operator in the plant noticed smoke coming from the overhead exhaust duct from the oven. He quickly hit the master "off" switch. As he did, there was a muffled explosion from the rooftop 20,000 acfm precipitator pollution control system. The resulting fire caused four alarms for the fire department and resulted in a loss costing $28,000 to correct.

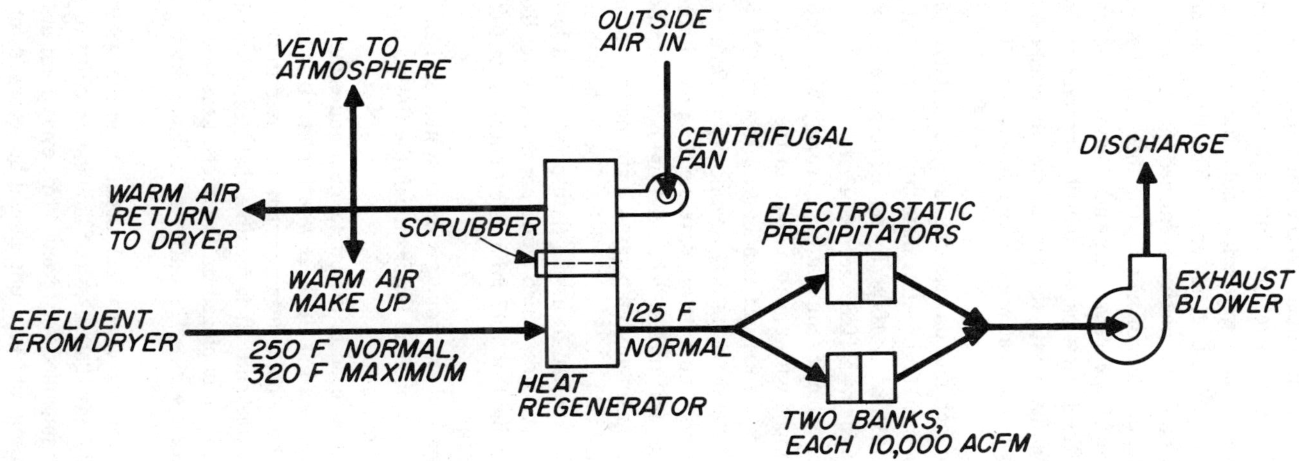

Figure 22-2. System diagram—air pollution control system utilizing heat wheels and–electrostatic precipitators.

All 10 of the individual 2,000 acfm precipitator units were burned black. Although the modularized precipitator framework and the potted high-voltage power supplies were salvageable, the ionizer sections and collector plates were damaged beyond repair. The 10-ft-diameter heat wheel was gutted. Its aluminum mesh was burned into a smaller volume, containing many melted droplets of metal. The sheet metal structure housing the units was fire blackened and expanded outward on one side with a tear in one seam. The hottest part of the fire appeared to be in either the right bank of precipitators (in the first of two in-series modules and at the middle level) or in the heat wheel. The blower was blackened but otherwise all right.

The manufacturing process used approximately 925 gal/week of oil in the spraying operation. After installation of the heat wheel and precipitator system, approximately 100 gal/week of oil were recovered from the precipitator drain system—a 10.8% recovery rate.

The dryer exhaust temperatures normally were 260°F. When the production line stopped with a material gap in the oven, the burners throttled down, but not before the exhaust temperature rose to 324°F. The heat wheel (upstream of the precipitators) was, however, designed to drop the temperature 125°F maximum before entering the precipitator. There was evidence that the wheel was turning when the fire occurred. So even if the material stopped, the maximum temperature entering the precipitator should not have exceeded 324 - 125, or 199°F. The high-flash combustible oils in the process were reported to have a flash point of not less than 360°F. The dry catch particulates, which could conceivably act like a wick, were within the allowable limit of 1.47 lb/hr in an emission test conducted two months prior to the fire.

The direct-fired two-zone gas dryer system had an interlocked start-up and shutdown procedure (even though the fire occurred during normal burner "on" operation).

One sprinkler head in the intake duct and eight in the oven were actuated, along with the head at the heat wheel inlet. It is known that a common solvent-free household cleaning agent was used in cleaning the heat wheel the previous weekend. It is possible that debris (consisting of oil-soaked fibers) could have impinged on the ducts, heat wheel, or the precipitators and acted like a wick. The outlet air is damp, having a wet bulb temperature of about 133°F.

Based on the two fires which occurred in the system under discussion and on similar installations, the following checklist of safety controls (Table 22-1) is recommended for review. Not every item in this list is needed on each installation, but each one should be reviewed for its applicability to precipitator air pollution control systems.

Table 22-1. Checklist for Safety Controls on Electrostatic Precipitators Installed on Combustible Emission Sources

	Item
1A.	An interlocked static flow switch to confirm air flow before the precipitator will operate. If exhaust air flow fails, it shall shut down the process and the control system.
B.	An automatic purge of system for at least two minutes prior to energizing precipitators.
2.	Temperature sensors in the inlet plenum with an indicator and adjustable over-temperature alarm at the control panel to automatically actuate sprinkler heads at the heat wheel and the precipitators.
3.	Temperature sensors in the outlet plenum with an indicator and adjustable over-temperature alarm at the control panel to automatically actuate sprinkler heads at heat wheel and precipitators.
4.	Sprinkler head water sprays in the inlet plenum with automatic actuation from a temperature sensor–and shutoff from either the control panel or the rooftop.
5.	Sprinkler head water sprays in the outlet plenum with automatic actuation from a temperature sensor–and shutoff from either the control panel or the rooftop.
6.	Sprinkler head water sprays in the plenum between the hot side of the heat wheel and the precipitators (when a heat wheel is in the system) with automatic actuation from a temperature sensor–and shutoff from either the control panel or the rooftop.
7.	Fuseable link fire bypass drop gate at roof openings (preferably near the inlet to the precipitator), which will drop to divert the exhaust flow to atmosphere and shut down the fan and precipitators and actuate sprinkler heads at heat wheel and precipitators.
8.	Sprinkler head water sprays in the inlet ducts.
9.	Direct reading dial type temperature indicators (with 6-in. probes) located in the inlet and outlet plenums and visible from outside the housing.
10.	Complete operation and verification of the oven purge system prior to turning on oven burners.
11.	Blowout panels and/or latches to be provided on all equipment enclosure access doors.
12.	Self draining floor area (with steel walking grid) in precipitator (and heat wheel, if applicable) equipment enclosure.
13.	Float system to keep sprinkler and/or spray cleaning water from going through drain and overflowing onto the production floor.
14.	Provide heat wheel rotation which puts hottest wheel temperature at bottom for more even heat distribution.
15.	Assure all temperature sensors are dry type and do not contain combustible liquids or gases.

Table 22-1, continued

	Item
16.	The process to shutdown if the dryer temperature exceeds 325°F.
17.	Pitch all ducting at least 6 in./10 ft toward drains.
18.	Provide sprinkler heads in duct lines every 10 ft. Provide cleanout/access panels at each sprinkler head.
19.	Differential pressure sensor with remote indicator light to indicate clogging of the heat wheel.
20.	Rooftop vent for equipment enclosure.
21.	Approved combustibles analyzer recorder to monitor the oven (or kiln) exit gas stream.
22.	Interlock the analyzer from Item 21 to sound an alarm at a 1% concentration of combustible gas and remove power from the precipitator at 2% concentration.

The safety controls listed in the table are illustrated in Figure 22-3. In this diagram an energy-saving heat wheel is included.

REFERENCES

1. Rossnagel, W. B. "Condensing/Precipitator Systems on Organic Emissions," *Proc. 3rd Annual Industrial Air Pollution Control Seminar*[(May 1973).
2. Factory Mutual Insurance. "Electrostatic Precipitators," *Hazardous Materials and Processes Handbook*, Section DS, 7-2, Norwood, Mass. p. 66-11.
3. Factory Insurance Association, Hartford, Conn.

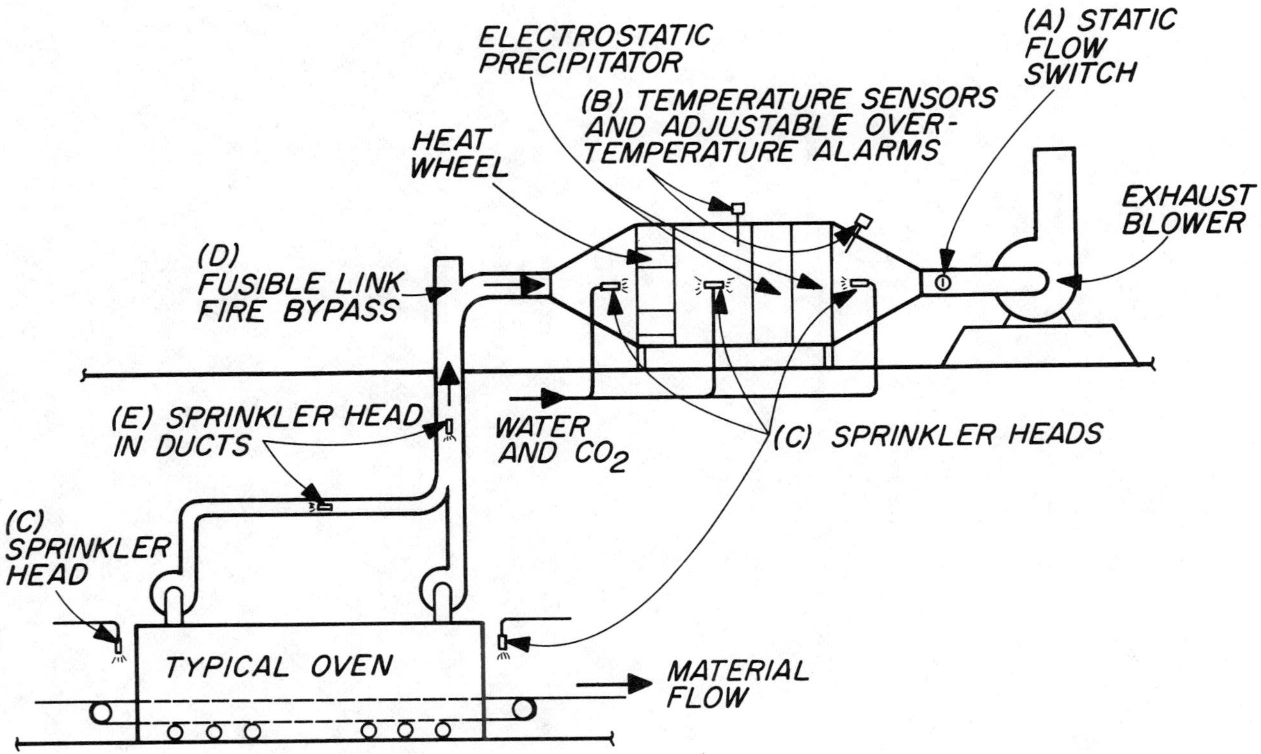

Figure 22-3. Schematic showing safety controls on electrostatic precipitators installed on combustible emission sources.

CHAPTER 23

INDUSTRIAL ODOR CONTROL BY OXIDATION WITH OZONE

Carl Nebel, Ronald D. Gottschling

Welsbach Ozone Systems Corp.
Philadelphia, Pa. 19100

Ozone is a powerful and useful oxidant. It occurs naturally in the upper atmosphere, at heights of 60,000 to 90,000 ft above sea level, in concentrations of 10 to 12 ppm by volume.[1] This ozone layer is of inestimable benefit as it filters out certain frequencies of ultra-violet light known to be lethal to all forms of life. At ground level, however, ozone usually exists in a greatly diluted and completely harmless concentration of less than 0.01 ppm. Under smog conditions in places like Los Angeles, the concentration may rise as high as 0.9 ppm, at which level it can be mildly irritating though far from toxic. However, such conditions are exceptional and represent ozone that is *not* under control. What we, as engineers and scientists, are concerned with is ozone that is *always* under control–in which connection it should be noted that no known human fatality has ever been caused by ozone, whether generated by nature or by man.

Like chlorine, fluorine, potassium permanganate, hydrogen peroxide, etc., ozone is an oxidant, but it differs from other oxidants in some very important ways, such as its potency and the facility and economy with which it can be generated on-site from air, electricity, and cooling water.

Despite the allegedly novel and superior methods of generating ozone publicized from time to time, the most practical and economical method employs a glass tube dielectric, around which is produced a corona discharge. The corona discharge converts a portion of the oxygen in the air feed gas to ozone. Ozone can be generated directly from ambient air, or preferably from ambient air that has been filtered, compressed, cooled, and dried (the latter is by far the most consistent and economical process). Packaged ozone generators are available that include not only the ozonator but all the electrical gear and air pretreatment equipment interconnected, thus requiring minimal installation (Figure 23-1). Figure 23-2 is a flow diagram for the generation of ozone from air.

Figure 23-1. Welsbach Model CPL-51-D19L.

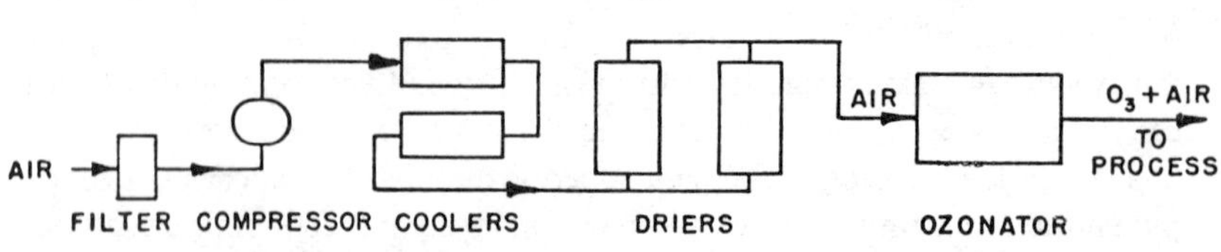

Figure 23-2. Ozone from air.

When ozone leaves the generator, its concentration in the air stream is approximately 1% by weight or 6,100 ppm by volume. In this state ozone is used commercially for potable water treatment, municipal and industrial water pollution control, chemical processing, food preservation, and deodorization. Indeed, the greatest number of installations to date is in sewage odor elimination.

Once installed, the ozonators are very simple to operate. They are automatic, and push-button controlled, and production can be varied over a wide range. Ozonators can, if necessary, be interlocked with other control equipment. Maintenance is minimal, and consists of a simple cleaning operation twice a year. The operating cost, which is negligible, is mainly the cost of electricity. In the larger ozonators, producing from 4 to 750 lb of ozone per day from an air feed gas, the electrical consumption is about 10 kwh/lb of ozone. In many odor control applications, one pound of ozone per day is adequate to treat 5,000 or 6,000 cfm of exhaust gas.

Ozonators are easy to install and simple to maintain. The ozone generation unit need only be set in place and connections made to sources of cooling water and electricity. A drain should be provided for the effluent cooling water and pipe or tubing should be run to conduct the ozone to the diffusion point. With larger-size ozonators, the same routine is followed, except that the dielectrics must be installed first, as in shipment they are packed separately because of their fragility. Their ozone output can be readily varied up to an 80% turndown by a variable-voltage transformer.

Not long ago, when most sewage treatment plants were located outside residential areas, the odor problem was not critical. However, many plants have become surrounded by rapidly expanding suburbs, and others soon will be. It is therefore essential to include odor control equipment in the design of new plants. In this application, ozone is very successful and commensurately popular.

The principle of ozone treatment of exhaust gases involves the addition of ozone diluted in air to the exhaust system containing odor-forming compounds. Elimination of the odor problem is accomplished by chemical oxidation of the odor by ozone. Ozone is not a masking agent; it does not replace one odor with another. Many industrial odor treatment methods use masking agents which replace one odor with another odor that is stronger but perhaps more pleasant. Experimental data from actual installations and pilot evaluations prove that ozone truly eliminates odors by chemical oxidation.[3,4]

Most odors are associated with molecules which have centers of high electron density, such as amines, sulfides, and unsaturated hydrocarbons. Ozone most often reacts chemically, as if it were an electron-deficient molecule. Thus a molecule having a site of excess electrons (*i.e.*, the odoriferous molecule) is attracted to a molecule which is deficient in electrons, ozone. The two molecules interact chemically to produce compounds, usually oxides and oxygen, which do not possess any odor.[5,6,7] Examples of these reactions are given below.

$$\mathrm{CH_3\text{-}S\text{-}CH_3 + O_3 \rightarrow CH_3\text{-}\overset{\overset{\displaystyle O}{\|}}{S}\text{-}CH_3 + O_2}$$

Dimethyl sulfide → Dimethyl sulfoxide

$$\mathrm{CH_3 - SH + O_3 \rightarrow CH_3 - SO_3H}$$

Methyl mercaptan → Methyl sulfonic acid

$$\mathrm{(CH_3CH_2CH_2\text{-}CH_2)_3\ N + O_3 \rightarrow (CH_3\text{-}CH_2\text{-}CH_2\text{-}CH_2)_3\ NO + O_2}$$

Tri-n-butylamine → Tri-n-butylamine oxide

$$\mathrm{R_2C{=}CRH + O_3 \rightarrow R_2C = O + R - CO_2H}$$

Unsaturated hydrocarbons → Ketone + Carboxylate

Note that the products of oxidation are always less toxic than the original odoriferous product.

The criteria to consider in designing any odor control system utilizing ozone are (1) type of odor, (2) concentration of the odor, (3) temperature of the exhaust gases, (4) humidity of the exhaust gases, (5) retention time within the exhaust system, and (6) the distribution of ozone within the exhaust system.

In many instances the feasibility of ozone oxidation of various odors can be established in the laboratory by duplication of the actual odor. Laboratory results can only answer the question of feasibility; on-site pilot testing must be undertaken to establish design parameters. Such application data can be developed empirically through a pilot test installation on the plant site. Equipment can be sized and economics established by treating a small portion of the actual exhaust gas with ozone. Figure 23-3 illustrates a typical evaluation technique for field test determinations. An ozone pilot contact chamber (as shown) can be readily constructed from 6½ sheets of 4 ft x 8 ft x 3/4 in. plywood. All variables previously mentioned are evaluated and their relative importance is determined. It should be noted that the variables may or may not be important for each specific type of application. Thus the variables are altered as required to solve the problem.

The effectiveness of the ozone dosage is determined at the exhaust point of the pilot test facility by standard threshold odor techniques. In some cases chemical tests can also be run on the amount and type of contaminants remaining after exposure to a given amount of ozone. Theoretically, at long contact times and high ozone dosage levels the ultimate products of ozone

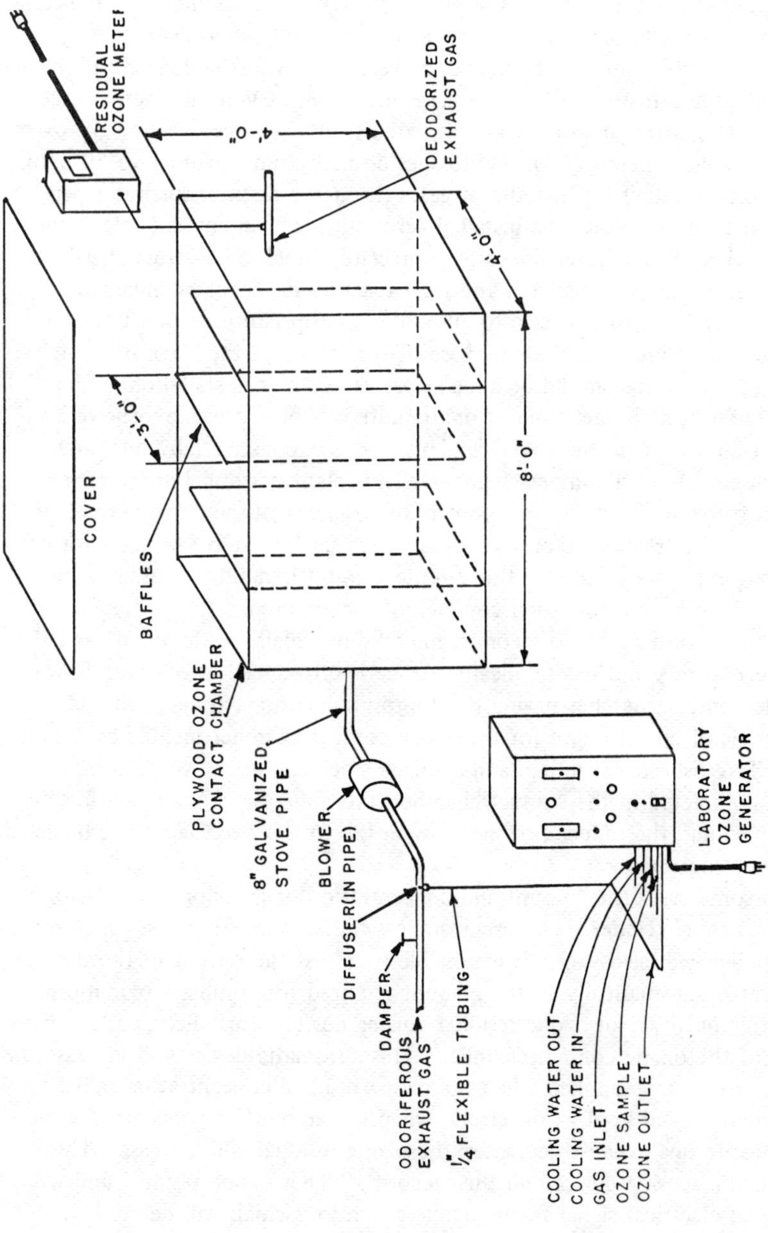

Figure 23-3. Field ozone pilot testing equipment.

oxidation of hydrocarbons are carbon dioxide and water. However, in most practical cases the oxidation products are harmless and inoffensive oxygenated materials. Once the pilot testing has been completed, full-scale equipment can be sized and the installation designs established.

It should be noted that the ozone dosage refers to the concentration of ozone applied to the odoriferous gases before any chemical reaction takes place. The latter should be differentiated from the concentration of ozone in the ozone generator. Ozone dosages normally range from 1 to 50 ppm.

Ozone is added to the exhaust gases through a diffuser such as a perforated pipe, which can be installed across the duct in any suitable shape or pattern. The diffuser does not appreciably block or obstruct the duct. No pressure drop is involved, and fan capacities need not be enlarged. If the stack is of sufficient size to allow the appropriate retention time, the ozone can be introduced at the base of the stack. If the stack is not tall enough, the ozone should be added upstream. In an installation in the Middle West, a contact time of one minute is achieved in inexpensive horizontal ductwork on the roof. A plant in the East uses a labyrinth in an old basement for a contact chamber. Thus plant personnel are never exposed to ozone. With proper control of ozonator output, the amount of ozone in the treated exhaust gas is less than 0.01 ppm—a safe level for continuous exposure—and even this minute amount is rapidly diluted in the atmosphere, where it is soon completely decomposed.

A given time (3 to 60 seconds) must be provided for the ozone and the exhaust to mix and react. Ideally, ozone is introduced into a plug flow ozone contact chamber having the appropriate contact time. Suitable materials of construction for the ozone contact chamber include concrete, fiberglass, asbestos sheeting, and stainless steel. After the oxidation has been completed in the contact chamber, the nonodoriferous oxidation products and the reduced ozone, now in the form of oxygen, are exhausted to the atmosphere.

In some cases the amount and concentration of an odor changes from time to time. Under these circumstances the amount of ozone which must be applied will also vary. It is possible to adjust the output of the ozone generator automatically to the amount required for complete oxidation. For this purpose, an ozone residual sensing device is installed at the exhaust vent of the ozone contact chamber. This automatic device will increase the amount of ozone applied when the ozone residual concentration falls below a predetermined level. Conversely, when the amount or concentration of an odoriferous exhaust decreases, the ozone residual will increase. The automatic sensing device will then record a higher ozone residual and will relay an electrical signal to the ozone generator, which will decrease its output. Normally, a residual level in the treated exhaust of 0.01 ppm is chosen as a safe operating level.

Exhaust systems that are treated with ozone include the following:

1. Fish processing plants
2. Rubber compounding plants
3. Sewage treatment plants
4. Rendering plants
5. Commercial kitchens and cafeterias
6. Paper mills
7. Chemical plant exhausts
8. Paint spraying exhausts
9. Food processing plants
10. Fermentation odors
11. Phenolic odors
12. Compost odors

RENDERING PLANTS

The rendering industry has long been plagued by odor problems from the reclamation of fatty acids and oils from the offal and carcasses of dead animals. Rendering of these materials is performed in large steam-jacketed cookers. Off gases from the rendering cookers normally contain condensible and noncondensible odors composed of acrolein, allyl amine, allyl mercaptan, ammonia, dibutyl sulfide, ethyl mercaptan, hydrogen sulfide, oils, skatole, and sulfur dioxide. The off gases are drawn out by a slight draft and passed through a water scrubber. Exhaust gases from the scrubber contain a noncondensible odor source, the scrubber having removed the condensible materials. Ozone dosages below 10 ppm by volume applied to the scrubber exhaust have proved satisfactory in the control of objectionable clinging odors (see Figure 23-4). In this example the building exhaust is approximately

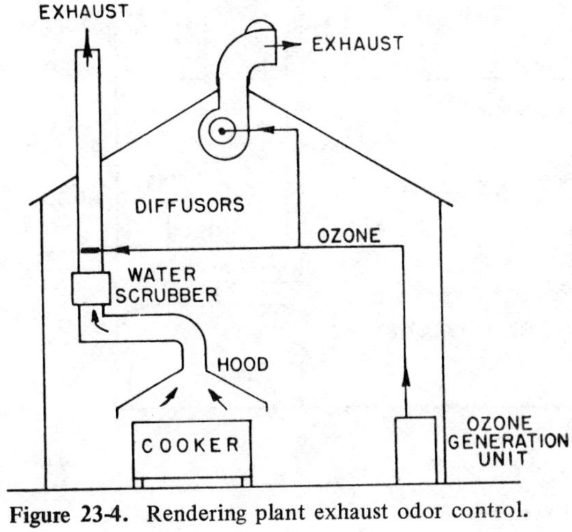

Figure 23-4. Rendering plant exhaust odor control.

25,000 cfm. The building and cooker exhaust was successfully treated with 42 pounds of ozone per day and a retention time of 5 seconds.

In a similar application, chicken feathers are normally treated in a rotary dryer and then put through a hammermill. Exhaust from the hammermill is sent to a cyclone separator where the solids are dropped to a packaging operation, while the exhaust gases are vented from the top. Ozone dosage, once again below 10 ppm/v, has been found ample for this operation.

COOKING ODOR CONTROL

A 44-story skyscraper, built in the early 1960s, reflected the latest in modern architecture. Before construction was completed it was decided to add a tenant restaurant. A glaring problem faced the consulting engineers on odor control. The New York City code prohibited the discharge of offensive odors at street level, yet the building could not feasibly be redesigned at such a late date. An exception to the code had to be made, with the clear understanding that there would be no trace of odors either from the restaurant waste gas or the deodorizing agent. Activated carbon was considered but was rejected due to economic considerations and increased duct resistance creating large horsepower requirements. Ozone was installed in conjunction with a gas scrubber, as shown in Figure 23-5. This provided the best overall odor-elimination scheme and the least amount of maintenance.

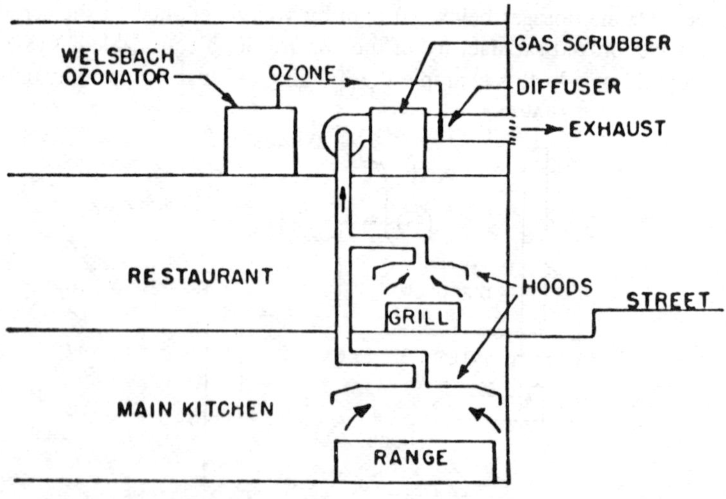

Figure 23-5. Ozone installed in conjunction with a gas scrubber.

A similar New York City ozone deodorization application is shown in Figure 23-6. In this case the 22,000 cfm required 1.5 ppm ozone for deodorization. The specific amount of ozone required in this application varies widelv with the time of day, the day of the week, and general weather conditions such as humidity and wind velocity.

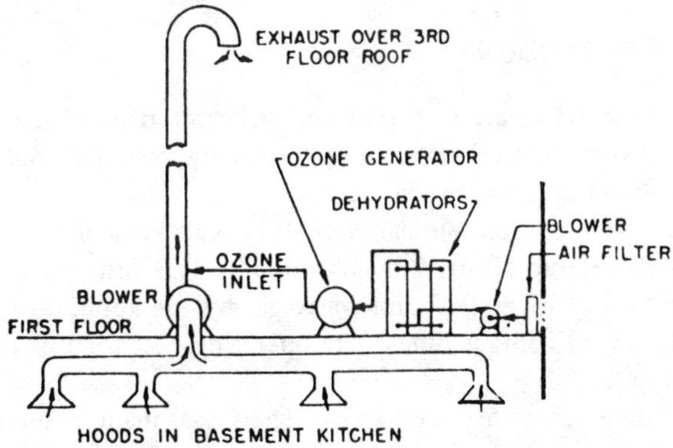

Figure 23-6. Ozone deodorization application.

In the design of the heating, ventilating, and air conditioning system of a New York City University dormitory building the engineers found that they were limited in the choice of locations for the fresh air intake and the kitchen exhaust. The only feasible locations were within 10 ft of each other, so that the possibility of bringing in air containing kitchen odors appeared probable. In addition, and perhaps more important, the discharge location was at ground level and would prove a nuisance to the campus. Air movement was at 55,000 cfm from the kitchen hoods.

The problem was rectified by treating the air with 1 ppm/v of ozone in two places–at the hood and after the fan (Figure 23-7). Grease filters are used to remove grease vapor and fumes from the exhaust system.

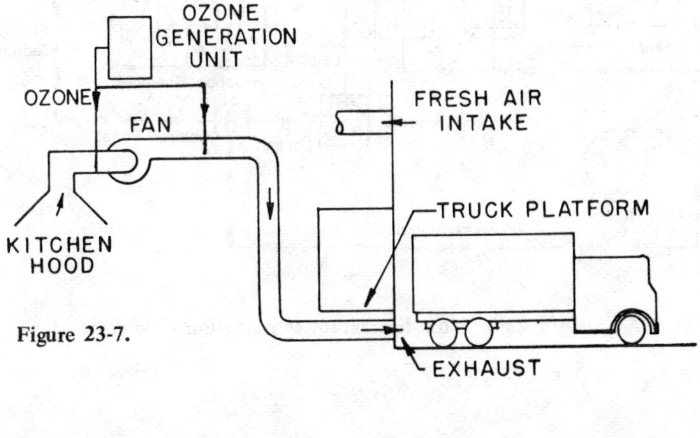

Figure 23-7.

An ozonator which produces 10 lb ozone/day was necessary to give a dosage level of three ppm. The odor character is no longer noxious, and even on fish cooking days the people walking nearby are not offended. The fresh air intake, above the discharge, does not reintroduce any malodorous air into the building.

FERMENTATION ODORS

Fermentation odors are very pungent. Characteristic of many food processers and pharmaceutical firms, these odors are very insidious and difficult to rectify by conventional processes.

A Chicago area pharmaceutical laboratory was particularly plagued since it was located near an affluent neighborhood. This firm eliminated the offensive nature of its 80,000 cfm waste gas by the application of ozone during a 10-second contact time. The operating cost for this is approximately 30¢/hr.[8]

Large volumes of air are used by an East Coast manufacturer in the aerobic fermentation of molasses to produce baker's yeast. Eight approaches for the elimination of this odor problem were considered: water scrubbing, alkaline solutions and oxidizing solutions, gas phase oxidation, thermal oxidation, condensation, dilution, dispersion, and masking. Gas phase oxidation with ozone was chosen as meeting all requirements of sanitary design, ease of cleaning, resistance to bacterial growth, avoidance of by-products, product contamination, and economy.[9] Figure 23-8 portrays the flow diagram for this oxidation system.

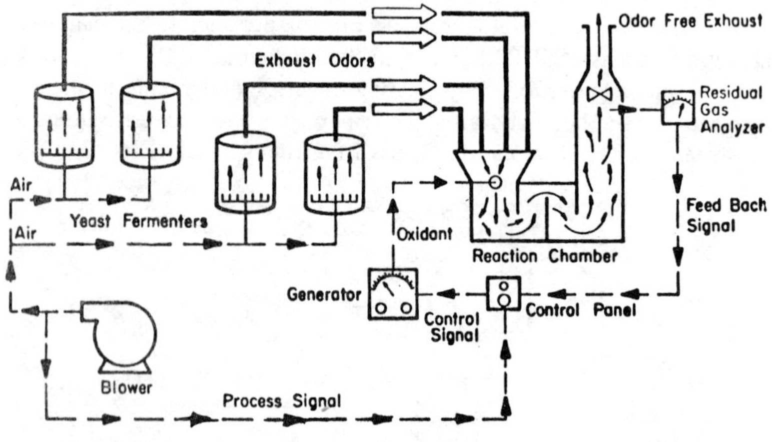

Figure 23-8. Flow diagram for an oxidation system.

RUBBER PLANT ODORS

A producer of hard rubber rollers for the printing press industry compounds thiokol rubber with carriers and other blending agents. Exhaust from the compounding and blending areas passes to a cyclone separator where particulate matter is removed (Figure 23-9).

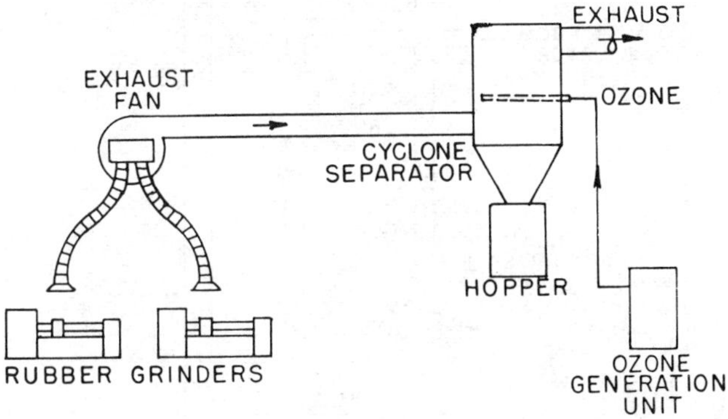

Figure 23-9.

The gaseous discharge from the separator contains mercaptans and other sulfide odors which produce a serious out-plant problem. The total gas flow of 2,000 cfm is treated with four ppm ozone. Ozone is added through a diffuser at the top of the separator in such a manner that it will not be consumed by the particulate matter containing unsaturated elastomers, but will be effective directly on the gases. Furthermore, the addition of ozone at the top of the separator increases contact efficiency by providing longer residence time.

FISH PROCESSING ODORS

Menhaden fish, caught off the New Jersey coast, are processed to produce an oil for the manufacture of paints and for a fish meal that is used as a source of protein in poultry and animal feeds. During the process in which this plant cooks, presses, and dries fish, relatively large amounts of fumes are produced. Under certain wind and/or atmospheric conditions, these fumes were carried to nearby communities.

The components of the plant's stack discharge proved to be made up primarily of effluent from the rotary stream driers, emitted in the form of water vapor, air, and a very small percentage of particulate matter. Amines were identified as the chief malodorous constituents of the effluent. Sulfides and mercaptans in lesser amounts were also detected. All of these compounds are known olfactory offenders, making the move toward more stringent control both understandable and justifiable. Scrubbers operating on the ejector-venturi principle were installed to remove the particulate matter. The off gases from the scrubber are treated with 10 ppm ozone in the exhaust stack (Figure 23-10). This system has been shown to be a totally successful operation for many years.[10,11,]

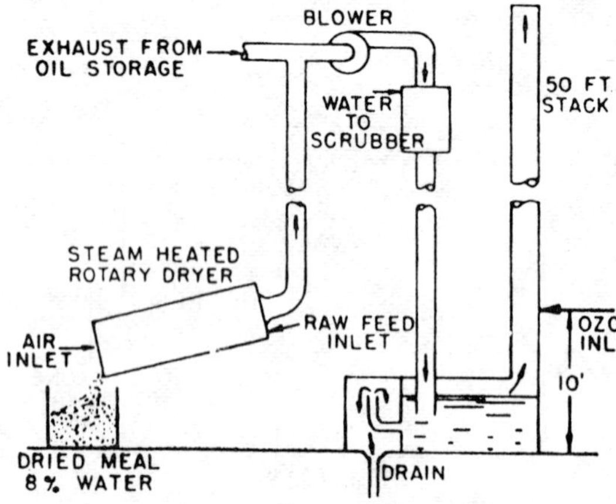

Figure 23-10.

PHENOLIC ODORS

A producer of felt products, used in the manufacture of mattresses, automobile seats, etc., employs a phenolic resin as a binder for the cellulosic materials. The product is subject to humidity variations and, when damp, develops an obnoxious odor. A test run was made so that results could be evaluated and design criteria established (*e.g.*, ozone production, contact time, etc.). The trials were successful and the felt lines now have ozone treatment as the final part of their system. The ozone generator is operated automatically and provides ozone to the exposure chamber (Figure 23-11). The gas infiltrates the felt, passes through, and is exhausted to the atmosphere.

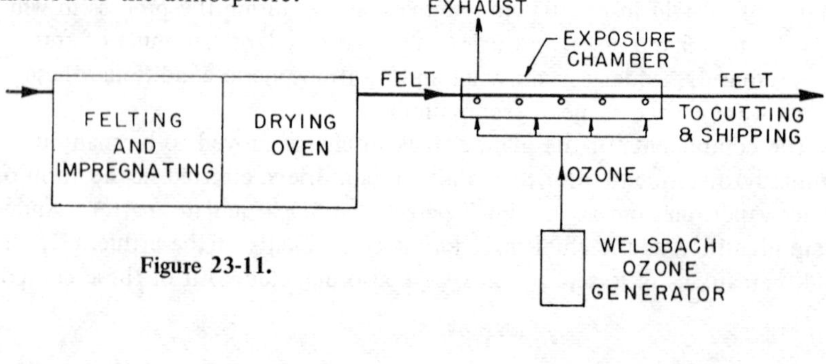

Figure 23-11.

PULP MILL ODORS

Pilot treatment of the effluent stack gases from a sulfate recovery system of a paper pulping mill in Canada has been successful.[12,13] In this plant the black liquor from 260 tons/day of wood pulp is burned in two recovery furnaces. A venturi apparatus disperses the liquor into the scrubbing gas. Entrained solids are separated in a cyclone from which 17,500 lb/day of salt cake, or sodium sulfate, are produced. Approximately 2,500 lb/day of salt cake are recycled to the venturi, leaving a net recovery of about 15,000 lb/day. The scrubbing gases from the cyclone separator are discharged to the stack, which has an inside diameter of 6½ ft and a height of 100 ft. The flow rate of these effluent gases is about 50,000 cfm. The stack gases contained both hydrogen sulfide and sulfur dioxide. Retention time in the stack is approximately two seconds with respect to the mixing of ozone with the other gases (Figure 23-12).

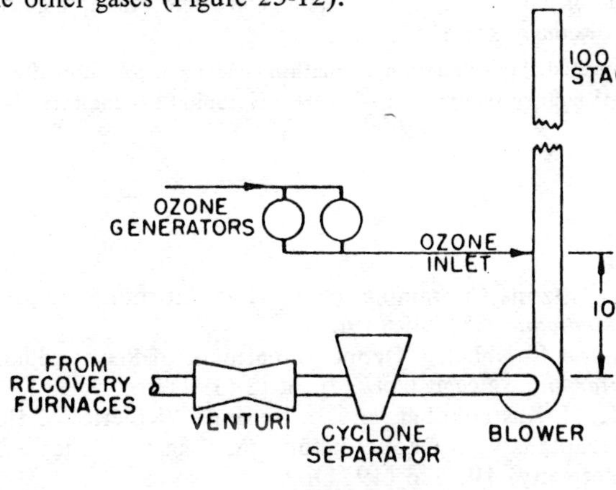

Figure 23-12.

Ozone is applied to the stack at a point 10 ft above the ground. The ozone generator has a capacity of 300 lb ozone/24 hr, or a dosage level of 80 ppm.

COMPOST ODORS

An integral part of the production of mushrooms is a compost process plant wherein a rich alluvial compost is developed for use as mushroom soil and a nutrient mulch. Timothy and alfalfa are subjected to conditions conducive to the bacterial growth required for transportation into the final compost. Gases enriched with ammonia and sulfur-containing compounds are released and treated with an acidic water scrubber to remove the ammonia. Unfortunately, the scrubbing process does not lower the odor level appreciably.

Attempts by one manufacturer to alleviate the problem with deodorants (masking agents) were unsuccessful. The odor was finally removed by treating the 5,000 cfm exhaust gas with 5.5 ppm ozone (4.85 lb O_3/day) in a contact chamber having a 40-second retention time. The operational costs, including power, cooling water, and maintenance, are 92¢/day.

The advantages of an ozone exhaust odor control system are:

1. Low operating cost; only power and cooling water are required for the ozonator.
2. Low maintenance; cleaning periods are required only every six to nine months.
3. Ease of operation; push-button control and automatic operation.
4. Nonrestriction of exhaust duct or stack; ozone equipment does not require higher fan capacity nor cause dirt and grime buildup requiring frequent cleaning.
5. When requested, preinstallation evaluation field testing ensures the user of satisfactory requirements before any capital expenditures are made.

REFERENCES

1. Bennett, G. "Ozone Contamination of High Altitude Aircraft Cabins," *Aerospace Medicine,* **33**, 969 (1962).
2. Unangst, P. and C. Nebel. "Ozone–Treatment of Sewage Plant Odors," *Water and Sewage Works,* p. R42 (1971).
3. Kirchner, K., H. Kastenhuber, and L. Biering. "Kinetics of the Reactions of Mercaptans with Ozone in the ppm Region," *Chem. Eng. Technol.* (Germany) **10**, 626 (1971).
4. Schmitz, L. R., R. E. Erickson, and L. M. Yates. "The Ozonation of Methyl Mercaptan," *Proc. 27th NW Regional Meeting, ACS* (June 17, 1972).
5. Maggiolo, A. and S. J. Niegowski. "Preparation of Tertiary Amine Oxides by Ozonation," *Adv. Chem. Series,* **21**, 202 (1959).
6. Maggiolo, A. and E. A. Blair. "Ozone Oxidation of Sulfides and Sulfoxides," *Adv. Chem. Series,* **21**, 200 (1959).
7. Bailey, P. S. "The Reactions of Ozone with Organic Compounds," *Chem. Reviews,* **58**, 925 (1958).
8. Sundberg, R. and G. H. Weyermuller. "Ozonator Operating Costs Only $2,600/yr," *Chemical Processing, Chicago*, **33**, 18 (1970).
9. Riggin, C. W. "Eliminate Fermentation Odor," *Food Engineering,* **44**, 145 (1972).
10. Anon. "Gas Scrubbers Solve Odor Problems at Fish Meal Plant," *Air Engineering,* **8** (11), 39 (1966).
11. Gelber, P. "One-Two Punch Controls Processing Odors," *Food Processing-Marketing,* **28**, 80 (1967).

12. U.S. Patents 2,830,869 and 2,830,870.
13. Cromwell, W. E. "Ozone in Air Pollution Abatement," *Industrial and Engineering Chem.,* **51**, 83A (1959).

CHAPTER 24

ODOR CONTROL WITH HYDROGEN PEROXIDE

Charles W. Raleigh

FMC Corporation
Princeton, N.J. 08540

Although hydrogen peroxide is an industrial chemical used in such diverse fields as textile bleaching, dye oxidation, ore and metal treatment, plastics manufacture, and food and starch treatment for the past 40 years, the application of hydrogen peroxide to waste treatment is a relatively new but rapidly growing use. Hydrogen peroxide has been found to be particularly useful in odor control in the oxidation of organic wastes, the correction of filamentous bulking, and the prevention of denitrification (and the resultant rising sludge) in secondary waste treatment systems.

Hydrogen peroxide adds no contaminants to the treatment system. Any unreacted hydrogen peroxide decomposes into water and oxygen.

MANUFACTURING WASTES

Sulfides, Mercaptans, Sulfur Dioxide

Hydrogen sulfide and mercaptans are waste products of petroleum refining, coal coking, the evaporation of black liquor in kraft pulping, viscose rayon manufacture, and the purification of natural gas. Hydrogen sulfide is often the cause of odor problems when organic waste products are collected and stored in nonaerated lagoons. Sulfide-containing gases are frequently scrubbed with alkaline solutions. These are also a source of liquid sulfide wastes requiring treatment before discharge into a sewer or stream. Sulfur dioxide and liquid sulfites frequently result from the burning of high-sulfur fuels, sulfuric acid manufacture, and chemical operations.

Hydrogen peroxide is used commercially to oxidize sulfur dioxide in sulfuric acid manufacture and to reduce the chemical oxygen demand (COD) of sulfite-containing chemical wastes. An economical peroxide system for treating sulfur dioxide in power plant emissions has not been developed.

Chemistry

Under neutral conditions, hydrogen sulfide reacts mole for mole with hydrogen peroxide to form sulfur and water.

$$H_2S + H_2O_2 \rightarrow S + 2H_2O$$

Under alkaline conditions sulfates are usually produced.

$$NaSH + 4H_2O_2 \rightarrow NaHSO_4 + 4H_2O$$

Mercaptans. RSH is oxidized with hydrogen peroxide according to the equation

$$RSH + 3\ H_2O_2 + OH^- \rightarrow RSO_3^- + 4H_2O$$

At room temperature, the reaction is slow. In a 3% hydrogen peroxide solution with Fe catalyst the reaction is complete in about 40 minutes. At 60°C the reaction is rapid. In acid solution it is complete in less than 15 seconds; in neutral and basic solution in less than a minute, even with only a stoichiometric amount of hydrogen peroxide.

Sulfides. $SO_3^{=2}$ is oxidized to sulfates. In acid solution the reaction is very rapid; in alkaline solution it is slower. The reaction is

$$SO_3^{-2} + H_2O_2 \rightarrow SO_4^{-2} + H_2O$$

The times necessary for 90 to 99% SO_3^{-2} oxidation at several pH levels (when initial SO_3^{-2} concentration was 7.2 mg/l and 10% excess of hydrogen peroxide was used) were

	Oxidation Time (Minutes)	
	% Oxidation	
pH	**90**	**99**
7.9	0.4	2
8.2	1.4	5
9.4	5.0	20
10.8	6.5	25
11.4	11.0	43

Table 24-1 summarizes the preferred operating conditions for oxidizing a number of sulfur compounds with hydrogen peroxide.

Table 24-1. Hydrogen Peroxide Oxidation Technology for Sulfur Compounds

Compound or Ion	Formula	S Conversion to Sulfate, H_2O_2:S		Preferred Oxidation Condition
		Molar Ratio	Weight Ratio	
Sulfide	S^{-2}	4	4.25	Alkaline solution or presence of catalyst
Polysulfide	S_2^{-2}	3.5	3.72	Alkaline solution
	S_3^{-2}	3.33	3.54	
	S_4^{-2}	3.25	3.45	
	S_5^{-2}	3.2	2.40	
Sulfur	S_8	3	3.19	Very alkaline solution from pH $>$12
Thiosulfate	$S_2O_3^{-2}$	2	2.13	Alkaline solution or presence of catalyst
Dithionite	$S_2O_4^{-2}$	1.5	1.59	Alkaline solution
Polythionate	$S_6O_6^{-2}$	2.17	2.31	Alkaline solution
	$S_5O_6^{-2}$	2	2.13	
	$S_4O_6^{-2}$	1.75	1.86	
	$S_3O_6^{-2}$	1.33	1.41	
Sulfite	SO_3^{-2}	1	1.06	Acid or slightly alkaline solution
Dithionate	$S_2O_6^{-2}$	0.5	0.53	Elevated temperature
Methane thiol	CH_3SH	3	3.19	Elevated temperature, presence of catalyst
Ethane thiol	C_2H_5SH	3	3.19	Elevated temperature, presence of catalyst
Dimethylsulfide	CH_3SCH_3	1	1.06	Elevated temperature, presence of catalyst
Dimethyldisulfide	CH_3SSCH_3	2.5	2.66	Elevated temperature presence of catalyst

Typical Applications

Cutting Oils. Hydrogen peroxide has been found effective in inhibiting bacteria growth in cutting oils and in preventing the disagreeable sulfide odors that result.

Cutting and grinding oil emulsions used in the metal machining industry are prepared from mineral or synthetic oil concentrates containing numerous additives. After emulsification, attack by bacteria and mold is rapid and additional preservatives are often needed. Even so, frequent replacement of these emulsions is required because disinfection is not complete. The bacterial activity produces disagreeable odors and, in extreme cases, the loss of necessary physical properties through demulsification.

In addition to the shortened storage time of the emulsions after the initiation of bacterial attack, the hygienic viewpoint must also be considered. Commercially used preservatives show not only limited bactericidal/fungicidal activity but also poor compatibility with the skin. An investigation was made to determine by bacteriological and chemical testing whether sufficient preservation of emulsions and resolution of other difficulties were possible with hydrogen peroxide.

Several emulsions from industrial plants were sampled and examined for the microorganism species present. The microorganisms were cultured and used for measuring the bactericidal performance of hydrogen peroxide in saline solution and later in oil emulsions. Performance in these laboratory tests was encouraging and tests were subsequently run in industrial plants with equally good results.

1. In saline solution, complete kills were effected with 0.1% hydrogen peroxide (Table 24-2).

Table 24-2. H_2O_2 Bactericidal Action in Hours

	% H_2O_2 in Saline Solution		
Microorganisms	**0**	**0.05**	**0.1**
Escherichia coli	$>$ 120	3	3
Bacteria of the coli group	$>$ 120	3	3
Pseudomonas aeruginosa	$>$ 120	24	6
Bacterium proteus	$>$ 120	3	3
Saprophytic staphylococci	48	3	3
Gram labile spore formers	$>$ 120	96	6
Gram negative spore formers	$>$ 120	3	3
Candida albicans	$>$ 120	$>$120	3

2. In laboratory emulsions, some species survived at the maximum test concentration of 0.05% hydrogen peroxide (Table 24-3).

Table 24-3. H_2O_2 Bactericidal Action–Organisms/Milliliter (x 10^7) at 24 Hours

Microorganisms in 1:50 Emulsion	Aral-Oil B 207		Esso Cutting Oil 35	
	Control	0.05% H_2O_2	Control	0.05% H_2O_2
Escherichia coli	18	2.5	16	0
Bacteria of the coli group	0.56	0	11	0
Pseudomonas aeruginosa	T.N.T.C.	T.N.T.C.	T.N.T.C.	T.N.T.C.
Bacterium proteus	36	15	7.0	3.0
Saprophytic staphylococci	1.4	0	0.68	0
Gram labile spore formers	0.20	0.03	0.04	0.03
Gram negative spore formers	180	0	94	0
Candida albicans	200	36	0.70	0.08

3. Emulsion stabilities were not affected by hydrogen peroxide treatment. Up to 0.2% hydrogen peroxide produced no visible changes in a 4% emulsion over a three-week period.
4. At the pH typical of emulsion (6.0-9.5), hydrogen peroxide concentration as high as 0.25% in water did not oxidize nitrite, a widely used but easily oxidized rust inhibitor.
5. Hydrogen peroxide stability in fresh emulsions with added catalytic iron presented no problem. Even in industrial tests, a half-life of two days at the 0.1% hydrogen peroxide level was noted.

This system is believed to be currently in commercial use in Europe and is under study by industrial cutting oil users in the United States.

Pulp and Paper Manufacture. The following two examples may be considered typical of the types of industrial applications where hydrogen peroxide has been helpful in controlling odor problems.

A Pacific Northwest pulp mill was treating approximately 50 mgd of wastewater through a 300-ft-diameter clarifier before discharge to the river. The sludge solids from the clarifier (300,000 gal/day) were sent to a rotary drum filter for thickening and the filtrate was returned to the clarifier. Sulfide levels as high as 75 ppm were measured in the liquid feed to the filter with atmospheric sulfides up to 600 ppm in the wet well.

When hydrogen peroxide was added to the filter feed line at a rate of 440 lb/day (100% basis), essentially all of the sulfide was destroyed. The 440 lb of hydrogen peroxide per day is equivalent to 2.3 parts hydrogen peroxide to 1.0 part sulfide.

A paper mill in the Northeast produces paperboard with a recycled waste furnish. The waste treatment system, handling 2.5 mgd, consists of a primary settling tank, aerated lagoon, and secondary clarifier. Sludge from the secondary clarifier is returned to the primary settling tank, combined with primary sludge, and returned to the paper mill for use as the board furnish.

Chlorine was added to the sludge system for control of odors, including hydrogen sulfide. Chlorine usage was 300-400 lb/day, or approximately 2,000 ppm based on sludge weight. A problem of short life on the disc refiner plates was blamed on corrosion caused by the high 80 ppm chloride content of the furnish at the refiners. Cost for a monthly change of refiner plates was estimated at $500. Handling the heavy chlorine cylinders was also a problem in cost and manpower.

A 35-day plant-scale trial demonstrated the effectiveness of hydrogen peroxide in controlling waste treatment plant sludge odors. Seventy-three lb/day of 50% hydrogen peroxide replaced 300 lb of chlorine for this purpose at an estimated annual savings of $1,000/year. Overall waste treatment plant performance, as measured by effluent BOD, effluent solids, and turbidity, was equal or superior to the hydrogen peroxide system as compared with chlorine treatment. Examination of the disc refiner plates after 35 days' operation on peroxide-treated sludge furnish showed no signs of corrosion. The plant now routinely uses hydrogen peroxide in place of chlorine for odor control.

The following table gives the estimated chemical cost of the hydrogen peroxide system compared with the chlorine system.

Chlorine System Costs	
Chlorine 300 lb/day at 5.04¢/lb	$15.12
Cylinder transfer cost $23/ton (300 lb at 1.15¢/lb)	3.45
Chlorine cost per day	$18.57
Chlorine cost per year (108,000 lb/yr Cl_2)	$6,685
Hydrogen Peroxide System Costs	
Hydrogen peroxide 73 lb/day (H_2O_2-50% @ 21.5¢/lb)	$15.69
Hydrogen peroxide cost per year (26,280 lb H_2O_2-50%)	$5,648
Approximate Net Annual Savings	
Hydrogen Peroxide vs. Chlorine	$1,037/yr

Three conclusions can be drawn:

1. Hydrogen peroxide controlled waste treatment plant sludge odors at an equal or lower cost than chlorine. Estimated annual savings are $1,000/yr.

2. Examination of the disc refiner plates after 35 days' operation on hydrogen peroxide-treated sludge furnish showed no signs of corrosion.
3. Overall waste treatment plant operation, as measured by effluent BOD, effluent solids, and turbidity, showed performance equal or superior to the hydrogen peroxide system, as compared with the chlorine treatment.

Soybean Processing. Wastewater from a soybean processing plant generated odors in a holding tank used to settle solids and decant oil. Subsequent discharge of the aqueous phase generated further objectionable odors when entering the municipal sewer.

The 0.3 mgd combined wastewater stream from the soybean processing operation contains soybean solids and oil, as well as significant sulfate from in-process neutralization of sodium soaps with sulfuric acid. Alum (0.4 wt%) is added to the wastewater and flocculated by adjustment of the pH to about 5.9. This facilitates settling of the solids when the mixture is fed into a 32,000-gal holding tank. The settled solids are removed, the oil that rises to the surface is decanted, and the aqueous layer is discharged to the municipal sewer. The aqueous phase has a high BOD of the order of 1,000 mg/l or more when discharged. Normal biological activity depletes the dissolved oxygen in the wastewater and high sulfate content encourages generation of hydrogen sulfide in the holding tank, and later in the municipal sewer.

Initial attempts to control the odor problem by aeration, or addition of up to 75 ppm chlorine to the wastewater, were unsuccessful.

Addition of hydrogen peroxide to the wastewater controls the odor problem. The hydrogen peroxide is added at a rate of 40 ppm (100% basis) to the wastewater before it enters the 32,000-gal holding tank. Supplemental hydrogen peroxide is added to the effluent from the holding tank on the rare occasions when some odor develops.

An ancillary advantage for hydrogen peroxide, when compared with chlorine for controlling odors in the holding tank, is recovery of soybean solids and oil with no objectionable chlorinated residues contamination.

Textile Plant Wastes. A Northeast textile mill treats its plant waste liquors with a combination of biological and chemical processes. In spring of 1973 the mill began experiencing odor problems at several points in their wastewater treatment system. These odors, described as "septic" and typical of dye house odors, were severe enough that the town authorities considered bringing legal action against the plant.

The treatment process at this plant is shown schematically in Figure 24-1. The plant effluent, which contains no domestic waste, varies from about 500M to 800 M gpd. The wastewater passes from the plant through

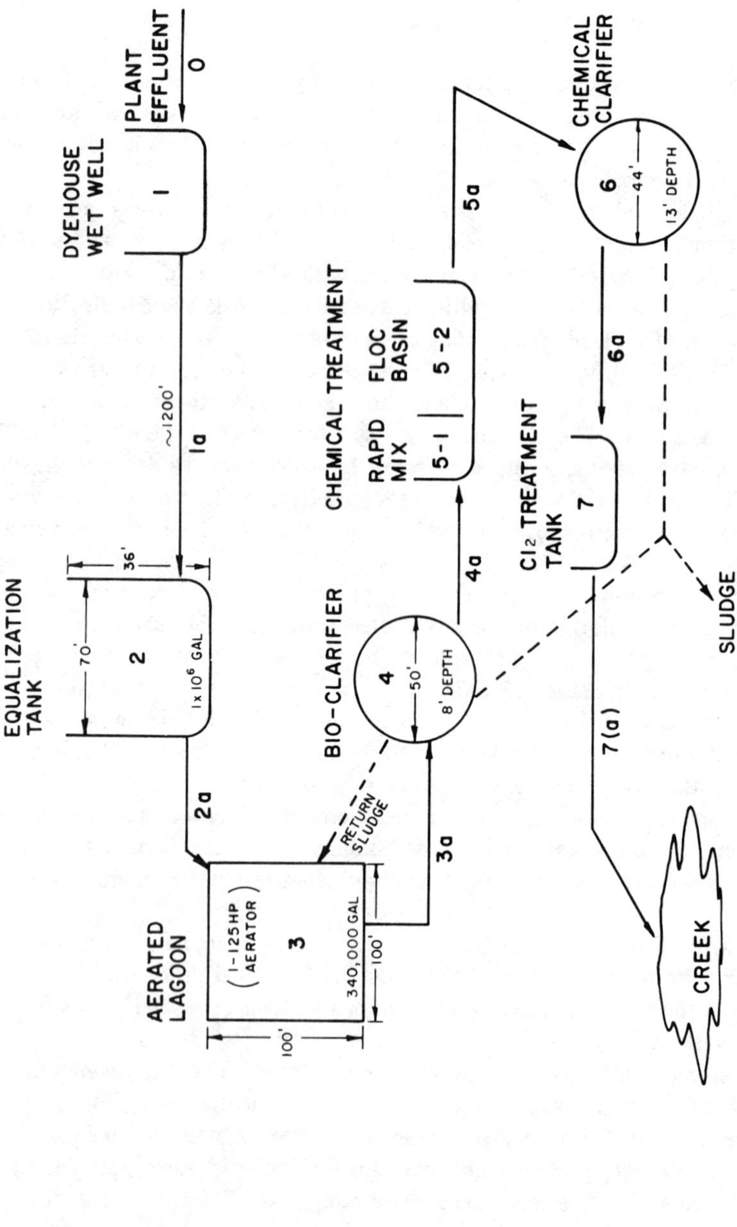

Figure 24-1. Textile wastewater treatment process.

the wet well in the dye house to the 1-million-gallon capacity equalization tank. After equalization for steady flow, the liquor enters an aerated lagoon where biological action occurs, reducing BOD and stabilizing the waste.

After approximately 10 to 12 hours of aeration, the mixed liquor enters the bioclarifier for solid-liquid separation. A portion of the solids (150 gal/min) is turned to the aerated lagoon and the remainder hauled away for ocean disposal. The supernate is treated in a chamber with alum to precipitate fine or colloidal particles and to improve color quality. Alum precipitation further reduces BOD and COD. A flocculant is added to aid in settling and caustic is used for pH adjustment.

Solid-liquid separation is next carried out in the chemical clarifier. The solids are combined with the waste solids from the bioclarifier for disposal. The treated liquor is then chlorinated or discharged directly into a local creek. The creek passes through the center of town and ultimately feeds into the river.

The first location in the treatment process, where odors were severe enough to cause numerous complaints from neighbors, was in the area of the equalization tank and the aerated lagoon. In both cases raw waste was involved, and thus was considered as one problem area. Part of this problem could be attributed to the failure to maintain a dissolved oxygen (DO) level in the raw waste sufficient to prevent septic conditions from arising.

Odors were also noticeable in the effluent of the bioclarifier and in the influent and effluent of the chemical clarifier. Again, the septic portion of the odor was attributed to an insufficient DO concentration in the liquor.

A laboratory study was initiated to evaluate hydrogen peroxide as a means of eliminating or reducing the odors. In the laboratory study, hydrogen peroxide with catalytic amounts of iron was found to be effective for controlling both the septic and dye house odors of the waste liquors after biological treatment. Hydrogen peroxide treatment of the raw waste, before biological action, was effective in reducing only the septic portion of the odor. Based on these laboratory studies, plant testing was initiated.

Hydrogen peroxide as a 50% solution was pumped directly into the flowing waste liquor. Evaluations to control the odors from the equalization tank and aerated lagoon were carried out by adding hydrogen peroxide directly to the raw waste in the wet well (Figure 24-1, location 1) in the dye house.

To control the odors of the final plant effluent, several addition points were tried. Initially, hydrogen peroxide was added to the liquor after chemical clarification (Figure 24-1, location 6a). The next steps were to add at locations prior to chemical clarification.

Treatment of the raw waste with only 11 ppm hydrogen peroxide sustained a DO level in the equalization tank effluent of about 1 ppm.

Treatment of the chemical clarifier effluent with 20 to 30 ppm hydrogen peroxide was more than sufficient to eliminate all hydrogen sulfide odors and provide DO in the final plant effluent liquors.

A hydrogen peroxide bulk storage and feed system has been installed and the system is now in year-round use.

Pharmaceutical Manufacture. A pharmaceutical manufacturer of vitamins and proprietary drugs uses an activated sludge process to treat chemical wastewater from its pharmaceutical plant. This treatment process is shown schematically in Figure 24-2. The plant effluent varies from 1.6 to 2.2 mgd, with an additional 1 mgd of waste being treated due to sludge recycle from the final clarifier to the inlet of the aeration tanks. Total average aerated waste is 2.9 mgd.

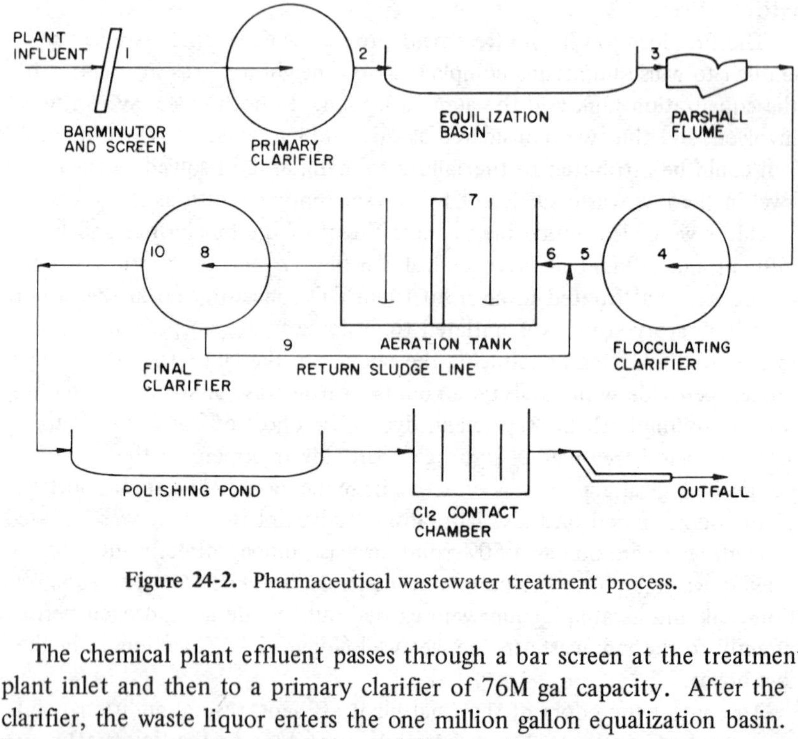

Figure 24-2. Pharmaceutical wastewater treatment process.

The chemical plant effluent passes through a bar screen at the treatment plant inlet and then to a primary clarifier of 76M gal capacity. After the clarifier, the waste liquor enters the one million gallon equalization basin. Retention time is approximately one-half day. From the equalization basin the chemical waste liquor passes through a Parshall flume to the floc clarifier for solid-liquid separation. At the floc clarifier exit, return sludge from the final clarifier combines with the effluent from the floc clarifier before entering the aeration tank. The waste liquor splits as it enters the two sections of aeration tanks, sized at 720M gal each. Here biological action occurs, reducing BOD and stabilizing the waste.

After approximately two hours of aeration, the waste enters the final clarifier for liquid-solid separation. The supernatant from the final clarifier is treated with chlorine before going to one of two polishing ponds. Chlorine is added at a rate of 200 lb/day. The combined capacities of the two polishing ponds is 2 mgd, with a waste retention time of approximately a day. Post chlorination follows the polish ponds before the wastewater is discharged in a stream adjacent to the treatment plant.

Pungent chemical odors could be detected in the floc clarifier area of the plant. The septic odors could also be detected around the aeration tanks. Here the return sludge from the final clarifier and wastewater from the flox clarifier are combined. A septic odor was attributed to insufficient DO in the return sludge line. Pungent odors from the floc clarifier were a result of insufficient DO in the plant wastewater.

Hydrogen peroxide was tested at three points in the waste treatment plant. Applications were made at a point 50 ft ahead of the treatment plant inlet, in the effluent of the Parshall flume, and in the return sludge line from the final clarifier.

Addition of hydrogen peroxide at rates of 30 ppm and 50 ppm at the plant inlet showed no significant effect in controlling chemical odors, which were primarily of a "solvent" character.

Hydrogen peroxide added to the effluent of the Parshall flume at concentrations of 20 to 25 ppm effectively reduced the pungent chemical odors and provided additional benefit in higher DO concentration through the floc clarifier.

Hydrogen peroxide treatment of the return sludge from the final clarifier, at a concentration of 20 ppm, was effective in eliminating the hydrogen sulfide and septic odors at the aeration tanks.

As a result, a bulk hydrogen peroxide storage and handling system has been installed and hydrogen peroxide is now routinely used for this application.

Aldehydes

Aldehydes, such as formaldehyde, are frequently found in wastes from plastics, leather, photographic, and antibiotic manufacture. At levels of 130-175 ppm, formaldehyde is reported to be lethal to bacteria in sewage, and even at lower concentrations will "shock" a conventional activated sludge system unless carefully activated over a long period of time. Pretreatment of formaldehyde wastes, therefore, is generally desirable.

Formaldehyde is oxidized rapidly with hydrogen peroxide to formic acid (within 10 minutes).

$$CH_2O + H_2O_2 \rightarrow H \cdot COOH + H_2O$$

The formic acid formed is then slowly oxidized further to carbon dioxide:

$$H \cdot COOH + H_2O_2 \rightarrow CO_2 + 2H_2O$$

The overall reaction is

$$CH_2O + 2H_2O_2 \rightarrow 3H_2O + CO_2$$

Thus 2.26 parts of hydrogen peroxide are in theory required to completely oxidize 1.0 part of anhydrous formaldehyde. The commercial formalin solution as used in photography contains 37% formaldehyde and would require 0.84 parts of hydrogen peroxide for oxidation. One gram per liter of 37% formaldehyde has a BOD_5 equivalent of 0.37 g/l.

Tests have shown that a formaldehyde solution having an initial BOD_5 of 23,000 mg/l after treatment with a stoichiometric amount of hydrogen peroxide had a BOD_5 of 2,000 mg/l--a reduction of 91.3%. The reaction is best conducted at a pH of 9 or above.

This reaction of hydrogen peroxide and formaldehyde has been used as a method of formaldehyde analysis.

Amines

A recent study by the St. Louis, Missouri, Metropolitan Sewer District indicated that amines are one of the major odor-causing components in sewage sludge. Amines usually impart a fishy odor to waste, which may be detected at extremely low levels. The odor threshold value for trimethyl amine, for instance, is 0.0002 ppm.

Primary amines are decomposed by hydrogen peroxide; secondary amines react to form hydroxylamines; and tertiary amines are oxidized to amine oxides.

Phenols

Phenolic wastes are frequently produced in the chemical, plastics, coke, and petroleum refining industries. Since phenols are toxic when chlorinated and impart a disagreeable taste or odor to water, there are strict limits on the phenol content of waste discharges. These limits may be as low as 15 ppb. Biological systems such as trickling filters and aerated lagoons have been partially effective in reducing phenol levels, but they may be inactivated by shock loading and low temperatures. Hydrogen peroxide in the presence of trace amounts of iron salts, such as ferrous sulfate, will rapidly oxidize phenol over a wide range of temperatures and concentrations.

Starting with pure phenol, the initial oxidation products formed are catechol and hydroquinone. These are further oxidized to dibasic acids and ultimately to carbon dioxide and water.

Studies were conducted using pure phenolics on a number of industrial and municipal wastes. Starting with a solution containing 2,000 mg/l of phenol and using 0.02% Fe^{++} as a catalyst, the relationship between H_2O_2/COD and phenol reduction was clearly shown (Table 24-4). All of these tests were carried out at room temperature. The initial pH of the test solutions was 6.5. Phenol levels were determined by the 4-amino antipyrine test.

Table 24-4. Hydrogen Peroxide Treatment of Phenol Wastes*

H_2O_2/COD Wt Ratio	Reduction in Phenol %	Reduction in COD %
0.3	86	28
0.4	94	32
0.6	99	40
0.7	99.8	44
0.8	99.9	52
1.0	100	69

*Initial phenol concentration 2000 mg/l.

A second series of tests was carried out on a mixture of phenol, isopropyl phenols, and xylene to determine if all were equally oxidized by hydrogen peroxide. The samples were treated with hydrogen peroxide and a ferrous sulfate catalyst at 90°F (32°C). Reaction was complete in approximately five minutes. The resulting mixture was then extracted with ethyl ether and analyzed by vapor phase chromatography (VPC), using diisobutyl adipate as an internal standard. The hydrogen peroxide weight ratios varied from 0.25 H_2O_2/1.0 phenol to 3.0 H_2O_2/1.0 phenol. The results are plotted in Figure 24-3. At a ratio of 1.0 H_2O_2/1.0 phenol, all of the phenol phenol and approximately 90 to 95% of the substituted phenols had been oxidized. At a ratio of 2.0/1.0, no phenols were shown on the VPC.

While iron salts are the most widely used catalysts for the oxidation of phenols by H_2O_2, other cations may be used. In a 1961 report of a study conducted in the U.S.S.R., the following order of catalytic activity was given: Fe^{++}, Fe^{+++}, Cu^{++}, Al^{+++}, Cr^{+++}.[11] Aluminum was said to enhance the catalytic activity of iron salts. Other studies, however, have not shown any substantial improvement when aluminum and ferrous salts are used in combination, as compared to the use of Fe^{++} alone.

Certain enzymes may also be used as catalysts for phenolic oxidation. Peroxidase, an enzyme found in nearly all plant cells, is very specific for this oxidation. Its optimum is pH 6. Starting with a mixed industrial

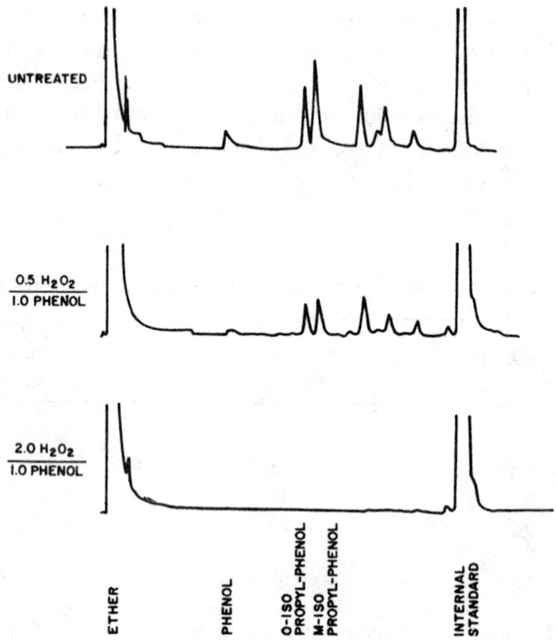

Figure 24-3. Vapor phase chromatographs—effect of H_2O_2 treatment on mixed phenolic wastes.

waste primarily containing phenols and low molecular weight alcohols, a 0.1 H_2O_2/1.0 COD ratio reduced the phenol level by 92% (604 to 48 mg/l) when peroxidase was used as a catalyst. Under normal conditions, however, ferrous salts appear to be the most practical catalyst for phenol oxidation.

Nitrogen Oxides

Nitrogen oxides found in the atmosphere and effluent gas are mainly NO (nitric oxide) and NO_2 (nitrogen dioxide). Both of these oxides are hazardous to health and corrosive. The American Conference of Government Industrial Hygienists has proposed threshold limit values of 25 ppm for nitric oxide and 5 ppm for nitrogen dioxide. Adverse health effects in children have been reported with nitrogen dioxide levels of less than 0.1 ppm.

NO_x Sources

Combustion of Fuels. At high temperature, the oxygen and nitrogen of air combine to form nitrogen oxides. Typical flue gas contains 100 to 1,500 ppm of nitrogen oxides.

Nitric Acid Manufacture. Waste gases which cannot be economically recovered in the final absorber usually contain 2 to 3% of nitrogen oxides, based on weight of nitric acid produced.

Metal Finishing Operations. Many metal-surface treatment operations which use nitrates, nitrites, or nitric acid evolve nitrogen oxides. Examples of such operations include bright dipping, phosphating, and pickling.

Chemical Processes. Many processes where nitric acid, nitrates, or nitrites are used as reagents evolve nitrogen oxides. Examples include the manufacture of explosives, plastics, and dyes.

Treatment

There are several practical procedures for the removal of nitrogen oxide contaminants. Scrubbing with a hydrogen peroxide containing solution is an effective and simple method of removing and destroying these contaminants. It is extensively used in Japan, where nitrogen oxide regulations are more stringent than in the U.S. In Japan the maximum allowed ambient concentration is 20 ppb. Although hydrogen peroxide in water can be used, it is advantageous to use strongly alkaline or acid solutions. The removal of the contaminants is more complete than with other methods, such as air oxidation.

In Kanto Kenka's process, an aqueous scrubbing solution which contains 0.2% hydrogen peroxide and 10% nitric acid is used. Scrubbing an air stream containing 1,000 ppm NO and 1,000 ppm NO_2 with this solution reduced the total NO_x content to 50 ppm.

Test Procedure

In lieu of a test procedure that is officially recognized by EPA, interim use of ASTM test D-1607 for NO_2 and ASTM test D-1608 for both NO and NO_2 appears suitable.

BIOLOGICAL TREATMENT PLANTS

Biological treatment plants and their associated collection systems offer a number of potential sites for the generation of odors. Long gravity mains at low sewage flows, and particularly in warm climates, create septic sewage and hydrogen sulfide. This is a problem even with new plants, where the design capacity greatly exceeds the sewage flow and hence leads to long detention times.

Force mains with long residence times will invariably lead to septic and sulfide odors. In this instance odors usually first appear at the discharge of the force main, where the sewage enters a wet well or connects to a gravity main. The addition of acidic industrial waste will greatly aggravate this problem. As may be seen in Figure 24-4, decreasing the pH from 7.0 to 6.0 will almost double the amount of free hydrogen sulfide. Going from a pH of 9.0 to 6.0 will give a ninefold increase in free sulfide.

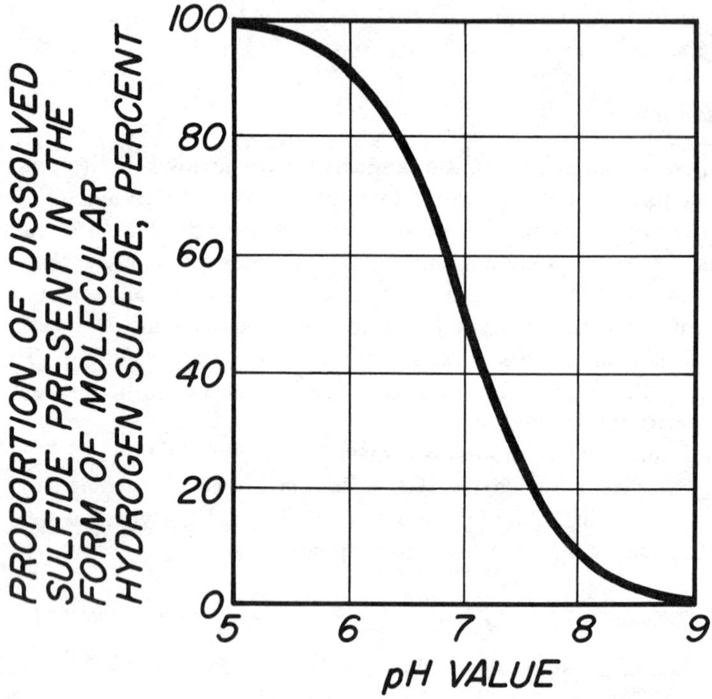

Figure 24-4. Relative concentrations of dissolved molecular hydrogen sulfide.

Excessive retention time or high BOD wastes will frequently lead to odors in the primary settler or trickling filters if they are used for secondary treatment. Sludge handling systems are another potential source of odors.

Hydrogen peroxide is used in the treatment of all the above problems. When hydrogen peroxide is used to treat sulfide odors in gravity or force mains, a high dosage is initially added (30 to 50 ppm) to reduce the odor-generating slimes on the pipeline wells. Dosage is then gradually reduced to a level equivalent to 2 to 3 parts hydrogen peroxide to 1 part hydrogen sulfide. Under neutral conditions, the sulfide is oxidized to colloidal sulfur, which settles out with the primary sludge. The hydrogen peroxide also adds a small amount of dissolved oxygen to the system, retarding the system from going anaerobic.

AGRICULTURAL WASTES

Food Processing Wastes

Food processing wastes are frequently a source of odor problems, both at the processing plant and in subsequent waste treatment systems.

Since much of the food processing business is traditionally seasonal, elaborate waste treatment systems are seldom employed by a cannery. Three widely used systems for handling food wastes are settling ponds, spray irrigation, and transfer to a municipal secondary waste treatment plant. There are two problems in treating ponds: (1) getting good mixing of the oxidant and waste and (2) the sludge layer of the pond is frequently a continued source of hydrogen sulfide generated by anaerobic bacteria, making oxidizer treatment expensive.

In spray irrigation one source of potential odors has been long feed lines leading to the irrigation fields. One example where hydrogen peroxide helped with this problem was a potato processor. The irrigation field was almost 3.5 miles from the processing plant. Anaerobic conditions led to a buildup of dissolved sulfides to a level of 6 to 7 ppm. Fifteen ppm of hydrogen peroxide added at the pump discharge eliminated essentially all of the dissolved sulfides and resulting objectionable odors.

Heavy seasonal loadings of cannery wastes can lead to several problems when they are discharged to a municipal treatment plant.

1. If the distance from the cannery to the treatment plant is far, the heavy oxygen demand of the wastes may lead to the generation of sulfides and odors at intermittent manholes, lift stations, or the plant entrance.
2. Odors may appear at the primary clarifier for the same reasons listed above. Or if a trickling filter is used, the sprays may not be able to keep up with the oxygen demand of the waste and odors may develop.
3. If an aeration basin is used for secondary treatment, irregular flows from the cannery may lead to shock loading and overloading of the aeration system. Odors and/or bulking problems may result.

Hydrogen peroxide has proved useful in all of the above conditions. If odor in the waste distribution system leading to the plant is the problem, the peroxide should be added upstream of where the odor first appears to give a 15 to 30 minute reaction time. Care should be taken not to add the hydrogen peroxide to concentrated canning wastes, if enzymes are present in quantities, which might decompose the peroxide and cause excessive use.

If the odor problem is at the primary clarifier, adding hydrogen peroxide to the plant influent may be desirable. Adding a small amount of hydrogen peroxide prior to a trickling filter may also reduce odors. Care, however, should be taken that excessive peroxide is not added to prevent lifting the slime from the filter bed.

Livestock and Poultry

Among the significant contributors to air pollution is animal manure. The odors are a problem particularly when manure slurry accumulations are disturbed for transfer or field spreading, when feedlot accumulations on the ground are wetted, or when waste is being treated in anaerobic lagoons.

Liquid-manure handling systems are among the most economical and practical of the many storage and disposal techniques, but the operations are proving to be increasingly unacceptable because of the offending odors. Rural locations in the past were sufficiently far from population centers that dilution by prevailing winds was adequate. Now, high animal densities are required by modern feeding methods, and urban sprawl extends into what were once rural areas. Dilution is no longer a solution to the pollution problem.

Despite the difficulties and experimental constraints of odor measurement, it has been possible to conduct meaningful tests on chicken and dairy-cow manure slurries with good odor control. The most common malodors from anaerobic manure decomposition in slurries are sulfurous in nature, *i.e.*, hydrogen sulfide, mercaptans, and sulfides which are susceptible to oxidation. Properly applied, hydrogen peroxide can do much to effect nearly complete oxidation of these materials and provide economical odor control.

Field trials were conducted at a large poultry farm in Pennsylvania that regularly received complaints from neighbors about the extremely foul and unpleasant odor when chicken manure slurry was spread on the fields. Sufficient hydrogen peroxide (to attain levels of 100, 150, and 175 ppm, based on total slurry weight) was added to the discharge side of a pump used to transfer a 5 to 10% solids slurry from the collection pit under the birds to a 1,400-gallon tank truck equipped to field spread the manure.

An area 10 feet wide and a quarter of a mile in length was used for manure application from each level of hydrogen peroxide treatment. Each test strip was separated from the adjacent test zones by a distance of 100 feet. The farm manager and two extension service people from Pennsylvania State University determined the overall quality and type of odor by sense of smell. Hydrogen sulfide was specifically determined quantitatively with a test apparatus. Additional checks were made 18 hours later to determine the efficiency and persistence of the treatments.

All three hydrogen peroxide levels of 100, 150, and 175 mg/l virtually eliminated the hydrogen sulfide and reduced the odor of the chicken manure significantly. The lowest level, 100 mg/l, was sufficient for control, and brought malodors under control very quickly. The offensive chicken manure odor remained significantly reduced in the test plots compared to the control (non-treated) plots on the second and third day after application.

Other trials on dairy-cow wastes were conducted at an eastern university's cowbarn liquid-manure storage tank. Hydrogen peroxide, diluted to 10% concentration, was added to the liquid manure pit to provide a final hydrogen peroxide concentration of 100 mg/l. During the test, a heavy-duty circulating pump provided vigorous mixing and circulation of the entire 2,400 gal of manure in the system. The atmosphere at the exhaust fans were tested periodically for hydrogen sulfide. As in the previous work, exhaust quality and odor type were determined by the sense of smell of experienced department and extension service personnel.

Prior to agitation of the slurry, only a silage or animal odor was apparent—but malodors were released as soon as agitation upset the quiescent system equilibria (Figure 24-5). With 100 mg/l hydrogen peroxide, sulfide-type malodors were eliminated. The remaining ammonia odor was probably constantly present but masked by the sulfide odors. It too was eliminated with an additional 100 mg/l hydrogen peroxide in a follow-up treatment. Malodors, however, will redevelop within a few hours, due to bacterial activity, unless the slurry is applied to a field where natural aeration can take over.

Hydrogen peroxide brought the malodors under control very rapidly when added to either type of liquid manure. Since the odors evolve mainly during infrequent periods of agitation, such as flushing, transfer, or removal, a 100 mg/l hydrogen peroxide treatment is very economical. Some care must be exercised in the treatment to avoid the inefficiencies of foaming from decomposition. Foaming results from too rapid addition of hydrogen peroxide, too high a hydrogen peroxide concentration, or insufficient mixing. In any of the processing, the hydrogen peroxide by-products (water and oxygen) are environmentally acceptable and contribute no secondary contamination.

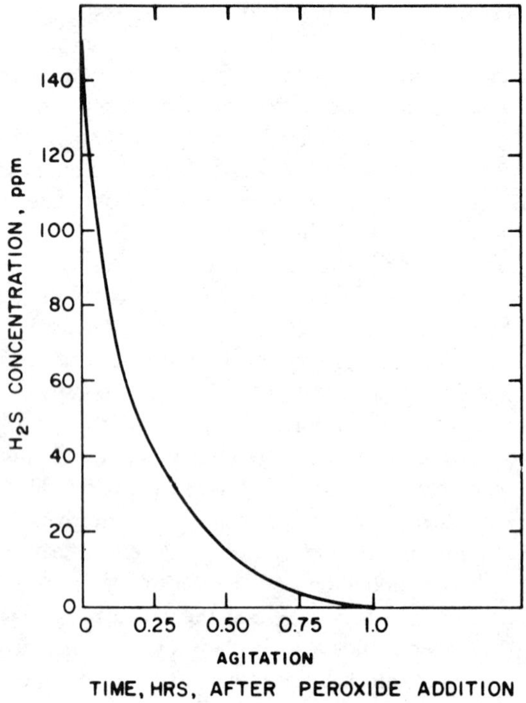

Figure 24-5. H_2S concentration versus time during hydrogen peroxide treatment (100 mg/l H_2O_2).

HYDROGEN PEROXIDE–GENERAL

Properties

Hydrogen peroxide is sold as a water solution. The concentrations most generally used for waste treatment are 35 and 50% by weight. Some of the more important physical properties are shown in Table 24-5.

The freezing point of 50% hydrogen peroxide is -62°F. Outdoor storage, therefore, does not present a problem. It is very important, however, that the purity of hydrogen peroxide be maintained during storage in order to prevent catalytic decomposition. The most generally used materials of construction for the storage and handling of hydrogen peroxide in concentrations up to 50% are high-purity aluminum, 300 series stainless steel, polyethylene, and Teflon. Materials such as mild steel, lead, copper, and copper alloys cause rapid breakdown of the hydrogen peroxide and must be avoided when handling the concentrated peroxide.

Table 24-5. Physical Properties of Hydrogen Peroxide

H_2O_2 Concentrations, Weight %	35	50	70
Active Oxygen Content, Weight %	16.5	23.5	32.9
Grams H_2O_2 per liter	396	600	903
Specific Gravity 20°C/4°C	1.13	1.20	1.29
Lbs per Gallon at 20°C	9.4	10.0	10.8
Boiling Point °C	108	114	126
Boiling Point °F	226	237	258
Freezing Point °C	-33	-52	-40
Freezing Point °F	-27	-62	-40
Viscosity 25°C (centipoise)	1.0	1.06	1.12
Total Vapor Pressure at 30°C (mm Hg)	23.3	18.3	10.1
Heat of Dilution cal/g mole of H_2O_2 @ 25°C & 1 Atm	-84	-178	-381

REFERENCES

1. Gas, Light and Coke Company. "Improvements in or Relating to the Removal of Hydrogen Sulfide from Gas and Liquids," British Patent 421,970 (January 1, 1935).
2. Parker, C. D. "Mechanics of Corrosion of Concrete Sewers by Hydrogen Sulfide," *Sewage and Industrial Wastes,* **23** (12), 1477 (December 1951).
3. Satterfield, C. N., R. C. Reid, and D. R. Briggs. "Rate of Oxidation of Hydrogen Sulfide by Hydrogen Peroxide," *J. Amer. Chem. Soc.,* **76**, 3922 (1954).
4. Pomeroy, R. "Generation and Control of Sulfide in Filled Pipes," *Sewage and Industrial Wastes,* **31** (9), 1082-1095 (September 1959).
5. Betz Laboratories. *Handbook of Industrial Water Conditioning,* 6th ed., (1962), pp. 55-59.
6. Eisenhauser, H. R. "Sweetening of Petroleum Distillates with Hydrogen Peroxide," *Industrial and Engineering Chemistry, Product Research and Development,* **4** (3), 185-189 (September 1965).
7. Sullivan, R. J. "Air Pollution Aspects of Odorous Compounds," prepared for National Air Pollution Control Administration, PB No. 188089 (September 1969).
8. Dresser Industries Inc. "Improvements in or Relating to a Process for Drilling a Well," British Patent 1,64,935 (September 24, 1969).
9. Chin, C. and M. G. Hicks. "Hydrogen Peroxide Studies of Oxygen Demand. *J. Water Poll. Control Fed.,* **42** (7), 1327-1342 (July 1970).
10. Friedman, L. "Method of Treating Sewage," FMC Corp. U.S. Patent 3,530,067 (September 22, 1970).
11. "Three Men Die from Fumes in Tannery," *Trenton Evening Times* (November 20, 1970).

12. Eye, J. D. and D. P. Clement, "Oxidation of Sulfides in Tannery Wastes," *J. Amer. Leather Chem. Assoc.*, **67** (6), 256-267 (1972).
13. Kibbel, W. H., C. W. Raleigh, and J. A. Shepherd. "Hydrogen Peroxide for Industrial Pollution Control," *Proc. 27th Industrial Waste Conference*, Purdue University (May 1972).
14. Chen, K. Y. and J. C. Morris. "Kinetics of Oxidation of Aqueous Sulfide by O_2," *Env. Sci. and Technol.*, **6** (6), 529 (June 1972).
15. Turk, A., R. Haring, and R. Okey. "Odor Control Technology," *Env. Sci. and Technol.*, **6** (7), 602 (July 1972).
16. von YuBishie, H. "Deodorization Method for Human and Animal Excreta," Japan Patent 1972-31025 (August 11, 1972).
17. Shepherd, J. A. and M. F. Hobbs. "Control of Sewage Hydrogen Sulfide with Hydrogen Peroxide," *Water and Sewage Works*, 67-71 (August 1973).
18. Stecher, P. G. *Hydrogen Sulfide Removal Processes.* (Park Ridge, N.J.: Noyes Data Corp., 1972).
19. Cheremisinoff, P., Y. Habib, and K. B. Rao. "Hydrogen Sulfide Pollution Control—A Review," *Ind. Wastes*, **20**, 22-28 (November/December 1972).
20. Shepherd, J. and M. Hobbs. "Sewage Treatment with Hydrogen Peroxide," U.S. Patent 3,705,098 (December 5, 1972).
21. Air Pollution Control Assoc. *Control Technology: Gases and Odors.* (Pittsburgh, Pa.: APCA, 1973).
22. Cross, F. L. *Air Pollution Odor Control Primer.* (Westport, Conn.: Technomic Publishing Co., 1973).
23. Geisler, C., C. Chin, and G. Hicks. "Hydrogen Peroxide Treatment of Industrial Wastes," *Eng. Bull., Purdue Univ. Eng. Ext. Serv.*, no. 140, pp. 897-901 (1973).
24. Chin, C., G. Hicks, and C. Geisler. "Bio-Oxygen Stabilization Using Hydrogen Peroxide," *J. Water Poll. Control Fed.*, **45** (2), 283-290 (February 1973).
25. Shepherd, J. A. and E. C. Shreve. "Odor Control with Hydrogen Peroxide," *Deeds and Data*. (Water Pollution Control Federation: April 1973).
26. Air Liquide. "Flue Gas Purification," U.S. Patent 3,733,393, (May 15, 1973).
27. Bethea, R., B. Murthy, and D. Carey. "Odor Controls for Rendering Plants," *Env. Sci. and Technol.*, **7** (6), 504-510 (June 1973).
28. Hammond, M. G. "High Strength Acid Containing H_2O_2 to Scrub SO_2," U.S. Patent 3,760,061 (September 18, 1973).
29. Environmental Protection Agency. "Hydrogen Peroxide Cures Filamentous Growth in Activated Sludge," EPA Publication No. 670/2 73-033 (October 1973).
30. St. Louis (Mo.) Metropolitan Sewer District. "Odors Emitted from Raw and Digested Sewage Sludge," EPA Report 670/2 73-098 (December 1973).
31. FMC Corp. "Deodorization of Animal Excrement," French Patent 2187-729 (1974).
32. FMC Corp. "Deodorization of Animal Excrement with Hydrogen Peroxide," German Patent 2328532 (1974).

33. Caropreso, F. E., C. W. Raleigh, and J. C. Warner. "Attack Bulking Sludge with H_2O_2 and a Microscope," *Bull. Cal. Water Poll. Control Assoc.*, **10** (4), 44 (April 1974).
34. Beltran, M. "Engineering/Economics Aspects of Odor Control," *Chem. Eng. Prog.*, **70** (5), 57 (May 1974).
35. Leonardos, G. "A Critical Review of Regulations for the Control of Odors," *J. Air Poll. Control,* **24** (5), 456 (May 1974).
36. Kanto Kenka Kogyo Co. Ltd. "N-Oxide Removal from Exhaust Gas," (February 28, 1973).
37. Ozawa, T. "Absorbing Nitrogen Oxides for Exhaust Gases," Japan Patent 8083-069 (November 6, 1973).

CHAPTER 25

HYDROGEN PEROXIDE: PROPERTIES, STORAGE, AND HANDLING

Clifton W. Cooke

E. I. duPont de Nemours & Co.
Environmental Products Section
Newport, Del. 19804

Hydrogen peroxide is a clear waterlike solution that is produced and sold in large quantities (about 200 million lb per year) in the United States. Higher purity and improved methods of stabilization, shipping, and storage have improved the stability of hydrogen peroxide. Thirty-five years ago, refluxing a 35% hydrogen peroxide solution for one hour was likely to result in considerable loss of active oxygen. Today the same concentrations can be refluxed 10 to 15 hours with almost no loss in active oxygen.

The reactions of hydrogen peroxide (and its active oxygen) are separate and distinct from those of the classical inorganic oxidizing agents such as chlorine, "hypochlorite," etc. Many inorganic oxidizing agents are contaminated by their by-products and reduced states. Unlike these, hydrogen peroxide yields no noxious or polluting species of its own; its reduced state (from oxidation or reduction) is water.

Hydrogen peroxide is miscible in all proportions with water. Its freezing and boiling points are plotted in Figure 25-1. Other physical properties are summarized in Table 25-1. More information on hydrogen peroxide can be found in the American Chemical Society (ACS) Monograph 128, *Hydrogen Peroxide.*[1]

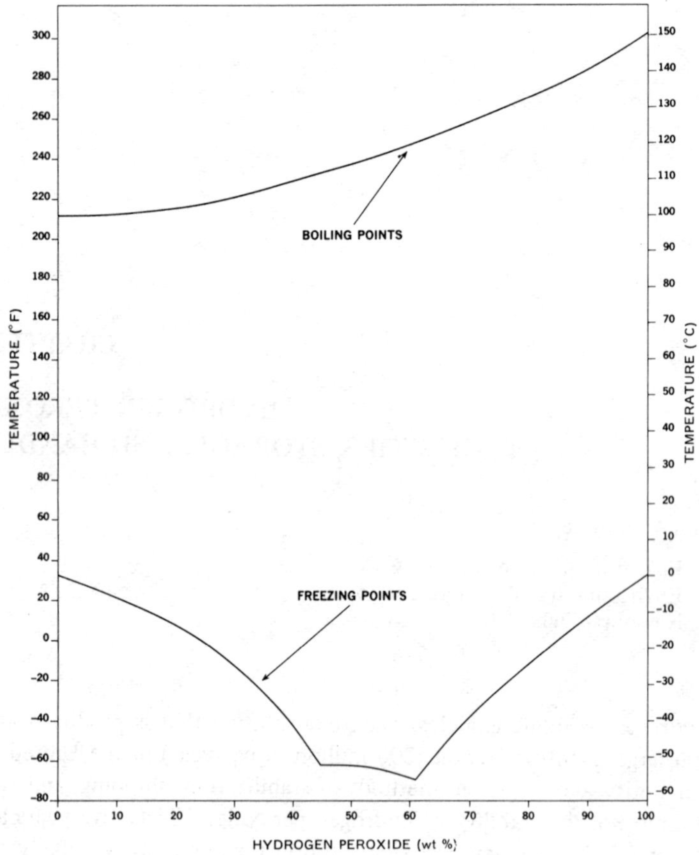

Figure 25-1. Boiling points and freezing points of hydrogen peroxide.

DECOMPOSITION

Pure solutions of hydrogen peroxide are inherently very stable. When they are carefully purified and kept in clean, unreactive containers, decomposition is extremely slow. In general, the stability of very pure hydrogen peroxide increases with concentration.

Decomposition is the most important reaction of hydrogen peroxide. Although the net reaction is simple, the several factors that contribute or cause this reaction are still not thoroughly understood—particularly those believed to involve chain reactions and heterogeneous catalysis. With some possible exceptions, hydrogen peroxide decomposes faster with catalysts,

Table 25-1. Physical Properties of Hydrogen Peroxide

Property	H_2O_2 Concentration, wt %		
	35	50	70
Active Oxygen (approx.) Wt %	16.5	23.5	32.9
Density (20°C, 68°F)			
g/ml	1.133	1.196	1.288
lb/gal	9.45	9.98	10.75
g H_2O_2 (100%)/ml	0.397	0.598	0.898
Apparent pH (approx.) by glass electrode	2.5	1.8	0.5
Freezing Point,			
°C	-33	-52	-40
°F	-27	-62	-40
Boiling Point			
°C	108	114	126
°F	226	237	258

alkali or acid, increases in temperature, and electrolysis. Decomposition must be controlled when hydrogen peroxide is manufactured, stored, shipped, and used.

CATALYSTS

Decomposition into water and oxygen is accelerated by a wide variety of catalysts—even a slight trace of catalytically active material (parts per billion). Obviously, under commercial operating conditions minute traces of catalytic materials, ever present in the surrounding atmosphere and in the container walls, cannot be kept from hydrogen peroxide solutions. All equipment and material must be thoroughly cleaned and passivated before use.

The enzyme catalase, halogens (except fluorine), or the heavy metal salts such as iron, chromium, copper, manganese, cobalt, lead, and silver can be used as catalysts if it is necessary to decompose hydrogen peroxide into water and oxygen.

RADIATION

The extent of decomposition caused by sunlight or other radiation is not completely established. Exposure to sunlight can increase the temperature;

this increases the decomposition and causes photochemical decomposition. Under laboratory or plant conditions, the effect of *diffuse* sunlight is probably insignificant.

TEMPERATURE

A 10°C temperature increase of pure unstabilized hydrogen peroxide solution will increase its decomposition 2.2 times. But adequately stabilized hydrogen peroxide solutions can be refluxed 10 to 15 hours without noticeable decomposition.

HAZARDS

Decomposition can effect the release of oxygen and heat that can create hazards and less efficient use of the hydrogen peroxide. Decomposition may occur in the vapor phase as well as in the liquid phase; vapor concentrations of hydrogen peroxide above 26 mole % can decompose with explosive violence.

PRECAUTIONS IN USE[2]

Hydrogen peroxide solutions obtained commercially are not explosive in the sense of propagating detonation. If handled properly to avoid contamination, hydrogen peroxide may be stored indefinitely and transported from storage area to chemical operation with little difficulty. Gross contamination of peroxide solutions, however, may result in violent rupture of the containers with possible injury to personnel from flying metal or glass fragments. Proper handling or ordinary (35 to 50% concentration) hydrogen peroxide solutions is achieved readily by teaching personnel concerned to abide by a few rudimentary principles of cleanliness and care. The ease with which such instruction is usually assimilated makes the occurrence of pressure ruptures in industry very rare.

High-strength hydrogen peroxide solutions containing over 52% by weight of H_2O_2 by definition also present few hazards when properly handled.

PERSONAL SAFETY AND FIRST AID

All persons who handle hydrogen peroxide should be thoroughly familiar with Chemical Safety Data Sheet SD-53 (revised 1969), published by the Manufacturing Chemists Association (MCA). Copies are available from the MCA, 1825 Connecticut Avenue, N.W., Washington, D.C. 20009.

HEALTH HAZARDS

Hydrogen peroxide (27.5 to 52%) solutions are strong oxidizers and cause irritation of the eyes, skin, nose, and throat. Contact with eyes may cause burns, although the effects may be delayed. If swallowed, injury may occur because of acute swelling of esophagus or stomach caused by sudden oxygen evolution. However, no chronic systemic effects have been reported. Short-term skin contact may result in a temporary whitening or bleaching, accompanied by a mild stinging sensation. Longer-term exposures may result in skin irritation and the possibility of blisters or burns. Repeated exposure may cause chronic skin inflammation due to thickening of the skin.

Hydrogen peroxide (52 to 90%) solutions are powerful oxidizers and cause severe burns to eyes and skin. Effects on the eyes may be delayed, and ulceration of the cornea and possibly blindness may result. If it is swallowed, injury may occur because of acute swelling of the esophagus or stomach caused by sudden oxygen evolution. No chronic systemic effects have been reported. Inhalation of vapor or mist may cause extreme irritation and inflammation in the nose and throat.

SAFETY PRECAUTIONS

All personnel who handle hydrogen peroxide should exercise care to prevent contact with solutions or vapors. Do not get in the eyes, or skin, or on clothing. Do not breathe mists or vapors. Adequate ventilation should be provided. Wash thoroughly after handling. All contaminated clothing, including gloves, shoes, aprons and coveralls, etc., should be promptly removed and thoroughly flushed or washed with water to eliminate traces of hydrogen peroxide before reuse.

PERSONAL PROTECTIVE EQUIPMENT

The following personal protective equipment is recommended whenever hydrogen peroxide is handled: gas-tight, impact-resistant glass or plastic chemical goggles; neoprene, butyl rubber, or vinyl gauntlet-style gloves; neoprene, polyethylene, or polystyrene aprons; neoprene or plastic boots; hard hat; and clothing of Dacron polyester fiber. Flammable clothing such as cotton, rayon, leather, or wool should not be worn; if they are splashed with high-strength peroxide, severe burns may result from the combustion of such materials. Protective clothing should be clean. If soiled, even polyester fabrics may burn if they are contaminated with high-strength hydrogen peroxides.

Because of the strong oxidizing nature of hydrogen peroxide, protective creams *should not* be used.

In emergencies or in performing work where there is a possibility of considerable exposure, a complete suit of neoprene and self-contained breathing apparatus should be worn. See MCA Chemical Safety Data Sheet SD-53 (p. 15, sec. 5.2.2.3) for additional information on the proper selection and use of respiratory equipment.

SITE FACILITIES

The following safety facilities should be readily accessible in all areas where hydrogen peroxide is handled or stored.

Safety shower, with quick-opening valves which stay open. Water should be supplied through insulated lines.

Water hydrant and hose: a means of flushing spills with large volumes of water under low pressure.

Eye-wash fountain, or other means of washing the eyes with a gentle flow of tap water.

FIRST AID

In case of contact, immediately flush eyes or skin with plenty of water for at least 15 minutes while quickly removing contaminated clothing and shoes. For eyes, promptly call a physician, preferably an eye specialist.

If swallowed, encourage vomiting. Give lukewarm water freely and encourage belching if there is any evidence of swelling of the esophagus or stomach. Call a physician.

If inhaled, remove the person to fresh air. If irritation of the throat or nose is apparent, refer the patient to a physician.

STORAGE AND HANDLING

Hydrogen peroxides should always be stored in the original, unopened drums. Fiber/polyethylene drums may be stored four high, but the vent holes must not be obstructed by drums or pallets. Drums other than the Weatherpak® should be kept out of the sun and away from heat in a dry, inside location.

Bulk storage installations are suggested when the annual peroxide consumption is 75,000 to 100,000 lb or more of 35 or 50% hydrogen peroxide.

Table 25-2. Drums and Bulk Carriers Used to Ship Hydrogen Peroxide

		Grades of Hydrogen Peroxide		
Drums (non-returnable)	**Capacity (Gal)**	**35 (lb, net)**	**50 (lb, net)**	**70* (lb, net)**
Weatherpak®	55	500	500	--
		275	275	--
		Load Net (approximate)		
Carrier	**(Approx. Gal)**	**(lb)**	**(lb)**	**(lb)**
Tank Cars	4,000	39,000	41,000	44,000
	6,000	58,000	61,000	66,000
	8,000	77,000	81,000	88,000
	17,500	166,000	175,000	193,000
	19,000	180,000	190,000	210,000
Tank Trucks**	1,000	99,500	10,000	--
	1,500	14,000	15,000	--
	2,000	19,000	20,000	--
	2,500	23,500	25,000	27,000
	3,000	28,000	30,000	32,000
	3,500	33,000	35,000	38,000
	4,000	38,000	40,000	43,500

*70% hydrogen peroxide is not packed in drums.
**Dependent on equipment available and road weight limitations.

PACKAGES

Labels on drums include the Department of Transportation (DOT) white label, which shows hydrogen peroxide is classified as a corrosive material. Tank cars and tank trucks use DOT markings. The freight classification is "Hydrogen Peroxide, Corrosive Material."

UNLOADING TANK CARS AND TANK TRUCKS

Details on unloading procedures and precautions are contained in MCA Chemical Data Sheet SD-53. A water hose should be on hand to wash down any spills. Avoid dropping articles into the carrier tank. Materials used in handling peroxide must be very clean.

A tank truck of hydrogen peroxide is unloaded by the driver. Tank cars of hydrogen peroxide can be unloaded either by pump or compressed air.

Caution: Tank car customers should not unload the first shipment without the assistance of a specialist to supervise and instruct plant personnel in safe, proper procedures. Dome fittings of hydrogen peroxide tank cars are usually color-coded.

COMPRESSED AIR UNLOADING

If compressed air is used to unload tank cars, it must be free of oil, water, and dirt to avoid contamination of the hydrogen peroxide. The pressure applied to tank cars must be kept below 15 lb and an emergency air shut-off valve should be located away from the area in case of leaks.

The equipment for air pressure unloading is shown in Figure 25-2. Black iron or galvanized iron pipe and fittings can be used up to the filter; stainless steel or aluminum pipe and fittings should be used between the filter and the tank car. The air hose is reinforced rubber.

These components should be checked at least every six months and maintained in good operating condition. The filter should be opened and the cartridge replaced at least once a year.

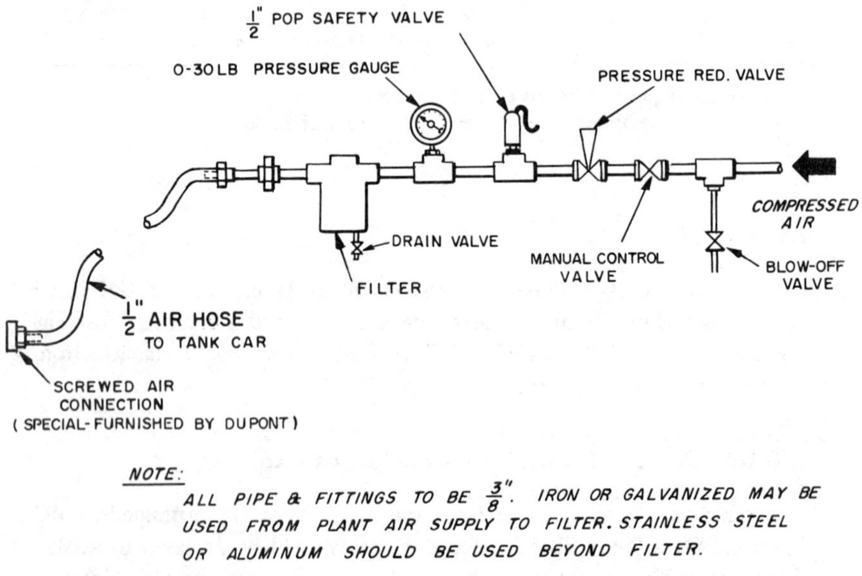

Figure 25-2. Equipment for air pressure unloading of tank cars.

STORAGE INSTALLATIONS

Figure 25-3 shows a bulk storage system where peroxide is transferred by low-pressure air from the tank car to the storage tank and by pump from the storage tank to the point of use. Figure 25-4 shows a single transfer pump with lines valved to direct the peroxide flow. Figure 25-5 illustrates a bulk storage installation receiving a tank truck delivery of 70% (dilution grade). The plant's dilution water is fed to the tank, then the 70% is added through the jet mixer inside the storage tank at a rate of 50 to 60 gpm. Figure 25-6 shows a storage tank for use with 35 or 50% hydrogen peroxide.

The services of a specialist are usually available upon request to assist in the design and installation of a satisfactory storage system for any grade and/or concentration of hydrogen peroxide. He will survey the mill, help choose the storage tank site, and prepare a general layout and material list. He will also ask a laboratory to check the quality of the mill water for dilution purposes.

After the tank is in place on the foundation, the specialist will oversee the installation of piping and the jet mixer by mill personnel. He will passivate the entire system with nitric acid, then check the system for serviceability.

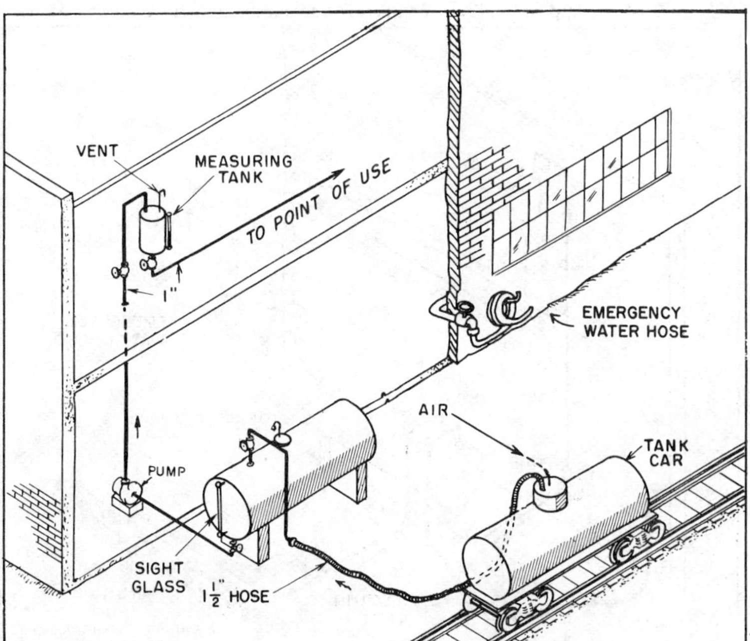

Figure 25-3. Air unloading of tank car and pump transfer to point of use.

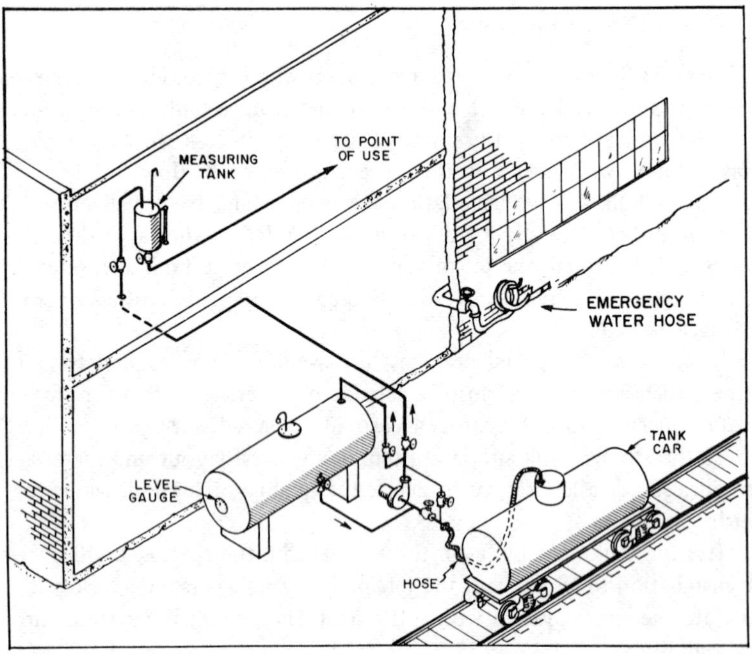

Figure 25-4. Pump unloading of tank car and pump transfer to point of use.

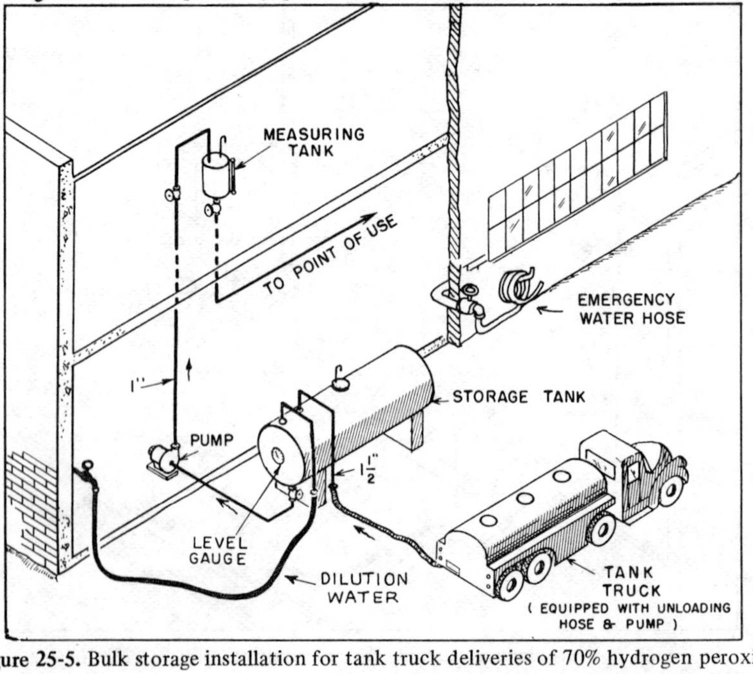

Figure 25-5. Bulk storage installation for tank truck deliveries of 70% hydrogen peroxide.

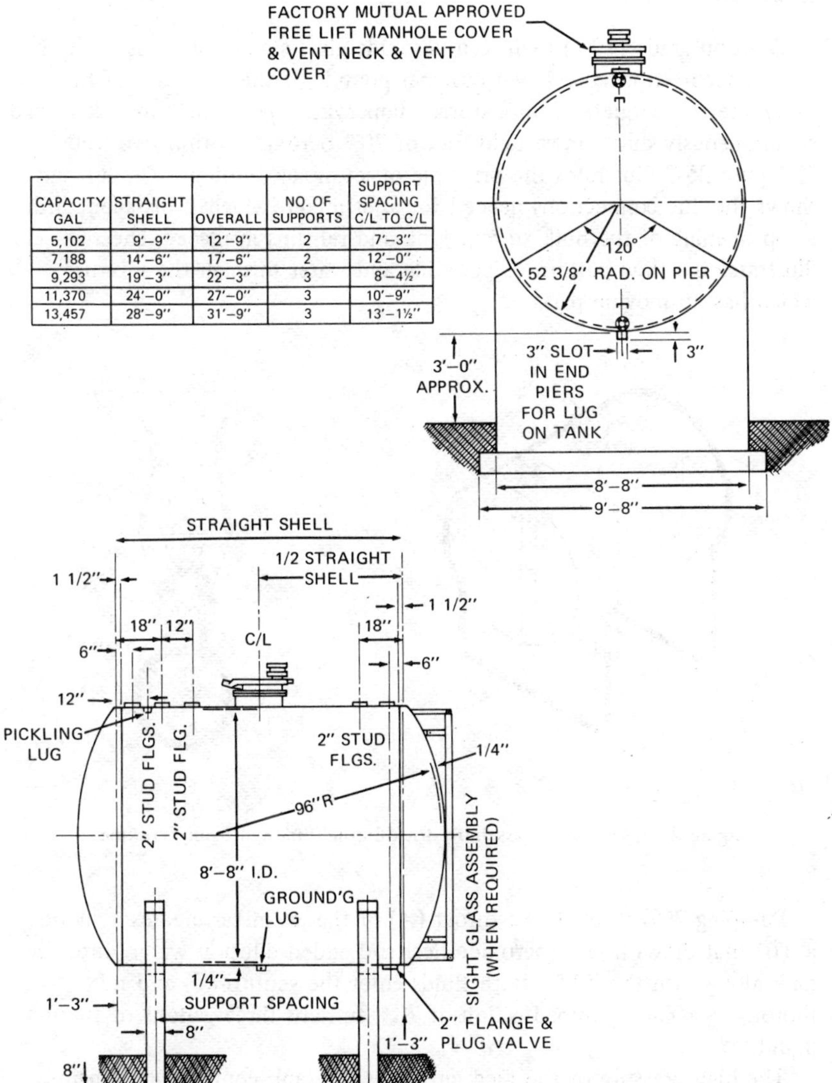

CAPACITY GAL	STRAIGHT SHELL	OVERALL	NO. OF SUPPORTS	SUPPORT SPACING C/L TO C/L
5,102	9'–9"	12'–9"	2	7'–3"
7,188	14'–6"	17'–6"	2	12'–0"
9,293	19'–3"	22'–3"	3	8'–4½"
11,370	24'–0"	27'–0"	3	10'–9"
13,457	28'–9"	31'–9"	3	13'–1½"

NOTE:

MATERIAL TO BE ALLOY 5652 ALUMINUM
DESIGN PRESSURE = 15 P.S.I.
TEST PRESSURE = 22½ P.S.I. AT BOTTOM
INSULATE BETWEEN TANK & PIER WITH TARRED PAPER, FELT, CLOTH OR SIMILAR MATERIAL.

Figure 25-6. Storage tank for 35 or 50% hydrogen peroxide.

BULK DILUTION

Dilution grade (70%) hydrogen peroxide gives stable solutions at 50 and 35% concentrations at a lower cost per pound. If the flow rate of the dilution water is adequate, a bulk storage tank with a jet mixer can receive and simultaneously dilute most deliveries of 70% peroxide within two hours.

Figure 25-7 illustrates the principle of jet-mixer dilution. The left cut shows the line connections needed for a jet mixer, which is installed through a top opening of the bulk storage tank and remains in place. The right cut illustrates the flow pattern of peroxide and water through the mixture, which has no moving parts.

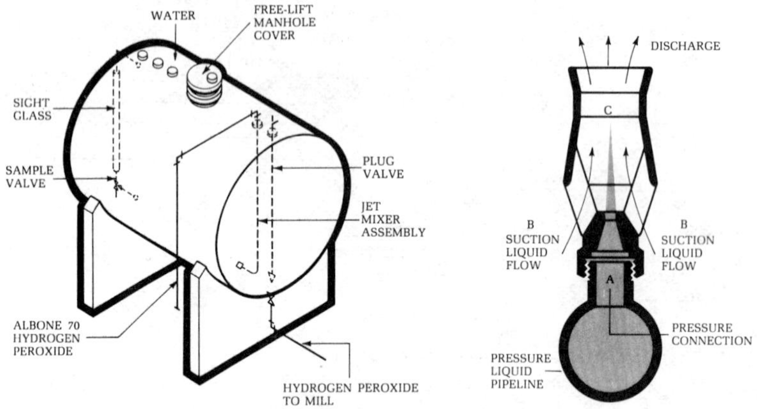

Figure 25-7. Jet mixer assembly for diluting 70% hydrogen peroxide.

Pumping 70% through the throat (A) of the jet mixer creates a suction at (B) that draws liquid (peroxide heel and added dilution water) into the tank along with the 70%. Both fluids enter the venturi (C) and mix thoroughly. Each pumped gallon of 70% entrains three gallons of suction liquid.

The high pressure of the feed circulates the tank contents and ensures uniform composition in the storage tank. Tests show the concentration of the diluted hydrogen peroxide immediately after delivery is within 0.2 to 0.5% of the calculated strength.

STORAGE TANKS

The storage tank capacity should be large enough to ensure an adequate supply of hydrogen peroxide between deliveries. Hydrogen peroxide storage tanks should be of class I aluminum alloy (Table 25-3), meet all applicable state and local requirements, be mounted on concrete piers or steel saddles, and be cushioned by asphalt-impregnated felt pads. Caution: Aluminum must be protected from the action of alkali, including that in masonry.

The storage tank will require an inlet, a bottom outlet, and a 20-inch manhole with a Factory Mutual Insurance Co.-approved free-lift manhole cover and vent with a dust filter attached. An opening for a liquid-level gauge connection or a sight glass assembly (Figure 25-7) will also be necessary. All connections to the storage tank should be flanged.

Small tanks (<350 gal)–molded especially for hydrogen peroxide–of virgin, unpigmented polyethylene are satisfactory for less than 52% hydrogen peroxide at 2 to 41°C (35 to 105°F), at atmospheric pressure, and out of direct sunlight.

LIQUID LEVEL GAUGE OR SIGN GLASS

A manometer-type level gauge or a sight glass is recommended for checking tank inventory.

MEASURING OR HEAD TANK

The measuring or head tank, if used, should have a capacity at least equal to the volume of peroxide required in an eight-hour, one-shift operation. The tank should be fitted with top and bottom outlets and a vent. To calibrate the tank, measured amounts of liquid can be marked on a gauge strip mounted behind the sight glass (Figure 25-8).

PUMP

A large and self-priming pump is required for unloading tank cars. Pumps used should be of class I or class II materials (Table 25-3).

A "hold-in" switch should be mounted at the measuring or head tank to control the pump operation during the head tank filling. If the pump is also used for unloading tank cars, a "start-stop" switch should be mounted at the pump location.

Braided pump packing of Teflon® TFE fluorocarbon resin without graphite is recommended. Nonlubricated braided resin of Teflon is suitable valve packing. Unpigmented polyvinyl chloride gaskets or envelope gaskets of Teflon give good service.

Table 25-3. Materials Suitable for Hydrogen Peroxide Service

Class I Materials

Class I materials are for unlimited contact with hydrogen peroxide.

Aluminum Alloys–Aluminum Association
Designations: 1060, 1100, 1160, 1260, 5254, 5652, 6063 (sheet and tubing) and B356 (casting alloy)
Chemical Stoneware
White Chemical Porcelain
Pyrex® glass
Kel-F® fluorocarbon polymer, pure unpigmented
Teflon® TFE fluorocarbon resin, pure unpigmented
Polyethylene, pure unpigmented (H_2O_2 to 52 wt % only and at ambient temperatures)

Class II Materials

Class II materials should be used only for short-time contact with hydrogen peroxide prior to storage or use.

Aluminum Alloys–Aluminum Association Designation 3003
Note: 6061 should not be used as it corrodes rapidly in peroxide service.
Stainless Steel Type 304, 316, 321, and 347
Koroseal® 700 polyvinyl chloride, molded
Tygon® 3604A polyvinyl chloride, unpigmented
Silicone SE 450, unpigmented (General Electric)
Silicone S-9711, unpigmented (Dow Corning)

Note: These metals should be in the nonporous state, free from blowholes or other casting defects. In addition they require cleaning, passivation and surveillance before use.

Caution: Do Not Use Ordinary Pipe Thread Compounds

Miscellaneous Materials

Joint Sealant
Gore-Tex® ribbon of Teflon TFE fluorocarbon resin

Grease
Fluorolubes® grease
Unpigmented silicone SE-450 or S-9711

Packings
Teflon fluorocarbon resin
Vitrium® fiberglass impregnated with fluorolubes

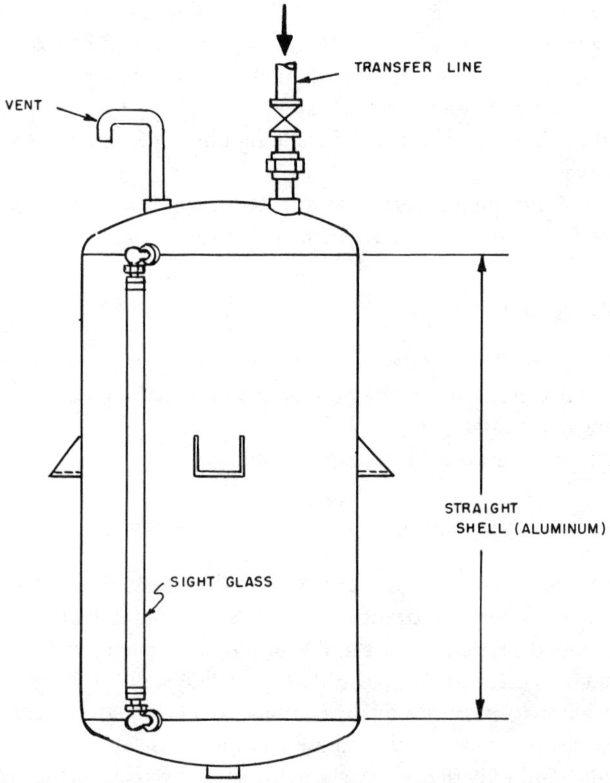

MEASURING TANK

NOMINAL CAPACITY, GALS	DIAMETER	STRAIGHT SHELL	CAPACITY STRAIGHT SHELL, GALS
10	12"	21"	10.39
20	12	41	20.09
30	18	30	31.68
50	18	49	51.74
75	24	42	82.24
100	30		
200	30		

Figure 25-8. Measuring tank for 35 or 50% hydrogen peroxide (E. I. duPont de Nemours & Co., Electrochemicals Department).

PIPING

Piping may be of either a class I or class II alloy, such as 6063. Either screwed aluminum pipe (schedule 40 or 80) with stainless steel fittings or welded piping with flanged joints is satisfactory. A flanged ball valve or a gate valve is recommended at the storage tank outlet; valve bodies may be of suitable aluminum or stainless steel alloy. The piping system should be so designed that there is no danger of trapping liquid peroxide between two closed valves.

Because aluminum piping must be protected against the action of alkalies, 304 or 316 stainless steel is preferred for underground pipe.

UNLOADING HOSE

A 1½-inch 304 or 316 stainless steel flexible metal hose is recommended for unloading tank cars. When the hose is not in use, the ends should be covered to prevent contamination.

Tank trucks carry a hose and pump for unloading.

PASSIVATION OF EQUIPMENT

As shipped, hydrogen peroxides can be stored without discernible loss of strength. Storage system construction materials are contaminated by dust, dirt, and embedded particles that can decompose hydrogen perocides, but when the system is cleaned and passivated, peroxide stability is ensured.

Do not attempt to passivate major equipment–inadequately passivated equipment causes peroxide to decompose. Proper passivation requires technical knowledge and experience. A specialist should passivate the storage system with nitric acid, and will prepare the system to receive hydrogen peroxide. He should be present when the first shipment is unloaded.

Acceptable peroxide equipment, properly passivated, should not require repassivation if the system remains uncontaminated, and it should become more passive with use. Tanks have been in service longer than 25 years and have not needed repassivation; but a specialist should periodically inspect the storage tank.

REFERENCES

1. Schumb, W. C., C. H. Satterfield, and R. L. Wentworth. *Hydrogen Peroxide.* (New York: Van Nostrand Reinhold Co., 1955).
2. Wallace, J. G. *Hydrogen Peroxide in Organic Chemistry.* (E. I. DuPont).

CHAPTER 26

POTASSIUM PERMANGANATE FOR ODOR CONTROL

Kenneth J. Ficek

Carus Chemical Co.
La Salle, Ill. 61301

The emphasis on odor control over the recent years has resulted in a great number of tests being conducted in pilot equipment and full-scale scrubbers to determine the most efficient and economical system to control odors from manufacturing facilities. Many of these tests included the addition of potassium permanganate to determine the areas in which it could be applied to help solve these problems. The purpose of this chapter is to try to "sort out" from past practical experience, and from a basic chemical understanding of oxidation, the following:

1. Chemically—the types of odor problems which $KMnO_4$ will and won't help solve
2. The concentration of contaminants which make $KMnO_4$ application uneconomical
3. The scrubbing systems found to be most effective
4. The methods of evaluating and testing
5. The proper operation and maintenance of a full-scale odor control system using potassium permanganate.

To better understand the process through which odor control is achieved using potassium permanganate, some basic permanganate chemistry will be reviewed.

Potassium permanganate is defined as a strong oxidizing agent with the chemical formula $KMnO_4$. It is used in aqueous solutions at various pH ranges to chemically change the structure of the chemical compounds which cause odors. This change in structure is due to the fact that potassium permanganate, in solution, provides highly reactive oxygen. In other words, it is a convenient form of providing oxygen to the system. A very simple reaction can be written as follows:

$$KMnO_4 + H_2O + \begin{array}{l}\text{Oxidizable}\\\text{Odor Producing}\\\text{Compounds}\end{array} \rightarrow MnO_2 + KOH + \begin{array}{l}\text{Oxidized Form of}\\\text{Odor-Producing}\\\text{Compounds}\end{array}$$

Obviously, if a compound is not capable of being oxidized, or can only be oxidized under very vigorous conditions, the potassium permanganate technique should not be pursued. Table 26-1 lists some compounds that are not oxidized by using potassium permanganate under the conditions present in a scrubber, and some which are highly resistant to chemical oxidation. If these compounds are the prime odorous compounds in an air stream, permanganate scrubbing should not be considered.

Table 26-1. Compounds That Do Not React Readily with Potassium Permanganate

Aliphatic Compounds	Nitrogen-Containing Compounds
Acetone	Nitrobenzene
Dipropyl Ketone	Pyridine
Methylisobutyl Ketone	
Diisobutyl Ketone	
n-Butanol	Sulfur Compounds
Methylethyl Ketone	Carbon Disulfide
Methylene Chloride	
Aromatic Compounds	Inorganic Compounds
Benzene	Ammonia
Toluene	Carbon Monoxide
Penta-chlorophenol	Carbon Dioxide

Generally, saturated or highly chlorinated compounds do not react with $KMnO_4$ at appreciable rates.

There is another group of compounds which are usually present in very high concentrations and in very high volumes of air. As a result of their very high corresponding potassium permanganate demands, the application of potassium permanganate is not only uneconomical but presents many maintenance problems. These include SO_2, NO, and H_2S. If these compounds were present in low concentrations, the application of permanganate might be practical.

Generally speaking, if the calculated permanganate consumption based on the permanganate demands in Table 26-2 exceeds 150 lb/day, a permanganate system *alone* should not be considered. Some pretreatment of the air with a less expensive chemical such as caustic, or acid, or water alone

Table 26-2. Compounds That Do React with $KMnO_4$

	Calculated $KMnO_4$ Demand lb $KMnO_4$/lb Substrate	
Aliphatic Compounds		
Formaldehyde	7.0	
Acetaldehyde	12.0	
Allyl Acetate	a. 12.6,	b. 6.3
Acrolein	a. 13.1,	b. 9.4
Aromatic Compounds		
Benzaldehyde	a. 15.9,	b. 1.0
Styrene	a. 20.2,	b. 4.0
Phenol	a. 15.7,	b. 12.3
o-Cresol	a. 16.6,	b. 10.7
o-Chlorophenol	10.7	
m-Chlorophenol	10.7	
p-Chlorophenol	10.7	
Nitrogen Containing Compounds		
Dimethylamine	a. 14.0,	b. 9.3
Trimethylamine	a. 16.0,	b. 10.7
Monoethanolamine	a. 8.6,	b. 1.7
Triethanolamine	a. 10.6,	b. 2.1
Putrescine	a. 13.1,	b. 4.8
Cadaverine	a. 14.4,	b. 4.1
Indole	a. 15.7,	b. 5.4
Skatole	a. 16.4,	b. 6.4
Sulfur Compounds		
Dimethyl Sulfide	3.4	
Dimethyl Disulfide	5.6	
Thiophene	16.3	
Diethyl Sulfide	2.3	
Diethyl Disulfide	4.3	
Inorganic Compounds		
Hydrogen Cyanide	3.9	
Hydrogen Sulfide (to sulfate)	12.4	
Nitrous Oxide	5.3	
Sulfur Dioxide	1.6	

Note: "a" and "b" represent two different hypothetical stages of oxidation that may be attained depending on conditions. In practice, the $KMnO_4$ demand will almost be less than the "a" value.

should be investigated to reduce the overall permanganate consumption. High concentrations of hydrogen sulfide can, for example, be substantially reduced by the use of a first-stage caustic wash. The use of a second-stage $KMnO_4$ scrubber would "polish" the air stream and remove any traces of hydrogen sulfide.

If an analysis of the air stream is available, a quick calculation using the $KMnO_4$ demands listed in the table should be run to determine the pounds of $KMnO_4$/day. (In some cases two demands are indicated, depending on the degree of oxidation achieved. Obviously, if the calculation using the lesser demand exceeds 150 lb/day, the possibility of using $KMnO_4$ economically is doubtful, unless a two- or three-stage system is installed.)

Example

Air stream analysis indicates 20 lb/day styrene; $KMnO_4$ demand is (a) 20.2 or (b) 4.0.

$$20 \text{ lb styrene/day} \times 20.2 = 404 \text{ lb } KMnO_4\text{/day}$$

Since $KMnO_4$ exceeds 150 lb/day, a single-stage system is not recommended. However, styrene is slightly water soluble, and if 75%, or 15 lb, could be removed with a venturi or other first-stage system, the remainder, or 5 lb, could be removed with a $KMnO_4$ scrubber as the second stage. $KMnO_4$ usage would be about 100 lb/day, or it could be as low as 5 x 4.0 (b demand), or 20 lb/day.

INDUSTRIES

Some of the industries which have odorous compounds in their effluent air stream are listed in Table 26-3. Generally, those industries in which $KMnO_4$ has been successfully applied to control odors include the following:

Rendering	Asphalt roofing
Foundry	Chemical manufacturing
Food processing	Municipal sewage treatment
Paint production	Industrial sewage treatment

Some of the applications which have *not* been successful include:

- Boiler flue gas–because of high concentrations of SO_2 (~3000 ppm)
- Oxides of nitrogen–tests do not indicate good removal due to absorption problems
- Some paints and coatings–many organic solvents do not react with $KMnO_4$.

PARTICULATES

Some particulate matter has a high $KMnO_4$ demand, which should be removed before the gas enters the permanganate scrubber. If this material is not cleaned from the stream, the permanganate consumption could be excessive and uneconomical. Usually a first-stage venturi or other particulate removal device can be utilized prior to the $KMnO_4$ scrubber. A test on coffee bean chaff indicated that each pound of chaff consumes approximately 10 lb $KMnO_4$.

Table 26-3. Manufacturing Processes which Produce Odors

	Aldehydes	Amines	Phenols	Organic Acids	Mercaptans	Ketones	Organic Sulfur Compounds	Organic Nitrogen Compounds	Hydrogen Sulfide
Foundry	X	X	X						
Rendering	X	X		X		X	X	X	
Pulp and Paper					X		X		X
Food Processing	X	X		X	X			X	
Municipal Waste Treatment	X	X		X			X	X	X
Paint and Varnish			X	X		X			
Grain Drying	X	X		X					
Fats and Oil	X	X		X					
Asphalt Roofing			X			X	X		X
Refining			X				X		X
Plastics	X	X	X			X			

ODOR CONTROL

Equipment used for the absorption of gases can be placed in five general categories, namely (1) packed towers, (2) plate towers, (3) spray chambers, (4) venturi scrubbers, and (5) jet scrubbers. The most successful results using $KMnO_4$ have been obtained with packed tower scrubbers containing saddles.

In the practical application of potassium permanganate for odor control, an absorber should have approximately one second dwell time for the gases to be transferred to the liquid phase. Many of the reactions with potassium permanganate take place very quickly, so that the contact time is required for the absorption rather than the oxidation. As was mentioned earlier, the two-stage system has been generally more effective than the single stage.

PILOT TEST

Many scrubber manufacturers offer pilot equipment to determine the effectiveness of their units to solve odor problems. If a pilot test is to be conducted, it is suggested that the customer

1. Contact a qualified chemical consultant for technical assistance in handling the $KMnO_4$ test and to make sure that the system has a chance chemically.
2. Make sure that the air stream being drawn into the pilot unit is representative of the problem air stream.
3. Try to obtain a two-stage pilot unit if possible, especially if the air stream contains particulates.
4. Use good techniques for determining odor levels, chemical concentrations, air flows, etc., or hire a testing firm. The costs of operating a full-scale unit can be far beyond expectations if sloppy techniques are used in the pilot tests.
5. Make sure any pilot unit is equipped with a good demister, so that $KMnO_4$ solution is not sprayed over the area.

EVALUATION PROCEDURE

In the actual pilot test or full-scale start-up a 2% $KMnO_4$ solution at neutral pH should be the first trial. If satisfactory odor control is obtained, it may be necessary only to add caustic, borax, or acid to make up for fluctuations in solution pH due to absorbed gases.

If satisfactory odor control is not obtained, the $KMnO_4$ concentration should be raised to 4% (1/3 lb/gal) and tests run. If still not satisfactory, raise pH first to 8.4 with Na_2CO_3 or borax, then ultimately to 12 using caustic.

If ammonia or some of the short-chain amines are present, an acid pre-scrub or an acid permanganate solution (pH 5-6) should be tried.

Care should be taken to keep the pH above 2 and below 12.5. Below pH 2, $KMnO_4$ is reduced by the acid; above the pH 12.5, $KMnO_4$ deoxidizes to become less efficient.

OPERATION OF A POTASSIUM PERMANGANATE SCRUBBER

After a full-scale scrubbing system has been installed, it should be run with water only to ensure that all interior components, such as packing, recirculation lines, etc., are thoroughly washed, before adding any chemicals to the system. While the system is running with water, the mist elimination system should be checked before $KMnO_4$ is added. A $KMnO_4$ solution being carried out of the scrubber can cause brown stains (manganese dioxide) on automobiles, clothing, etc., in the vicinity of the scrubber.

After the system has been thoroughly checked for leaks, the water should be flushed from the system and the recirculation tank should be filled with fresh water.

The pH should be adjusted to its proper level, and the permanganate should be added to the recirculation tank and made up to the strength determined from the pilot testing. (If no pilot testing was conducted, a 2% solution of permanganate should be made up [1/6 lb/gal] and the solution should be mixed for approximately 15 minutes to ensure that all the chemical has gone into solution.)

Using a Carus Chemical Co. Test Kit, CX-1, the $KMnO_4$ concentration can be very quickly determined. When the solution is at the correct strength, the recirculation pump and fan should be turned on to start the deodorizing process.

Samples of permanganate solution should be taken at different time intervals to determine the actual consumption during the start-up. In addition, constant checks should be made on the pH to determine if there is any fluctuation which may indicate a need for a buffering agent. *The solution level in the recirculation tank should be kept as constant as possible during the trial to accurately determine the permanganate consumption.*

If consumption is extremely high, it may indicate that the first-stage particulate removal system is not functioning properly. Often this particulate has a very high $KMnO_4$ demand; and if there is any carryover between the first-stage and the second-stage, permanganate consumption will be excessive.

If effective odor control is being achieved, constant checks on the chemical concentration should be run for the first several days to stabilize operating conditions. After the most effective concentrations of permanganate and the proper pH have been determined, it will only be necessary to make minor adjustments in the addition of chemicals.

CHEMICAL ADDITIONS

Permanganate can be replenished by two different methods: batch or continuous. In the batch process, a known quantity of $KMnO_4$ can be added daily to raise the concentration above the optimum level. Throughout the course of the day the concentration will decrease due to the oxidation reaction. If there is a minimum $KMnO_4$ concentration below which the odor control efficiency is questionable, batch additions may have to be made during the day to maintain a predetermined minimum concentration.

Permanganate can be added continuously during the operation of the scrubber at a rate equal to consumption, using either a dry feeder or proportioning pump. Selection of the type of feeding system depends on the rate of $KMnO_4$ reduction. If it is low, a proportioning pump can be used to meter in a known volume of $KMnO_4$ solution. If the $KMnO_4$ reduction rate is high, a dry feeder can be engineered into the system to feed a known weight into the solution recirculation line.

CLEANING THE SCRUBBER

As $KMnO_4$ reacts, manganese dioxide (MnO_2) is produced. The MnO_2 will be either in the form of a precipitate or can remain colloidal for some time, depending on the constituents in the scrubbing solution. In either case the packing can eventually become plugged or coated and severely restrict the air flow through the unit.

The manganese dioxide should be removed from the scrubber packing and other internal components on a regular basis, using a solution of a reducing chemical such as sodium bisulfite, preferably at a pH between 3 and 5 using a pH buffer such as citric acid. The cleaning schedule will depend on the amount of $KMnO_4$ used and the type of system.

In a system that does not operate 24 hours a day, a sloping bottom with a drain is recommended. The MnO_2 will settle while the unit is not operating; and before it is started up the next day, the drain should be opened to remove the sludge from the system. In a continuous operation, a filtering system can be installed to remove the MnO_2 while it is running.

Maintenance is involved in a $KMnO_4$ system, and if a schedule of cleaning the scrubber is not practiced, some serious problems can occur and cause downtime and needless expense in removing packing and cleaning it outside the scrubber, or possibly even replacing it.

SUMMARY

Potassium permanganate can be used to control the odors caused by many different manufacturing processes. It is not the answer to all of the problems—as is true of *all* pollution control techniques. If the chemistry is right, and if the proper system is installed, $KMnO_4$ can be used very economically and effectively to control odors. And with proper maintenance of the system, operating problems can be kept to a minimum.

CHAPTER 27

ODOR MODIFICATION

Oscar Bud Lauren

Environmental Odor Consultant
Countryside, Ill. 60525

Odor modification is the process whereby, when two substances of given concentrations are mixed, the resultant odor of the mixture may be far less intense and objectionable than that of the separate components, and may not even be perceptible. This decrease in olfactory intensity (modification) is thus recognized as an abatement of the malodor pollutant. An odor modifier is a specifically designed formulation of aromatic chemicals to achieve interaction of odorous molecules in a gaseous state which will result in the desired decrease of olfactory intensity.

Thomas V. Getchell[1] observed:

> We call substances odors physiologically if, when they are presented to the nose at or above some threshold level, they cause change in the activity of the receptor cells and cells higher in the olfactory pathway. Psychophysically, we call substances odors if we perceive them as an odor, or if we can demonstrate behavioral responses to the substances when they act on the olfactory receptors. As the interaction of the stimulant molecule and the chemoreceptive membrane cannot be monitored on the molecular level, knowledge of the extramental odor world is necessarily based on introspective psychophysical judgements or behavioral responses of experimental animals. *Thus from the human point of view a substance has an odor if one can smell it.*

Professor A. Dravneiks[2] stated:

> In a still rather elusive way the nature and strength of odor are related to some selected molecular or bulk properties of odorants, which presumably are accessible to object measurement.

Dr. H. Stone[3] further states:

> The complex mechanism of man's response to an odorous stimulus is not yet well understood. It is known that both physiological and psychological changes occur. Some of these changes are transitory and others, such as adaptation, are more lasting and influence response in not readily predictable ways.

One of the more important drawbacks is the lack of an exact terminology concerning odor. Degrees of intensity of smell, names of smells, and other classifying details vary from investigator to investigator, and a comparison of values obtained from different laboratories is often a difficult and complex matter. Sensations of sound or sight can be exactly defined because they can be exactly measured, but in the study of olfaction this, at present, has not succeeded.

True, all chemical and physical characteristics of an odor can be measured by standard or specially developed techniques, but this does not mean that any of these measurements allow an organoleptic interpretation. The mental background of the person experiencing an odor supplies the higher critical centers of the brain with the decisive information. A person with a pleasant memory of a particular smell will classify this odor as pleasing, while another person with a different background may take a conflicting view.

Masking is based on the premise that people perceive a mixture of smells as a single odor. Masking, therefore, is nothing other than the attempt to superimpose a pleasant odor on an unpleasant odor, which effort raises the total odor level and creates an overpowering sensation (hopefully more pleasant but, as it often works out, usually objectionable).

For example, a sewage plant operator might try masking the malodor of a sludge lagoon by spraying a vanilla fragrance in the immediate area. Unfortunately, several things might happen:

1. He might find that the sewage odor was so intense that the amount of vanilla required to mask the malodor would be so great that it would be hopelessly uneconomical.
2. The vanilla fragrance could become pronounced enough to prove objectionable to nearby citizens.
3. The sewage and vanilla odors might combine in such a way as to produce a completely new odor which could well prove more obnoxious than the original malodor.

A more subtle and scientific method of odor abatement exists, based on the principle of modification (compensation). This involves lowering the physiological odor intensity of a malodor—mixing it with selected chemical modification gases or vapors. Much work has been accomplished and reported in this area. E. P. Koster[4] in his paper "Intensity in Mixtures of Odorous Substances" related the following:

> Zwaademaker[5] and Backman[6] showed that when two substances of given concentration are mixed in a given ratio the resulting odor of the mixture may be far less intense than that of each of the separate components, and that eventually it may not be perceptible at all. This phenomenon, which is essentially a decrease of olfactory intensity, can be described as compensation (modification).

Koster further states:

> The main conclusions of our own work, in which we investigated the relative intensities of mixtures of a considerable number of pairs of odorous substances, were that compensation (modification) or decrease in intensity occurred frequently, especially in which the mixing ratios of the components were more extreme, such as 1:6 or 1:8. Conclusions: the results of these experiments show that intensity reduction or compensation (modification) occurs in the majority of cases in which two substances are mixed in uneven proportions.

Further, Moncreiff[7] reported:

> Nevertheless, when two air-streams carrying respectfully the unpleasant sweaty butyric acid and pleasant gin-like oil of juniper were mixed in certain proportions and an observer was asked to say when he could smell (not knowing if the air mixture contained either butyric acid or juniper oil or both) the answers were often negative, such as "pleasant neutral smell," "faint, neither recognizable." Another pair that was found was pyridine (rank tobacco) and oil of wintergreen. There is no doubt at all that odor compensation (modification) does take place so that neither of two odorants, present in sufficient quantity to be easily recognized alone, can be recognized in the mixture. What is the significance? A pleasant smell and an unpleasant smell will sometimes neutralize (modify) each other so that neither can be recognized. The only possible explanation is that the compensation (modification) is internal, that it takes place in the central nervous system, in the olfactory nerves, at the nerve junctions or on the cortex. It is an important observation that such internal compensation (modification) may take place in our ordinary life as an odor measurement, but rather to keep the animal as comfortable in his environment as is consistent with the warning function of the smell sense. According as one or the other sensation predominates, either the one or the other smell is noticed, and when they are exactly counterbalanced, then either no precise sensation is experienced or only a very weak impression, indefinite, which was noticeable only by great effort and corresponded with neither of the components.

Further work has uncovered many other modifying substances. Recent studies by Arthur D. Little have reconfirmed the well-established fact that two or more odorous materials may interact to produce a new odor complex.[8]

For example, in roasted coffee beans the following are found:

Diacetyl	$CH_3 \cdot CO \cdot CO \cdot CH_3$
Methyl-ethyl carbinol	$CH_3 \cdot CH_2 \cdot CHOH \cdot CH_3$

Furan	C_4H_4O
Furfuraldehyde (furfural)	$C_4H_3O \cdot CHO$
Furfuryl alcohol	$C_4H_3O \cdot CH_2OH$
Acetaldehyde	$CH_3 \cdot CHO$
Pyridine	C_5H_5N
Hydrogen sulfide	H_2S

Odors may range in a wide variety and there are many instances where a relatively small change of chemical structure swings the verdict of the observer from agreeable to highly objectionable. W. Summer[9] relates the following:

> Even once agreement about the method of odor classification will have been reached, the actual grouping will still cause considerable difficulties, not because of the method involved, but because of the duality (if not plurality) of odor sensations caused by various concentrations of one and the same substance. Such materials may at one degree of concentration cause repulsion whereas at another degree, usually high dilution, [they] will contribute a pleasant, even highly attractive odor. This dualistic effect becomes the more interesting and remarkable when it is applied to give a variety of foods their specific flavor or odor. Indole, in concentration a highly objectionable repulsive fecal odor, is used in dilution to give chocolate its "kick." It also occurs in the scent of orange blossom and jasmine. Hydrogen sulfide at concentrations sufficient to produce harmful physiological effects apparently loses its smell of rotten eggs and produces a pleasant odorous sensation. This is a dangerous characteristic.
>
> Indole is similarly interesting in that it seems to be a combination of two odors, although this is probably a statement which has a psychophysiological, rather than a chemical justification. In low concentrations it smells of jasmine and has a low threshold of perception. In high concentrations it has a strong odor of faeces and a-naphthylamine, and a considerably higher treshold of perception. Thus, only the jasmine scent is perceived when indole is highly dilute.

The understanding and implementation of the concept of modification has led, therefore, to positive solutions for problems of nuisance malodors. Pragmatically speaking, there can be no objectionable odor unless there is someone to smell it and complain. This fact had led to the development of a concept of bathing malodorous air with gaseous odor-modifying chemicals before it reaches the community. This procedure involves setting up a screen of modifying gas between the source of a malodor and the source of complaints. Odoriferous air must therefore pass through the screen and mix with the proper modifier gas. Many of these installations are operating successfully, in varied applications, throughout the country. Waste disposal plants, industrial facilities, feedlots, lagoons, etc., have eliminated nuisance odor complaints with the use of this proven, unique approach.

An alternative method of odor modification was applied in the following instance. A major processor of soya oil products was faced with complaints from the nearby community. The problem was an emission of malodorous vapors from the cooling towers. Samples of the liquid waste material were sent to the odor laboratories for analysis. A prescribed method of converting the liquid into a gaseous state was used with established odor-panel testing techniques and the following conclusions were drawn.[10]

The malodor was described as fatty, glue-like, rendering type. The Odor Recognition Threshold (RT) for the waste material was established at 0.084 ppm in air (0.63 mg waste in a chamber of 256 cf volume containing some 7,500 g of air; volatilization was achieved by spraying 1,260 mg of a 0.05% solution of waste in aerosol form into the chamber). The plant exhaust air fan has an output of 570,000 cfm. The flow is, therefore, 570,000/256 = 2,227 times. The volume of the odor chamber is equal to 1.4 g waste per minute at the RT level. This corresponds to 84 g or 3 oz/hr.

The preferred odor modifier has an established RT of 0.006 ppm, which indicates an odor effectiveness of 0.084/0.006 - 14-fold over the waste material. The odor modification efficiency was determined in a series of tests which gave the following results.

The odor modifier was sprayed into the odor chamber at a level of 0.024 ppm, or four times the RT level (1.8 g of a 0.01% solution = 0.18 mg modifier in 7,500 g air = 0.024 ppm). By simultaneously spraying 0.48 ppm waste into the chamber, the malodor was effectively modified (7.2 g of a 0.05% solution = 3.6 mg waste sprayed into 7,500 g air = 0.48 ppm).

The conclusion is that 0.024 ppm modifier successfully modifies the malodor of 0.48 ppm waste, a ratio of 1:20. This odor modifier chemical is incorporated directly into the liquid waste effluent and allowed to vaporize in the same air stream as the malodorous gas effluent.

Odor modifiers can be applied to contained and uncontained sources of malodor pollutants by one of four methods:[11] (1) direct addition, (2) spraying, (3) metered feeding, and (4) dispensers.

Direct Addition

This method adds the odor modifier directly to the material which is causing the odor problem. It is normally used only in batch processes and is the simplest method of application as no equipment is needed. The odor modifier must, however, be carefully chosen, since it will normally undergo the same processes as the raw material and, therefore, must withstand similar conditions. Such factors as stability in alkaline or acidic conditions, inertness in any chemical reactions which take place during the process, and ability to withstand processing temperatures must all be taken into account. Physical properties such as boiling range and vapor pressure are also important.

Spraying

This method can be adapted to a wide variety of problems and is probably the most useful means of applying odor modification materials.

Hand Spraying

Spraying with any simple form of hand spraying device, and in particular the type of knapsack sprayer frequently used in agriculture, is both effective and convenient for small areas. Such problems as small sludge drying beds, open heaps of kitchen waste, municipal refuse piles, the open tops of tanks containing offensive liquors, etc., are all readily treated by this means. When this method is used, the spray should be coarse in character in order to ensure sufficient retention of the odor modifier on the surfaces of the odor source.

Stack Spraying

Where the malodor is due to the gaseous discharge from an exhaust chimney, the odor modification material must be atomized as finely as possible and injected into the base of the stack. This can be done by means of compressed air or steam, both motivators being equally effective. The choice between them depends largely upon the availability of either service on the site. The odor-modifying solution can be fed to the spray nozzle or nozzles from a pressurized container or a gravity-feed header tank. Capital cost is usually low, and the system can include automatic controls based on time or process cycles, thus coordinating the introduction of the odor modifier into the process.

There are numerous processes in gas, petroleum, chemical, animal byproducts, rubber, and many other industries where this type of spraying equipment provides a suitable means of applying the odor control compound.

Aerial Spraying

This is a specialized system which can be used to control the odors from large open areas of waste material. When this method is adopted, a predetermined number of emission points are set up along the relevant dimension of the odor source, each emission point carrying a special spraying nozzle which provides a fine mist dispersion of the odor control compound. This sets up a "barrier" of the odor modifier which will modify any odors carried away from the affected area. Operation of such a system can be carried out either by liquid pressure or compressed air. Installations of this type are fully instrumented and usually completely automatic in operation, with facilities for individual spray-rate adjustment and sectional isolation provided.

Large sludge beds in sewage works and waste lagoons from processing factories are typical examples where this method has given excellent results. In addition, animal by-products factories, storage warehouses, and the petroleum industry are some other examples where similar results have been obtained with this versatile technique.

Metered Feeding

Where a liquid effluent stream is the cause of malodor, it is often possible to apply a continuous feed of odor-modification materials to the liquid. This technique is particularly suitable where the effluent stream is fed into a waste storage lagoon or open tank.

Due to the very low concentration of odor modifier required, it is important that the feed rate is accurate and reliable. Three types of units can be used for this method: a drip-feed apparatus (which must have a constant head device), a flow inducer, or a metering pump of the reciprocating type. This last is essential when considering the addition of a modifier to pressurized effluent pipework.

The choice between the water-miscible and oil forms of the odor modification material and the concentration used depends on the circumstances of the problem. The water-miscible form will mix with and modify the whole effluent, whereas the oil version will form a film over the surface. The "spreading power" of the oil version can be increased by using kerosene as a diluent before application, but it should be noted that an oil film may not be effective in turbulent conditions.

The economics of this form of treatment must be considered for each individual problem, but in many instances it provides a convenient and simple solution to malodor difficulties.

Examples of metered feeding applications are condenser effluents in animal by-products factories and effluents from chemical factories and gas works.

Dispensers

Where the odor control compound has to be mixed with an air stream, it is sometimes feasible to use some form of dispenser.

Where large volumes of odorous air are concerned, use can be made of small humidifying units of the spinning disc type. The water-miscible version of the appropriate odor modifier is simply added to the humidifier water reservoir by direct addition. Thus the odor modifier is vaporized and carried into the offending gases with the water vapor.

If the air volume to be treated is fairly small, a wick-type dispenser can be used. The wick is placed in the air stream and, at the bottom, dips into a supply of the odor modification material. This saturates the wick and presents a large surface area for evaporation.

Large lagoons can be treated by injection of an additive into the waste streams. One technique, which has been used in sewage treatment plants in the proximity of residential locations involves the generation of an air curtain around the source. A line of stovepipe 6 to 12 inches in diameter is laid along the periphery of the offending units, as close as possible to the offended receptors. Air containing the volatilized odor modifier is injected into the line. The volatilization of odorant may be controlled by passage of air over a pan of odor-modifying chemicals held at the proper temperature.

Another useful technique uses a windmill-type arrangement whose rotor partially dips in a reservoir of odor chemicals and is rotated under the impulse of flowing air. Through the use of properly spaced orifices in the duct, a curtain of material some 20 feet high is formed, which serves as a source of odor-modifying chemicals to modify the offensive odors before they can diffuse into the neighboring vicinity.[12]

One of the advantages of odor control with odor modification techniques is that where equipment is required for its application, capital outlay is usually low. Nevertheless, in order to obtain maximum benefit from these odor chemical compounds some care is necessary in the selection of equipment. In addition, many malodorous operations are temporary in that—due to a surge in the manufacturing process, the time of year, high temperatures, and the fact that the most adequate means of controlling odors are in the design stage—this type of control can be considered emergency relief until some of the other modifications of the process take place.

It has been well established that odor modification materials have their proper place in the abatement of malodor nuisances. There are, however, important considerations to think about before selecting this route.

1. *No attempt should ever be made to modify or otherwise disguise any gas or vapor that may possibly be toxic or harmful.*
2. Odor modification should not be used as a substitute for good housekeeping.
3. Where it is feasible and economical, the malodor should be eliminated or destroyed at the source.
4. Odor-modifying chemicals must conform to the specifications and regulations of EPA, OSHA, and other regulatory agencies, whether on a federal, state, or local level.

5. Odor-modifying materials should effect a lowering of olfactory intensity.

Odor control techniques are sophisticated and specialized, and only trained and experienced personnel should be involved. Admittedly, odor pollution is the least directly harmful to health of the environmental nuisances, but it can seriously affect mental attitudes, destroy the peaceful enjoyment of our homes and neighborhoods, and interfere with proper working conditions.[13]

REFERENCES

1. Getchell, T. V. *Proc. 3rd International Symposium, Olfaction and Taste.* (New York: Rockefeller University Press, 1969).
2. Dravnieks, A. *Theories of Odors and Odor Measurement.* I.I.T. Research Institute, Chicago, Ill. Necmi N. Tanyoiac (1968).
3. Stone, H. *Theories of Odors and Odor Measurement.*
4. Koster, E. P. *Proc. 3rd International Symposium, Olfaction and Taste.* (New York: Rockefeller University Press, 1969).
5. Zwaardemaker, H. "Uber die Proportionen der Geruchs Kompensa tion," *Arch. Anat. Physiol.* (Leipzig), **31**, suppl., bd. 59, p. 70 (1907).
6. Backman, E. L. *Experimentalla Under Sukninger Over Luktsinnets Fysiologi Upsal.* (Forhandl: Lakare Forenings, 1917).
7. Moncreiff, R. *Theories of Odors and Odor Measurement.*
8. Leonardos, G., D. A. Kendal, and E. Rubin. *The Role of Background Odor in Affecting the Perception of Odorant Chemicals.* (Cambridge, Mass.: Arthur D. Little).
9. Summer, W. *Method of Air Deodorization.*
10. Lauren, O. B. "Odor Modification," *Pollution Engineering* (July 1973).
11. Lauren, O. B. and S. E. Elinsky. "State-of-the-Art Report of Odor Control Technology," Air Pollution Control Association (March 7-8, 1974).
12. Lund, H. F., ed. *Industrial Pollution Control Handbook.* (New York: McGraw-Hill, 1971).
13. Lauren, O. B. "Odors and Malodors," 29th Annual Purdue Waste Conference (May 7-9, 1974).
14. Burton, M. *The Sixth Sense of Animals.* (New York: Ballantine Books).
15. Lauren, O. B. "Odor Modification," presented at Purdue University (November 1972).

CHAPTER 28

INDUSTRIAL ODOR CONTROL CASE HISTORIES

Paul N. Cheremisinoff

New Jersey Institute of Technology
Newark, N.J. 07102

Richard A. Young

Pollution Engineering
Barrington, Ill. 60010

The following case histories of industrial odor control installations are a typical cross-section of applications technology utilizing various equipment techniques.

1. LIQUID WASTE INCINERATOR OXIDIZER SYSTEM

The largest packaged and portable liquid waste incinerator ever produced in the world was designed and operated at 50 million Btu's per hour. The rate of firing necessitated bringing waste liquids to the plant test site by tanker truck.

The incinerator is a self-contained process with pumps (hot and cold waste, kerosene), natural/propane gas, temperature controls and recorders, safety bypasses, and flame safeguards–all with compatible materials of construction for the corrosive materials. This self-contained process arrangement allows for testing under actual operating conditions.

The unit tests included particulate, using the U.S. EPA test procedures and test train, hydrocarbon analysis, NO_X, SO_X, HCl, and carbon monoxide.

The liquid waste incinerator is designed to dispose of high Btu wastes (hot), such as solvents (or a combination of hot wastes and oil or kerosene), to use as the fuel to dispose of aqueous low Btu (cold) wastes. This system is used to dispose of waste liquid when conventional wastewater treatment systems cannot be economically or physically adapted for this use. This series of incinerators is available as "packaged" units in sizes from 3.5 million to 70 million Btu/hr.

The unit is designed for operation in the stack mode (as shown in Figure 28-1), *i.e.,* the natural draft or scrubber mode, for removal of toxics or particulate discharges. The scrubber that will be used in this application is a refractory-lined scrubber with a packed column.

The wastes (hot and cold) are injected into the refractory-lined incinerator simultaneously, with all controls and functions monitored from the feed end of the incinerator. A partial list of wastes used in the tests described above included methylene chloride, ethyl acetate, ethyl alcohol, acetonitrile, water, and sodium chloride. Pressure injection was used to feed these waste streams into the system. The incinerator is built with twin guns, each identical in capacity of hot and cold wastes and auxiliary fuel. However, when large suspended solids are in the waste stream, or the waste stream has a high viscosity, the equipment can be converted to steam injection by a change in nozzle. This nozzle change can be made in a matter of minutes without the necessity of shutting down the entire process.

The incinerator has a high turbulence in the firing chamber to assure complete oxidation by intimate contact of air with waste. The turbulence is created by the combustion air that is forced between the outer steel shell, where it is heated by transfer from the inner refractory-lined firing chamber and forced at a tangent to the circumference, into the firing chamber through combustion air tubes at the point where liquid injection takes place.

2. POULTRY PROCESSING ODOR CONTROL

The farmyard smell, a traditional part of the country scene, has acquired economic significance in Britain in recent years.

Not only have poultry units become much larger and more concentrated but towns have expanded and brought into the countryside people whose noses tend to be much more sensitive than those of farmers.

One source of trouble can be a manure drier, particularly if it is the sort that can cope with 100,000 birds or more. Some driers have even been closed down by court order because they were creating an offense in the neighborhood.

The difficulty with poultry manure is that in reducing its moisture content, from perhaps 75 to 10% in the final product, the water driven off into the atmosphere as steam can contain some of those volatile substances which give the manure its characteristic smell.

Figure 28-1. Liquid waste incinerator oxidizer system designed for operation in the stack mode, *i.e.*, natural draft or scrubber mode, for removal of odors, toxics, or particulate discharges.

In the early days of drying it was thought the problem could be overcome simply by lengthening the drier chimney in the hope that this would dissipate the fumes sufficiently. Unfortunately, the gases do not always disperse but are often carried on in a narrow band, according to the direction of the wind. Therefore, a tall chimney may get rid of smell in the vicinity of a drier but it can well create a real nuisance a kilometer or so away.

Another solution that is not always successful is to fit a drier with deodorizing equipment. This is a more difficult operation than it appears because only a few parts of a malodor per million can cause offense.

The British experience is that a more reliable method is to oxidize or burn the volatile substances by installing an afterburner in the drier chimney, its efficiency being dependent on the amount of heat it produces: the hotter the burner, the greater the reduction in the smell. This is because nearly 50 different gases are generated in the drying process, and although some can be oxidized at relatively low temperatures, others cannot be completely destroyed at less than 700°C (1,292°F).

Unfortunately, an afterburner can significantly increase the running costs of a drier. Not only can destroying the obnoxious substances consume more fuel than the actual drying–an increasingly important consideration in these days of high oil prices–but the high temperature to which the exhaust gases are subjected can cause the chimney to corrode.

To reduce the quantity of fuel, various design features can be incorporated, including substances which increase the burning of the odorous materials. But these can be costly and there comes a point when the sophistication of the burner involves more capital cost than it saves in fuel.

One of the latest British driers has overcome this problem by incorporating a heat exchanger which extracts heat from the hot exhaust gases to warm the incoming air to 343°C (650°F) before it enters the furnace.

Not only does this cut running costs by raising the temperature of the incoming air to nearly 70% of that required for drying the manure but it also reduces the temperature of the exhaust gases and therefore the corrosive effects on the chimney.

Trials of this drier over a number of years have shown that obnoxious fumes are destroyed so effectively that the manufacturer–Douglas Rownson Ltd. of Basingstoke in southern England–now virtually offers a guarantee that it is completely nuisance-free.

Another manufacturer–Phoenix Organics Ltd. of Newark in the midlands–claims to have developed a drier that produces so little dust that the afterburner needs to consume only 1¾ gallons of fuel per hour to make the exhaust gases free from smell.

Now that the public nuisance problems have been overcome, sales of manure driers are expected to rise quickly. One manufacturer sold 48

Figure 28-2. One of the most popular manure driers in Britain is this machine from Phoenix Organics. Forty-eight have been sold in the past year, including some for overseas.

during the past year (including some overseas), and as world demand for both fertilizers and protein increases, poultry industries have less excuse to regard the manure as a waste product rather than as a valuable by-product that can make a significant contribution to the profitability of egg production.

3. FOOD PROCESSOR ODOR CONTROL

A ketchup and chili sauce plant of a major food processor had a problem because its processing kettles were venting high rates of superheated steam containing objectionable amounts of vinegar and spice odors. The residential community in which this plant was located had registered strong complaints against acrid acid odors, and the state EPA had ordered abatement.

Because of the sensitive nature of the processing operation, the specifications on the odor and acid emission control equipment were extremely stringent. The requirements were:

1. Scrubbers were to be mounted directly on the kettle stacks, were not to interfere with processing operations at any time, and were to be "fail-safe" with respect to food contamination.
2. The scrubbers were to have a minimum efficiency level of 95% for acetic acid removal from 220°F superheated steam.
3. Water was stipulated as the scrubbing liquor, and because the supply was limited, minimal steam condensation and water usage were desired.
4. Because food-flavor characteristics were affected by operating pressure, a scrubber backpressure of less than 1 in. W.C. was required, and no auxiliary blower could be used.

A number of conventional venturi and packed gas scrubber pilot units had been tested and had failed to meet requirements. The plant was faced with a six-month deadline on a court-ordered shutdown if compliance was not achieved.

The solution proposed the use of a spray scrubber as a prototype installation on a single stack with a guarantee of meeting the acid removal efficiency requirement with water scrubbing. Following installation, the prototype scrubber (the unit at the far right in Figure 28-3) was brought up to design operating conditions, and proved the performance guarantee. Initial pressure drop across the scrubber was in excess of the 1 in. W.C. specification, but modification of the spray system to yield a "push-pull" condensation effect on the steam reduced the backpressure to the 0.7 in. level.

Using the design modifications established in prototype operation, three additional stack scrubbers were installed in July, 1972. Tests on these scrubbers showed that at water usage rates corresponding to about 25% steam condensation, acetic acid removal was substantially 100%, both by mass balance and by odor. Even with an inlet temperature of 60°F and maximum water inlet rates, the exit liquor from the scrubber was invariably at its boiling point, indicating a heat transfer efficiency near 100% and unusually effective contacting.

The spray scrubber water is "reuse" water, taken from bottling line waste cooling and rinse water, which is now collected for scrubbing purposes in a 1,200-gallon storage tank. To minimize the load on the waste treatment facilities, scrubber water is used only during the cooking and steam venting period, and is controlled by solenoid valves operating off each cook cycle control panel. The effluent scrubber liquor was found to be so corrosive and odorous that it became necessary to discharge the liquor through a closed fiberglass-reinforced polyester (FRP) pipe drain system and to seal the plant yard sewer vents.

Figure 28-3. 100% operating efficiencies have been achieved by in-line spray scrubbers in the removal of acetic acid and spice odors from superheated steam exhaust from food processing kettles. Flavor preservation required scrubber operation at back-pressures not exceeding 1 in. W.C. and following prototype (far right) operation, new scrubber design gives 0.7 in. W.C. loss. Installed in mid-1972, the plastic units shown are mounted directly in-stack and have provided total emission control at this ketchup/chili sauce plant of a major food processor.

Despite the high acid and organic content of the scrubber effluent liquor, the existing treatment facilities were found to be adequate to handle the additional load.

Following installation and testing, the scrubbers were approved by the state EPA, and five-year operating permits have been issued.

4. TRUCK TERMINAL ATMOSPHERE MADE SAFER BY WET SCRUBBER

When a wet fume scrubber solves air pollution/health and fire hazards at a chemical tank truck terminal, and does so with low first costs and low operating costs, the operator enjoys a combination of good housekeeping and good economics that is hard to beat.

An FRP wet fume scrubber with Dynel shield uses 320 gpm of 5% caustic solution to process 8,400 cfm of noxious gases and vapors from the tankers and ventilating ductwork. The Dynel shield protects the scrubber from chemical attack when fluoride vapors are scrubbed.

The heart of the system is a 24 in. x 24 in. jet venturi scrubber. The scrubber creates the motivating force to entrain, scrub, and absorb the odors, chemical vapors, and gases. As shown in the schematic, no fans or blowers

Figure 28-4. 12 ft high, 24 in. inlet wet fume scrubber uses 5% caustic solution to remove noxious fumes and vapors from chemical trucks; clean air is vented from stack.

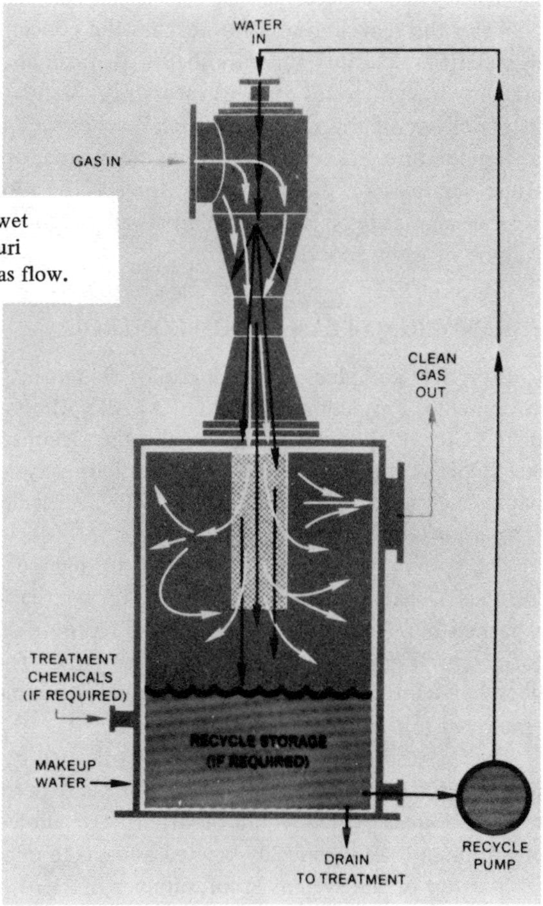

Figure 28-5. Schematic of wet fume scrubber shows venturi configuration and liquid/gas flow.

are required for most applications. Since there are no baffles or moving parts to become plugged or clogged, the scrubber is virtually maintenance-free.

In operation, the sequence is as follows. A 5% caustic solution is pumped from the 6,000-gal storage recirculation tank at 320 gpm and 60 psig to the top of the scrubber. The solution leaves the nozzle as a hollow cone spray, creating a draft that draws the gases and vapors into the moving stream, where they are continuously scrubbed and absorbed. The fume scrubber is also effective for the simultaneous removal of dust and particulate matter. The scrubbing solution is separated from the gas and returned to the recycling tank. Cleaned air that meets state and federal requirements is vented to the atmosphere.

When the recycling solution reaches the concentration limit, it is treated in a clarifier, and the tank is refilled with fresh absorbent. Consequently, not only air but water effluent meets EPA standards.

Efficiency of the scrubber, which reaches 95% and higher, varies according to temperature, concentration of contaminants, operating pressures, etc. Other variables are the capacity of the specific scrubber and the use of separators, clarifiers, and flocculators–all dependent on the nature and volume of gases and vapors.

5. RENDERING PLANT ODOR CONTROL

Complete compliance with the odor abatement regulations of the Environmental Protection Agency (EPA), the Air Pollution Control Division of the City of Cleveland (APCD), and the Occupational Safety and Health Act (OSHA) was assured by a new ventilating system. The system eliminates odors from rendering operations, in which reusable materials are reclaimed from a meat industry by-products.

The system, with a maximum-rated throughput of nearly 108,000 cfm, includes an exhaust fan and ductwork for ventilation of the process areas, a packed crossflow wet scrubber, and a chlorine-acid-caustic additions system for chemical neutralization of offensive odors. The entire system is of fiberglass-reinforced plastic construction for resistance to the chemicals employed in the odor abatement process.

Fresh outdoor air, drawn into the work area by the ventilation system, provides a comfortable environment for plant personnel. Off gases from rendering areas, which would be offensive if allowed to enter the outside environment, are chemically treated and made odor-free.

Start-up of the system in midsummer of 1974 marked accomplishment of a major goal. Much of the total cost–approximately $175,000–was for the structural modifications needed before the ventilation system could be installed on the plant roof.

The pollution problem was approached to meet requirements of EPA, OSHA, and APCD. The company has worked closely with APCD in defining and complying with odor abatement requirements. More than two years ago, the company was given a "good neighbor" award by Cleveland's Ward 2 Civic Club for its efforts toward abatement of air pollution.

A strong factor in decisions to install the system was the three-stage, packed-bed design of the crossflow scrubber. This approach, representing the newest technology in odor abatement systems for rendering operations, was accepted during the past year by the Fats and Proteins Research Foundation (FPRF). It is also the type of system recommended by the National Renderers' Association (NRA).

Figure 28-6. Rooftop view of inlet duct and cross-flow scrubber in the three-stage packed-bed design used to control rendering plant odors. Downtown Cleveland skyline is in the background.

The installation is the largest single unit (of any make) featuring the three-stage, packed-bed approach to odor abatement. Completion time for the system, from order date to start-up, was about six months.

The cross-flow scrubber, of modular design, measures 32 ft from end to end, 22 ft in width, and 11 ft 2 in. in height. All controls and recycling equipment are beneath the plant roof—with no outdoor sump, which would be vulnerable to freezing temperatures.

Intimate mixing of chlorine in the air stream before it enters the scrubber causes oxidation of odoriferous compounds which would not otherwise be removable by normal gas absorption techniques. In the three-stage system, gases pass through a Tellerette packed bed that is irrigated successively by sulfuric acid and caustic.

In the second stage of the scrubber, the air stream is reacted with recycled acid for absorption of odor-causing ammonia- and amine-based compounds as it passes through the packed bed. An integral entrainment separator prevents acid carryover into the third scrubber stage, where a recycled alkaline scrubbing liquid, flowing through the final packed bed, absorbs odorous hydrogen sulfide, mercaptans, lower organic compounds, and excess chlorine.

Automatic controls need no attention beyond normal maintenance. Chemical feeds are shut off immediately in the event of power or fan failure. For added safety, the chlorine system is under vacuum, preventing any escape of the gas if, for any reason, a feed line should be broken.

In business on Cleveland's West Side since 1876, the company processes meat industry by-products into tallow, greases, meat meal, and other materials for use in a wide range of industries. Typical end uses include cosmetics, farm animal foodstuffs, soaps and detergents, plastics, paints, concrete, fertilizer, lubricants, deodorants, and asphalt tile. Some 27 of the company's 55 employees are drivers of trucks which pick up bones, fat, offal, waste, restaurant greases, and other materials from slaughterhouses, stores, restaurants, trim operations, locker plants, and meat wholesalers.

The watchword of the industry is "recycle"—and companies in the rendering business convert 95 to 100% of raw material input into products for further use. That compares very favorably with recycling figures of 5, 20, and 50% for glass bottles, waste paper, and steel products, respectively.

6. IMPINGEMENT TRAY SCRUBBER CLEANS INDUSTRIAL SLUDGE INCINERATOR GASES

American Cyanamid uses an impingement tray scrubber to remove fly ash from combustion gases exhausted from its industrial sludge incinerator at Wallingford, Connecticut. The incinerator burns the excess activated sludge

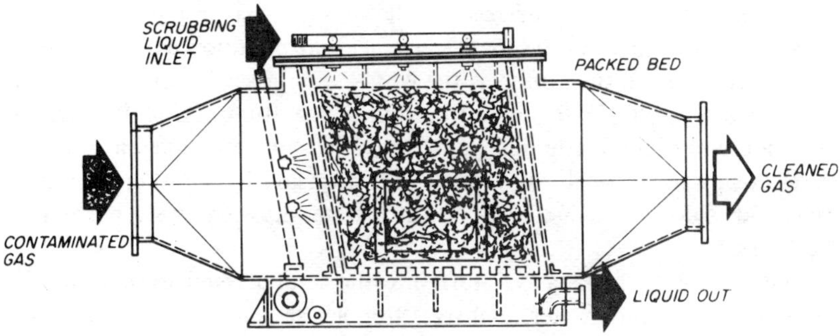

Figure 28-7. Crossflow packed scrubber.

Figure 28-8. Impingement tray scrubber (right) cleans incinerator exhaust gases before discharge to atmosphere at wastewater treatment plant. Industrial sludge incinerator is large unit on left. Combustion quencher is above elbow leading to tray scrubber

from a new $4.5 million water treatment plant. The plant manufactures a wide variety of plastics and resins. The scrubber was installed to enable exhaust gases to meet Connecticut's stringent air pollution control codes.

Prior to entering the scrubber, the combustion gases are quenched from approximately 1,400°F to their saturation temperature of 182°F. Eighty-two thousand cfm of saturated gases are cooled by water spray as they travel downward through the vertical inlet duct and by a curtain of water which flows down the duct wall. Quench water also removes a portion of the fly ash.

Quenched gases make a "U" turn into the scrubber and travel upward to the trays. Quench water discharges through an outlet drain.

The cylindrical scrubber houses three impingement trays. Water flows successively across each tray, starting at the top and continuing through downspouts to the other two. Gases travel upward through tray perforations. Jets, formed as the gases pass through the perforations, churn and agitate the circulating water, providing thorough mixing of gas and water. Ash particles become embedded in droplets of water and are carried away in the water. Small baffle plates, one above each tray perforation, increase turbulence and mixing by deflecting jets into the water. The baffles increase scrubber effectiveness roughtly 25%.

Gases are cooled to approximately 140°F as they pass through the scrubbing water. This condenses water vapor and decreases the chances of formation of a steam plume (which might be mistaken for air pollutants) as the gases reach the atmosphere. Moisture content is reduced about 400% by the cooling action of the scrubbing water.

Quench water and scrubber water return to the water treatment plant, where solids are removed. Treated effluent is used for both quenching and scrubbing.

Ash particles entering the scrubber range downward to one micron in size. Scrubber exhaust gases contain less than 0.016 gr/scf of particulate matter. The scrubber is constructed of 316 ELC stainless steel for corrosion resistance.

CHAPTER 29

ATMOSPHERIC TRANSPORT, DISPERSION, AND STACK DESIGN CONSIDERATIONS

Paul N. Cheremisinoff

New Jersey Institute of Technology
Newark, N.J. 07102

The transport and diffusion of odors in the atmosphere and the design of stacks for safe effluent release and odor dispersion are dependent on many factors. Meteorological effects such as wind and turbulence are highly significant. Local topography, including both man-made obstructions and natural features, also sets criteria for systems and stack design. Physical and chemical properties of gaseous effluents must be carefully considered, and adjacent stacks accounted for, or a multiple-source problem is created. Certain limiting factors may prohibit extremely tall stacks. Local ordinances often limit heights, as do economic considerations.

An accurate analytical expression for all these factors does not exist at this time. By assuming some of these factors to be negligible, equations have been used to predict atmospheric dispersion. Experimental models are often used to supplement calculations. Computer programs have been created to model proposed stacks and the conditions they will face. A bibliography is offered at the end of this chapter to enable the reader to refer to empirical derivations of some of the mathematical expressions that have been formulated by a host of able investigators in this highly speculative area.

METEOROLOGY

Some understanding of basic meteorology is a fundamental requirement of a successful solution in atmospheric odor dispersion. Atmospheric transport of pollutants is largely controlled by weather factors—the most important being wind.

Wind is large-scale pressure variations over the earth, which produce air flows. Wind distribution in an area is determined by the intensity of these pressure variations. Characteristic patterns can be developed for the movement of these pressure systems and diurnal heating effects in a given area. These patterns are plotted on polar diagrams called *wind roses.* Variations of seasonal wind speed and direction are readily apparent in this type of graph. The *prevailing wind* is defined as the predominant wind direction in a wind rose. If wind direction is constant over long time periods, it is called a *persistent wind.*

Horizontal wind speed and direction vary with height. At high altitudes, wind flow is primarily caused by the horizontal pressure gradient known as *gradient winds.* Friction effects nearer the ground slow the wind speed and change wind direction. *Angular change* of wind with height depends on a variety of factors, the most important being the thermal stability of the atmosphere and underlying terrain changes. Angular change is usually lower during the day than at night and lower over smooth surfaces than rough. The *mixing height* is the height above the ground surface at which relatively vigorous vertical mixing occurs. It is important that this height be reached by a stack plume if good dispersion is to be achieved. Mixing height also varies with diurnal periods, ground surface, and precipitation.

Several common terms are used in connection with winds. *Drainage winds* occur when a marked slope exists on the land. Winds developing close to the surface are cooled by radiation at night and flow down with gravity. *Valley winds* are caused by the nighttime flow of drainage winds toward the valley center, from which such flows tend toward the lower end of the valley. During the daytime the flows tend to travel up the center of the valley.

Land and sea breezes are caused by temperature differences between adjacent bodies of water and land. During the day the air over the land is warmer than that over the sea, and breezes circulate toward the land. At night the reverse happens, due to the warmer surface of the sea; and hence, temperature will determine prevailing wind directions.

Turbulence is local fluctuation in the wind flow. This is an important parameter since it has a significant effect on dispersion. Turbulence can be separated into two classes: mechanical and convective. Mechanical turbulence is caused by the influence of buildings or other objects on the surface which obstruct natural air flow. Turbulence increases with the irregularity or roughness of terrain. Convective turbulence is due to a difference in temperature between the surface and overlying air streams. These differences can be caused by a large variety of factors, including the radiant heating of buildings, water masses, or land features.

The *vertical temperature gradient* is an indicator of convective turbulence. Temperature gradients can be divided into three general categories: neutral, stable, and unstable. Mechanical and thermal influences generally occur simultaneously in differing ratios. One affects the other, and it is often difficult to distinguish the type of turbulence. In general, a neutral temperature gradient indicates primarily mechanical turbulence, while an unstable gradient adds convective turbulence. A stable gradient suppresses mechanical turbulence.

Stability is the ability of the atmosphere to resist vertical motion or to suppress or enhance turbulence. It is related to both wind shear and the temperature gradient. Measurement of stability is accomplished through the temperature gradient. As a standard, a small volume of air is considered to rise within the atmosphere. Assuming no heat exchange between this air and the environment, the rate at which the air cools is defined as the dry adiabatic lapse rate (-1°C/100 m).

Potential temperature is the temperature this small volume of air would have if its pressure were raised adiabatically to 1,000 microbars. A potential temperature increase with height implies stability. The actual vertical temperature distribution is called the environmental lapse rate.

Environmental lapse rates can be divided into five types. *Super-adiabatic* conditions exist on days of strong solar heating. These conditions favor instability and turbulence. When the environmental lapse rate is nearly identical to the dry adiabatic lapse rate, a *neutral* condition exists. A more gradual temperature decrease with height is known as *subadiabatic.* If air temperature is constant with height, *isothermal* conditions exist. When air temperature increases with height inside a stable atmospheric layer, dispersion is restricted. This condition, known as an *inversion*, can exist at surface level or, in the case of elevated inversions, as high as several hundred meters. Inversions can be caused by a number of factors, but they always result in suppression of vertical motion—which will trap a smoke plume or stack exhaust and have the effect of increasing ground-surface concentration of odors and pollutants.

PLUME TYPES

The geometrical configuration of plume dispersion in the atmosphere has been related to the vertical temperature gradient. Stack plumes may be classified into six types, each of which is associated with an environmental lapse rate, as shown in Figure 29-1.

In designing an exhaust system or stack the engineer should be familiar with the conditions causing each plume type. If particular conditions can be anticipated in a stack location, the design must include provisions to

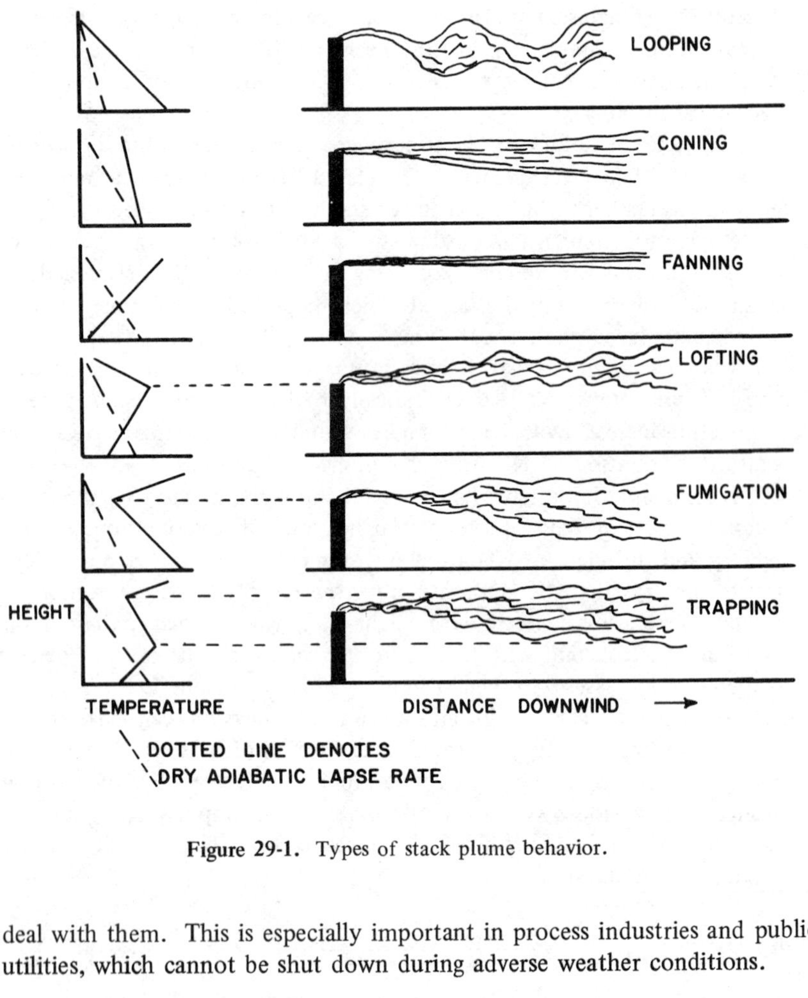

Figure 29-1. Types of stack plume behavior.

deal with them. This is especially important in process industries and public utilities, which cannot be shut down during adverse weather conditions.

EFFECT OF TOPOGRAPHY

Surface irregularities range from isolated projections or depressions to dense successions of features. Collectively, these features are called *ground roughness* and have a large effect on local turbulence. These irregularities can be caused by buildings or natural terrain, which can exhibit mechanical thermal or combined influences over air stability.

Buildings

When a moving air stream approaches an obstacle, air is displaced vertically and laterally. This air displaces adjacent streams of air. The mean velocity increases upward from the ground, rapidly at first but slower at higher elevations. This is called *streamline distortion* and is exhibited in Figure 29-2.

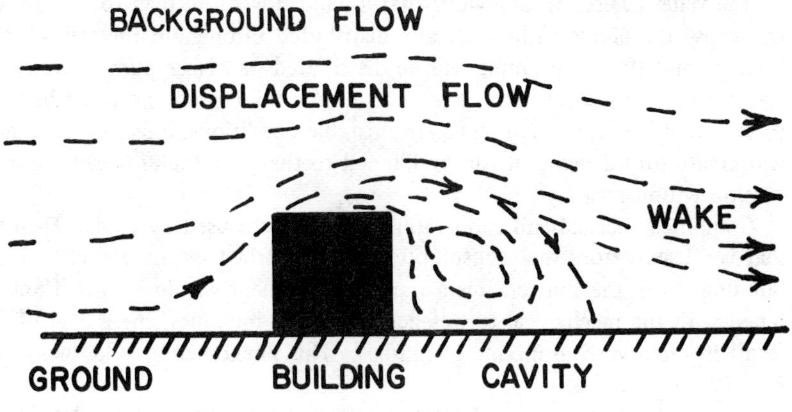

Figure 29-2. Mean flow around a building.

This flow disturbance produces a highly turbulent wake. Within the upper portion of the wake, directly behind the building, a cavity is formed. Within this cavity is a circular flow, which will trap effluents. Cavity size is influenced by building shape and wind directions, but is principally a function of the frontal area of the building presented to the wind.

A building's effect on plume dispersion will vary with stack location and height. A plume from a tall stack will clear the displacement zone but may not clear the wake, depending on atmospheric turbulence. A short stack will cause the plume to be caught in the displacement zone, cavity, and wake. A medium stack, whose plume centerline is in the vicinity of the cavity boundary, will cause the plume to enter the displacement zone and the wake, and it may enter the cavity. Effluent from a tall stack will clear the wake, while effluent from a medium stack will clear the cavity. A stack close to the roof will cause the plume to be distributed throughout the cavity and the wake. Any downwind stack will not be affected by the cavity, but lower stacks will disperse plumes into the wake.

Natural Topographic Effects

Temperature differences between water bodies and the surrounding land have been mentioned. Diurnal variations in temperature cause changes in wind direction and inversions.

The direction of wind flow in a valley will determine dispersion. A wind blowing parallel to a valley will result in normal dispersion, confined only by the valley walls. It is obvious that a wind flow from high to low ground will cause the plume to be vertically distributed throughout the valley; for reverse wind flow, the plume will be distributed on high ground. Under stable conditions, stack plumes tend to follow the terrain and resist being forced over hills and mountains. In unstable conditions, hills or mountains will create turbulence that can be likened to that of a building causing dispersion within a cavity.

Combined thermal and topographic effects are caused by cities. Heat that is released from fuel consumption and absorbed by streets and buildings from the sun, creates a warm environment within a city. Combined with the mechanical turbulence caused by buildings, these heat effects result in rapid vertical mixing of plumes. This phenomenon is known as a *heat island.*

There are no accurate analytical equations for predicting topographic effects. Estimates can be made using the general rules mentioned, but differences in topographical elements prove this method is only approximate. Experimental data, as provided by scale-model wind tunnel tests, give the most complete answers to the question of topographical effects.

EFFECTS OF INITIAL PARAMETERS

The two initial source parameters, stack effluent exit velocity and stack effluent temperature, are extremely important to good dispersion. They can be manipulated (after stack construction) to correct dispersion much more easily than other alternatives.

The exit velocity of a stack effluent must be high enough to prevent aerodynamic downwash–the tendency of a gas to creep down the stack. This speed will depend on local atmospheric conditions and can be determined in wind tunnel tests. Additional elevation of the plume is also supplied by the momentum of the exit gas. It is important to provide additional elevation before atmospheric turbulence becomes controlling. The stack gas is usually much warmer than ambient air, thereby allowing its reduced density to cause a buoyancy effect. The gas will rise until it is cooled by the atmosphere. Pollution control devices, such as wet scrubbers, can cool the gas and restrict this rise.

MULTIPLE-SOURCE EMISSIONS

The most obvious effect of multiple-source emissions is an increase in pollution. Each source alone may not present a problem, but combined effects can result in undesirably high pollutant levels. Legal problems arise in setting responsibility for corrective measures. Before a stack is designed, consideration should be given to possible combined effects from more than one source, as well as possible interaction between pollutants of differing chemical composition.

When two or more stacks are closely grouped, they tend to influence each other. Their plumes combine and will rise higher, but not as high as the plume of one stack that replaces them all. There are no accurate analytical equations for evaluating plume rise or the concentration of effluent from combined sources, but an approximation is often used for a very rough estimate:

$$X_{max(N)} = X_{max(1)} N^{0.8}$$

where

$X_{max(1)}$ = maximum concentration for a single stack
N = number of stacks

Multiple stacks are often in line, and when wind is unidirectional, the stack plumes will rise their highest. This can be explained by noting that the lifting factor will be at a maximum, since all stack plumes will merge. This effect is heightened when a line of stacks is placed on a long building that is in line with the wind.

EFFLUENT CONSIDERATIONS

The type of emission or odor that is emitted into the atmosphere requires consideration of the process source. Process sources can be divided into two categories: *batch* and *continuous.* Batch processes may be shut down under extremely bad weather conditions with little equipment damage. Continuous processes, as exemplified by electric power generation, cannot have an interruption of operations. The sensitivity of the local area to the effluent is another important factor. Local annoyance level may be such that the additional expense necessary for better dispersion may be warranted.

The physical properties of the effluent gas should also be considered. Stack effluent may be composed of gases or a combination of gases and particulates which require special treatment. This treatment has a pronounced effect on design. The gas density will have an effect on buoyancy rise. The density is dependent on the temperature of the exit gas. Care must be taken that the temperature of the gas does not fall below the dew point of the gas. This could result in serious corrosion problems within the stack or precipitation problems in the locality.

Chemical properties must also be considered. Many effluents will react in the atmosphere to cause secondary pollution problems. An excellent example of this is oxidation of sulfur dioxide to sulfur trioxide to sulfuric acid in the atmosphere. Particles drop out of the air onto buildings and cars. Deposition or settling of particles could be another problem area.

PLUME CHARACTERISTICS

Effective Emission Heights

Most analytical methods use a two-step approach. The first step calculates the height of the plume above the surface. Step two estimates the diffusion of gas and deposition of particulates. These equations are very limited, in that they treat only a single horizontal straight plume. Topography and atmospheric stability are not taken into account. The diffusion equations assume a point source of the plume which cannot be applied for small downwind distances. While these equations are of questionable accuracy, they provide a rough initial estimate of the stack plume.

Plume dispersion is assumed to begin at what is called the *effective stack height.* This height is the sum of the physical stack height and the plume rise after emission. The latter term is the sum of the rises created by the buoyancy factor and momentum effect.

The equation used to calculate the plume rise can be best determined under actual conditions. If the plume is in a turbulent wake of the stack or buildings near the stack, the exiting gases will be caught in an aerodynamic downwash toward the ground. If plumes are free from the aforementioned turbulent zones, meteorological factors influence the rise. These emission factors are exiting stack velocity (v_s), effluent gas temperature (T_s), and stack diameter (d). Weather factors that affect plume rise include wind speed (u), ambient air temperature (T_a), and the shear of wind speed with height and atmospheric stability.

A number of investigators have attempted to correlate formulas for computing effective stack height on an empirical basis. These include Davidson-Bryant (1949), Bosanquet-Cary-Halton (1950), Holland (1953), Bosanquet (1957), and Moses and Strom (1964). The Holland equation is

$$\Delta H = \frac{v_s d}{u} (1.5 + 2.68 \times 10^{-3} p \frac{T_s - T_a}{T_s}) \quad (2.9)$$

where

ΔH	=	the rise of the plume above the stack, m
v_s	=	gas exit velocity of the stack, m sec^{-1}
d	=	inside diameter of the stack, m
u	=	wind speed, m sec^{-1}
p	=	atmospheric pressure, μb
T_s	=	stack gas temperature, $^{\circ}K$
T_a	=	air temperature, $^{\circ}K$
2.68×10^{-3}	=	constant, having units $\mu b^{-1} m^{-1}$

Holland concluded that a value of 0.8 to 0.9 times the ΔH from the equation should be used for stable conditions and 1.1 to 1.2 times ΔH for unstable conditions.

Much theoretical work has been done on the diffusion of stack gases. For short-term dispersion of stack gas from a continuous source, there are only two main parameters: wind speed and turbulence. The wind acts directly to separate the gas molecules, while the turbulence mixes the pollutant with the ambient air. The turbulence is dependent on stability and topography. Since there are many possible combinations of these factors, no one method is completely accurate. The dispersion equations usually assume a Gaussian distribution of the plume.

In recent years, computerized models have been developed that are based on many of these equations. With the use of computers, many additional data on weather and topography can be included in the calculations. Even with this additional information, these programs still suffer from the shortcomings of the analytical equations.

Adequate weather and topographical data cannot be supplied since knowledge of these important parameters is limited. However, computer programs, coupled with experimental observations, provide a valuable tool in plume prediction and diffusion possibilities in the atmosphere.

EXPERIMENTAL PROCEDURES

For a specific air pollution problem, the available analytical methods may often be inadequate. Two experimental methods, which have been widely used, are full-scale field experiments and scale-model wind tunnel tests. In a field experiment, the actual conditions of stack operation, topography, and atmosphere are used. This may often be impractical in cases where buildings are in the design stage, or when infrequent atmospheric conditions are to be included in the design. Because of the cost involved, limited field experiments are often used in conjunction with other data. Wind tunnel tests may be advisable when dealing with a limited time factor, since conditions can be changed as desired.

In field experiments a desirable tracer is released under conditions of stack operation. The tracer should be easily identifiable, and easily dispersed, and should closely simulate atmospheric motions. Fluorescent particles and water-soluble fluorescent dyes are commonly used. At times, a gas is necessary to simulate the atmospheric effect on small particles. Smoke has been used primarily as a visual tracer, and nuclear techniques have also been developed.

The primary measurements taken of the atmosphere are those that are most important in dispersion: wind speed, wind direction, and wind turbulence. Surface-wind direction is measured by vanes, while wind direction aloft is measured by different types of balloons. Wind speed is also measured at ground level and aloft. An anemometer is used at ground level, while a balloon is tracked to obtain the speed aloft. Wind turbulence is estimated indirectly from the environmental lapse rate or directly through vanes and anemometers.

Secondary measurements include visibility, humidity, precipitation, and solar radiation. The equipment necessary to determine the concentration of effluent will depend on the tracer used. The ability to carry out wind tunnel tests on a scale model will depend on the availability of data. Detailed information is needed on plant configuration, topography, meteorological characteristics, and the sensitivity of polluted areas. Meteorological data are taken in the same manner as in the previous section. In the development of a new plant it is often advisable to obtain data on the tolerances of these items in the area sensitive to the pollutant. Plant blueprints and terrain maps should provide all necessary information for construction of the scale model.

Wind speed and direction are easily modeled in the wind tunnel. Air density and thermal and convective turbulence can also be scaled down. Usually the three main stack parameters (stack height and the velocity and temperature of the exit gas) are varied for a fixed set of conditions. In this manner an optimum design can be determined. Particulates can also be treated by this method.

The actual data to be represented in the wind tunnel must be scaled down to the model. This, and "scale up" of the optimization study, is done with equations derived through dimensional analysis.

Wind tunnel experiments can be valuable in designing modifications to existing plants. The addition of nozzles or stacks can be studied, as well as changes in operating conditions. As more experience is developed on the effect of building shape and similar factors, it may be possible to modify the theoretical analysis to account for these factors. Wind tunnel tests have been effective in investigating atmospheric dispersal of gaseous wastes, which do not lend themselves to analytical solutions.

GENERAL RULES INFLUENCING STACK DESIGN

Over the years, several rules of thumb in stack design have been developed from experience. These are general guidelines, and most concern stack height. The most widely known of these rules calls for stacks to be 2.5 times higher than the tallest building within 20 stack lengths. At this height the aerodynamic effects of adjacent buildings are no longer a factor in dispersion. In flat country with few large buildings, this factor may be reduced to 1.5. Local laws often govern the minimum height of stacks.

A few other general rules should be kept in mind. A gas ejection speed lower than the wind speed results in downwashing. It has been found that a gas exit speed of 60 fps is sufficiently high to overcome most downwashing tendencies. Through experience it has also become apparent that any structure added near the top of an otherwise plain circular stack will be detrimental, and a square stack will be inferior to a circular stack. Exit stack velocities as high as 100 fps are also recommended in areas or cases which may be affected by severe problems.

DESIGN FACTORS FOR STACKS

Ground-level concentration of pollutants can be reduced through modifications of design features in both new and installed stacks. Building height is an important factor because of its effect on turbulence, and a building's shape may be just as important as height. The influence of a building's shape will be determined by wind direction. For example, a long narrow building with a large surface "normal" to the wind will have a more adverse effect on dispersion than a square building of the same height. It may be possible to favorably modify an existing building, but this would usually require costly additions. Wind tunnel tests can be used to test the effect of any suggested modifications.

The most obvious method of decreasing ground-level concentration is by raising the stack height. In general, ground-level concentration is found to vary with the inverse square of the plume elevation. This assumption is based on uniform turbulence. Thus additional factors, such as buildings, will significantly increase this equation. While an increase in physical stack height may seem an ideal solution, this is often not the case, as will later be discussed.

An increase in gas exit velocity will increase the momentum of the gas, thereby raising the effective stack height. This can be done by increasing the forced draft equipment, if practical, or by decreasing the exit area of the stack. In a new stack, a gradual decrease in stack diameter can be designed into the stack. In existing stacks, a nozzle can be placed on the top of the stack. Care must be taken in nozzle design so that the reduction is not too sudden. The effect of the nozzle is also controlled by the effluent. Again, wind tunnel tests can be used to test designs.

A problem is created by processes with varying stack emission rates. In an effluent with large momentum and buoyancy effects, reduction in emission rates will cause greater ground-level concentration. A stack designed for a full load will not be satisfactory for a partial load. Special devices which adjust the exit area are used so that the full-load emission rate can be reached with a partial load.

The effluent gas may be physically modified to create lower concentrations at ground level. Heat can be added to a gas to increase its buoyancy. Dilution at effluent temperature may be desirable to increase the effluent exit speed. If air is added at ambient temperature, it may have little effect on ground-level concentration because it will lower the density and temperature of the effluent. Because it is costly, heat addition may be reserved for especially poor meteorological conditions.

Pollutants may be removed from the exit gas before emission. Before installing add-on control devices, effects on plume characteristics should be considered. An example of possible adverse effects on atmospheric dispersion is wet scrubbing to remove sulfur dioxide. Scrubbing will reduce effluent temperatures and introduce water droplets. Evaporation of these droplets, as the plume rises, causes further cooling of the gas, which becomes denser than the ambient air. The heavier gas will fall to the ground, carrying any remaining pollutants.

Combining two or more adjacent stacks is a potential solution to a multiple-source problem. The resultant stack will be larger and, at the same ejection velocity, will create a higher plume.

STACK HEIGHT LIMITATIONS

There are many limitations to the physical height of a stack. The most common of these are economic restrictions. Construction and material costs are high for tall stacks. Another factor may be height limitations due to air traffic in the area. Runway approaches have inclined limit lines, increasing from the end of the runways.

Wind pressure, another important factor in stack design, will increase with air density and the square of the wind speed. At higher altitudes, wind speed will increase, and stacks must therefore be designed for a higher lateral wind speed.

Wind forces also form breakaway eddies around stacks. These eddies form and break away with a regular periodicity. If this frequency corresponds to the natural vibration frequency of the stack, a resonant vibration is created. Without strong structural damping, the vibration may reach critical levels. This phenomenon occurs only in steel stacks.

CONCLUDING REMARKS

The design of stacks is a complicated matter, which requires knowledge of the factors affecting dispersion. It can be seen that much depends on meteorological conditions in the atmosphere. A thorough understanding of meteorology is desirable, as well as long-term weather data, for design. If possible, a form of meteorological control should be considered. This does not necessarily mean suspension of operation but rather, a scaling-down during adverse conditions, or dispersion of pollutants accumulated during adverse conditions when good dispersion is possible.

Analytical expressions for plume design are limited. They provide guidelines, but caution should be used in their application. Experimental methods, particularly wind tunnel tests, hold promise as they may provide additional data to be eventually used in conjunction with computer modeling.

Effluent considerations are also very important. Aside from necessary physical modifications to a stack, an effluent may necessitate the addition of devices to remove objectionable concentrations. Consideration must be given to the effect these devices will have on plume dispersion.

Many factors are involved in dispersion, and a large capital investment may be necessary for stack construction. An optimization study should therefore be made before a final stack design is accepted. The engineer must consider all present conditions and must account for future factors which will influence atmospheric dispersion. The designer should be familiar with all assumptions made before a method, analytical or experimental, is used to assist his design.

REFERENCES

1. Bosanquet, C. H., W. F. Carey, and E. M. Halton. "Dust from Chimney Stacks," *Proc. Inst. Mech. Eng.,* **162**, 355-367 (1950).
2. Bosanquet, C. H. "The Rise of a Hot Waste Gas Plume," *J. Inst. Fuel,* **30** (197), 322-328 (1957).
3. Busch, N. "Weather and Climate Factors in Industrial Site Evaluation with Respect to Air Pollution," paper presented at Kem-Tek 2 Congress, Copenhagen (November 4, 1971).
4. Davidson, W. F. "The Dispersion and Spreading of Gases and Dust from Chimneys," *Trans. Conf. on Ind. Wastes,* 14th Annual Meeting of Industrial Hygiene Foundation of America, 38-55 (1949).
5. Halitsky, J. *Wind Tunnel Model Test of Exhaust Gas Recirculation at the NIH Clinical Center,* Rep. No. 785.1, New York Univ. (1961).
6. Halitsky, J. "Diffusion of Vented Gas around Buildings," *J. Air Poll. Control Assoc.,* **12** (2), 74-80 (1962).

7. Halitsky, J. *Gas Diffusion Near Buildings: Theoretical Concepts and Wind Tunnel Model Experiments with Prismatic Building Shapes,* Geophysical Sciences Lab Report No. 63-3, New York Univ. (1963).
8. Halitsky, J. "Gas Diffusion Near Buildings," *ASHRE Trans.,* **69**, 464-484 (1963).
9. Hawkins, J. E. and G. Nonhebel. "Chimneys and the Dispersal of Smoke," *J. Inst. Fuel,* **28**, 530-546 (1955).
10. Holland, J. Z. *A Meteorological Survey of the Oak Ridge Area,* 554-559 Atomic Energy Comm. Report ORO-99, Washington, D.C. (1953).
11. Holzworth, G. C. "Estimates of Mean Maximum Mixing Depth in the Contiguous United States," *Monthly Weather Rev.,* **92**, 235-242 (1964).
12. Hosler, C. R. "Low Level Inversion Frequency in the Contiguous United States," *Monthly Weather Rev.,* **89**, 319-339 (1961).
13. Moses, H. and G. H. Strom. "A Comparison of Observed Plume Rises with Values Obtained from Well-Known Formulas," *J. Air Poll. Control Assoc.,* **11** (10), 455-66 (1961).
14. Moses, H., G. H. Strom, and J. E. Carson. "Effects of Meteorological and Engineering Factors on Stack Plume Rise," *Nuclear Safety,* **6** (1), 1-19 (1961).
15. Scorer, R. S. "The Behaviour of Plumes," *Int. J. Air Poll.,* **1**, 198-220 (1959).
16. Scorer, R. *Air Pollution.* (Oxford: Pergamon Press, 1968).
17. Sherlock, R. H. and E. J. Lesher. "Role of Chimney Design in Dispersion of Waste Gases," *Air Repair,* **4** (2), 1-10 (1959).
18. Smith, M. *Recommended Guide for the Prediction of Dispersion of Airborne Effluents,* ASME (1968).
19. Stern, A. C. *Air Pollution,* 2nd ed. (New York: Academic Press, 1968).
20. Strom, G. H. "Wind Tunnel Scale Model Studies of Air Pollution from Industrial Plants," *Ind. Wastes,* (Sept.-Oct. 1955, Nov.-Dec. 1955, Jan.-Feb. 1956).
21. Strom, G. H., M. Hackman, and E. J. Kaplin. "Atmospheric Dispersal of Industrial Stack Gases Determined by Concentration Measurements in Scale Model Wind Tunnel Experiments," *J. Air Poll. Control Assoc.,* 7 (3), 198-203 (1957).
22. Tadmor, J. "Consideration of Deposition of Pollutants in the Design of a Stack Height," *Atmospheric Environment,* **5**, 473-482 (1971).
23. Turner, D. B. *Workbook of Atmospheric Dispersion Estimates,* U.S. Public Health Service Publication 999-AP-26 (1970).

CHAPTER 30

PLANT SITE SELECTION FACTORS

Nicholas P. Cheremisinoff

Clarkson College of Technology
Potsdam, N.Y. 13676

Polluted air is a complex mixture of gases and a disperse phase of chemical compounds which may be odorous. Odor is always mixed with air, and usually in small concentrations. Many gases are odorless because they are not appreciably volatile or not capable of exciting the olefactory tract.

The object of odor control is not, as commonly supposed, to eliminate an odor by chemical action or other means. Solid and liquid particulates which are undergoing chemical changes or evaporating, are the sources of odor. Therefore, the guiding principle in the control of industrial odors is the removal of these solid or liquid sources.

The severity of odor pollution caused by a single industrial plant depends upon both the volume of production and its location. Ideally, plants should be far from residential areas because of the air pollution generated by smoke, odor, and toxic gases. However, it is also ideal for employees to live as close as possible to the plant to reduce transportation problems. Dilemmas such as these are difficult to resolve in selecting the best plant site. Additionally, numerous other factors such as meteorological, topographic, and orographic data must also be considered.

Specifically, planners should consider the following factors affecting air quality:

1. Specific pollutants produced, and their quality and quantity
2. Impact of the different pollutants on the surrounding environment
3. Federal, state, and local pollution control regulations
4. Available methods for the control of these emissions

5. Climatic conditions–*i.e.*, wind direction and speed variability, etc.
6. Effects of air pollution emissions from other sources on the proposed plant
7. Location with respect to residential and employment areas.

Initial decisions as to location are usually based on factors related to available utilities and ease of travel to and from the plant. In general, the size of the plant bears a strong relationship to basic materials, services, transportation, and labor.

The selection of a new site is a difficult and tedious job involving countless facts and figures, which are usually kept confidential during the initial studies. For expediency, many large companies establish a bank of air quality information that is gathered and administrated by a separate department.

AIR QUALITY INFORMATION BANKS

It is the function of this department to provide data on state and local air quality standards to planners and be available to various state and local organizations that promote industrial development. An engineer may be assigned within the group to act as a liaison between the public and industry. Public relations is an extremely important area that requires attention and can become a major deciding factor in final plant-site location. Public organizations must be assured that contaminated air will be treated properly, and that whatever odors remain will be sufficiently diluted before being discharged to the atmosphere.

The organization should be composed of members from various professions–engineers, chemists, statisticians, and economists. Other members might include a cross-section of employees and residents from various age groups for the purpose of getting citizen feedback.

PREPARING STANDARDS OF ODOR AND AIR POLLUTION

Air quality standards should be established so that accurate and meaningful evaluation of local information can be performed by the data organization. Terms related to odor classification, for example, require both qualitative and quantitative description. One of the standard methods for describing odor intensity should be adopted. Table 30-1 shows a common classification.

Old standards, such as Ringelmann numbers for air pollution, are becoming obsolete. Table 30-2 is one of the newer measures of air pollution. This type of index is useful for regional air quality management because it

Table 30-1. Odor Indices

Odor Index	Concentration of Odor (lb/cu ft)	Type Odor
0	10^{-3}	Imperceptible
1	10^{-2}	Threshold
2	10^{-1}	Noticeable
3	1	Strong
4	10	Overpowering
5	10^{2}	Dangerous, fatal

includes several parameters that present an overall pollution level for an area. Such an index does not, however, provide a good measure for individual plant sources. But by employing mathematical diffusion models, estimates of emissions from single sources can be obtained and related to the maintenance of the desired regional air quality.

All sources capable of generating odors should be specifically examined. In addition, the following factors should be evaluated:

1. Type of processes involved
2. Duration of various processes
3. Methods employed for the disposal of solid and liquid wastes, gases, and vapors
4. Methods employed for controlling gaseous and vaporous wastes
5. Determination of the specific particles causing odors
6. Maintenance of odor-controlling methods
7. Magnitude of the odor problem.

Table 30-2. Air Pollution Index
(Standard used by New York City)

Index = Carbon Monoxide x 1 +
Sulfur Dioxide x 5 +
Ammonia x 10 +
Aldehydes x 10 +
Dust Count x 20

Units–gas, ppm
dust, million particles per cu ft

PHYSICAL FACTORS INVOLVED IN PLANT SITE SELECTION

The following factors should be investigated when considering any site for locating a plant:

1. Terrain
2. Seasonal changes
3. Temperature variations
4. Variations in wind and wind direction
5. Vegetation
6. Presence and location of dumps
7. Open burning
8. Nature and location of other industrial activities
9. Nature and location of highway, rail, and water traffic.

Some of these factors cannot be fully evaluated. For example, only general conclusions can be made about climatic conditions. Weather conditions such as temperature, rain, wind, etc., greatly affect the distribution of airborne particulates, but it is difficult to exactly predict how they will affect emissions from a new plant.

Particulates can present a number of problems specifically related to vegetation growth. In general, three types of injuries occur to the leaves of plants:

1. Leaf tissue collapse
2. Color changes (chlorosis)
3. Alterations in growth.

The effect a manufacturing facility may have on surrounding vegetation is a major factor of consideration. Land use defines the type of vegetation that can exist. Farmers, forest preserves, nurseries, and homeowners may suffer sufficiently from a new industrial activity to bring suit to have the court enjoin the manufacturer from further production.

A detailed study should be conducted to include genetic susceptibility and time-concentration relationships for vegetation and pollutants. Again, climatic conditions must be carefully examined. Factors such as light intensity, temperature, carbon dioxide concentration, humidity, and wind must be examined in relation to conditions prior to exposure.

LEGAL ASPECTS

Planners should thoroughly investigate all local ordinances applicable to the plant processes, as well as legal provisions related to wildlife conservation and the common-law rights of citizens.

Most laws pertaining to odors classify the latter as nuisances. Nuisances are usually denoted as private or public. A thorough discussion of the legal aspects of odor enforcement is provided in the latter chapters of this book. It is important to note that in many states laws may not be applied until the nuisance exists. This means that although an odor nuisance is anticipated, no legal action may be taken until it occurs. After the plant or process is in operation and the nuisance is proved, the courts may take legal steps to shut down operations. Such difficulties can be obviated through proper legal zoning and city planning.

Allowable amounts of emissions should be carefully examined and compared to expected plant output. Furthermore, a quantitative study should be made of pollution from neighboring plants. As a general guide, the following relationship can be used between the pollution a plant generates, local air quality standards, and the background levels of existing pollution.

$$\text{Plant Emission} = \text{Local Air Quality Standard} - \text{Background Level Pollution}$$

This value can be compared with the estimated plant or production pollution to determine if the latter would be in compliance.

Thus far, selection factors related to odor and air pollution have been discussed. However, numerous other factors must also be considered in selecting a plant location. Primary considerations are usually production costs (which include plant construction costs) and transportation costs (which include transporting incoming raw materials and outgoing products). Table 30-3 lists some of the production cost and transfer cost factors that must be examined.

Table 30-3. Production and Transportation Cost Factors

Raw Materials	Fuels
Labor	Geographic
Water Resources	Labor Supplies
Transportation Rates	Markets

Raw materials may be simply defined as anything that is brought into a plant from an outside source for processing. Items such as mill supplies are included (lubricants, abrasives, machine parts, tools, furnace linings, etc.). It is rare that any one material has a locational influence on a plant—usually because of the volume and diversity of materials required. Nevertheless, it is a factor to examine. In general, however, the greater the number of different materials required, the less influential any one may be.

If the cost of raw materials is a large part of the finished product's value, the sources of raw materials play a great role in plant location. This is not true, of course, when materials are plentiful and can be obtained at equal cost. Companies with large expenditures for raw materials usually locate their plants near their sources.

Table 30-4 lists some of the major points to consider in examining regional labor markets. Many large plants require highly trained or experienced personnel, and problems may arise in the reluctance of some employees to leave their homes. Companies may have to provide an incentive to obtain sufficient skilled help to operate a new plant in some areas of the country.

Table 30-4. Labor Factors

Amount of Skilled Labor Available
Amount of Unskilled Labor Available
Average Intelligence of Workers
Industrial Experience Available
Current Wage Scales
General Efficiency and Quality of Work

GEOGRAPHICAL FACTORS

Physical features are important considerations in the plant location analysis. For example, rough terrain may make construction difficult and costly and present problems in transportation. Soil conditions are extremely important where heavy production machinery must be supported.

Probably one of the most important factors to consider is the water supply. Much of the water industry uses is not obtained from public water supplies. Water, however, is important for many processing, washing, scrubbing, and cooling processes in industry. Plants requiring large volumes of water should locate near a continuous supply. Although there may be surface or well supplies available, plants may be restricted in some cases from using them because of recreational, sanitary, or other regulations.

Once adequate water supplies are located, a careful examination should be made of the type and quality of the water that is available during all seasons of the year. This should be compared to the potential needs of the new plant. The water may contain too many minerals, or be heavily contaminated with bacteria or chemicals from neighboring plants. Table 30-5 lists other factors related to water that should be taken into consideration in determining plant site location.

Table 30-5. Water Considerations in Plant Site Location

Factors in Wastewater Treatment

1. Stormwater, cooling water, process water
2. Estimated wastewater discharge (total and individual processes)
3. Specific pollutants produced
4. Methods of control and cost
5. Neighboring waterways which may be used for disposal
6. Water quality
7. Wastewater effluent standards
8. Permit requirements for water use
9. Stream flow characteristics
10. Requirements for pretreatment, if discharged to public sewers
11. Percolation rates for onsite underground disposal systems
12. Electrical power dependability for emergency operations.

Factors in Water Supply

1. Estimated water requirements (process, cooling, fire protection, etc.)
2. Availability of water through public or private waterways, ground or surface water
3. Water properties (chemical and bacterial content, corrosive nature)
4. Specifications–amount of water available, variations in pressure and size of pipe lines required
5. Cost of water (cost per 1000 gallons)
6. Treatment requirements for process water, cooling, boiler-feed, etc.
7. Restrictions in use, if any
8. Future supplies outlook.

TRANSPORTATION

Transportation is a prime determinant in locating a manufacturing facility. Costs of transporting raw and finished materials to and from the plant largely depend on its location. Furthermore, prompt service to customers is vital.

An area that has railroad, truck, or water transportation freight competition is advantageous. Special shipping rates may be available. A detailed study should be made of all means of transportation and the existing rates.

AIDS IN PLANT SITE LOCATION

As we have seen, a plant site survey can be an exhaustive study; so it is helpful to have reliable short cuts. Some suggestions for quick approaches to plant site selection are summarized in Table 30-6, along with their disadvantages.

Table 30-6. Short-Cut Approaches to Plant Site Surveys

Approach	Disadvantages
1. Invite proposals from local chambers of commerce, power companies, real estate agents, and neighboring industries. The final decision can be based on a comparison of these proposals.	Not thorough and can lead to a great deal of confusion and misunderstanding. Information may also be biased and inaccurate.
2. Seek information through advertisements.	Not thorough coverage.
3. Rely solely on information provided by numerous federal, state, and local government departments.	These are excellent sources and information is quite accurate. However, factors such as public relations and effects on the environment can often be given too little attention in the study.

Some of the most valuable aid the planner can obtain is information that has been compiled by the various federal, state, and local agencies. Table 30-7 enumerates some of the federal sources.

Table 30-7. Government Agencies and Information

Department	Information
U.S. Department of Labor	Provides wage and hour statistics, cost of living and price data.
U.S. Department of Commerce	Provides regular censuses of population, manufacturing, distribution, and business.
U.S. Department of Interior	Prepares maps and data on topographical features, resources, etc.
Federal Trade Commission	Can provide data regarding economic and business conditions, and can give insight into changing legislation.
Interstate Commerce Commission	Can provide rate information on railway, express, and sleeping car companies, oil pipelines, and motor carriers. It also has information on freight forwarders and carriers operating coastwide, intercoastal, and inland waterway routes.

Regional and state agencies can also supply valuable, detailed information. For example, state and local planning boards compile data on resources and industrial opportunities, together with studies on unemployment, wage scales, costs and locations of plant sites, and taxes and laws affecting industrial operations.

Information from private sources like railroads and power companies should not be overlooked. Such organizations benefit directly from new plants and therefore they maintain departments for aiding industries seeking new locations. Railroads can be extremely helpful in answering questions related to local freight rates and shipping schedules. Power companies can answer questions about energy supplies. Both can give worthwhile opinions on plant location in their areas.

TOOLS USED IN PRELIMINARY SELECTION

There is a variety of tools that planners can use. For example, charts and diagrams can facilitate the survey and assure effective presentation. In addition, maps can be extremely helpful in visualizing the locations of competitors, competitive freight-rate areas, transportation routes, farm and residential areas, etc.

Other useful techniques include prepared questionnaires that can be submitted to the main engineering division or officials of the company. These can aid in determining which requirements the company considers most important in determining the plant site. They should not, however, be used as the sole criteria. The planner should be aware that respondents are not always aware of the full picture. Some may stress those items that are greatest in total cost, not realizing that economics can be influenced by a change in location. Furthermore, some respondents may feel that only essentials, such as water supply, deserve close attention.

The major advantages and disadvantages of each region should be listed. A quantitative rather than qualitative approach is the best; that is, items should be weighed on their overall importance. A systematic approach can be devised where factors can be assessed on a scale from 1 to 10 or 100 to 1000. For a particular case, power supplies may be valued at 2.5, public opinion 1.3, terrain 0.5, transportation 3, etc. Such an approach involves scoring each site for each factor. The area that has the highest total is presumably the best.

This is only one approach, and not always the best. A great deal of the final decision should be based on the competence of the planners' judgment, not on the type of approach used.

It is difficult to generalize all the factors that are important. Each plant is a separate problem with numerous considerations not mentioned here.

The emphasis in this chapter has been on odor- and air pollution-related factors, but these may not always be the most significant problems. Table 30-8 briefly lists other major environmental criteria for plant site selection.

Table 30-8. Other Environmental Factors in Plant Location

Solid Waste Considerations

1. State and local regulations on solid waste disposal.
2. Disposal facilities available (*i.e.*,iincineration, sanitary landfill, composting, recycling, etc.)
3. Local contract pickup and disposal rates (examines competition between haulers).
4. Methods of onsite disposal (*i.e.*, incineration, landfill).
5. Special handling facilities or practices that may be necessary for particular industrial wastes.

Ecological Considerations

1. The effects of gaseous, liquid, and solid wastes on the surrounding environment and wildlife.
2. Control measures available to minimize ecological effects.
3. Aquatic and wildlife propagation areas.
4. Tolerance levels of air quality, water quality, and noise levels.

The important point is that a systematic approach must be followed. Terminology must be defined and the major selection factors identified before any meaningful survey can be made. Figure 30-1 reviews the major points covered and presents a general outline of attack.

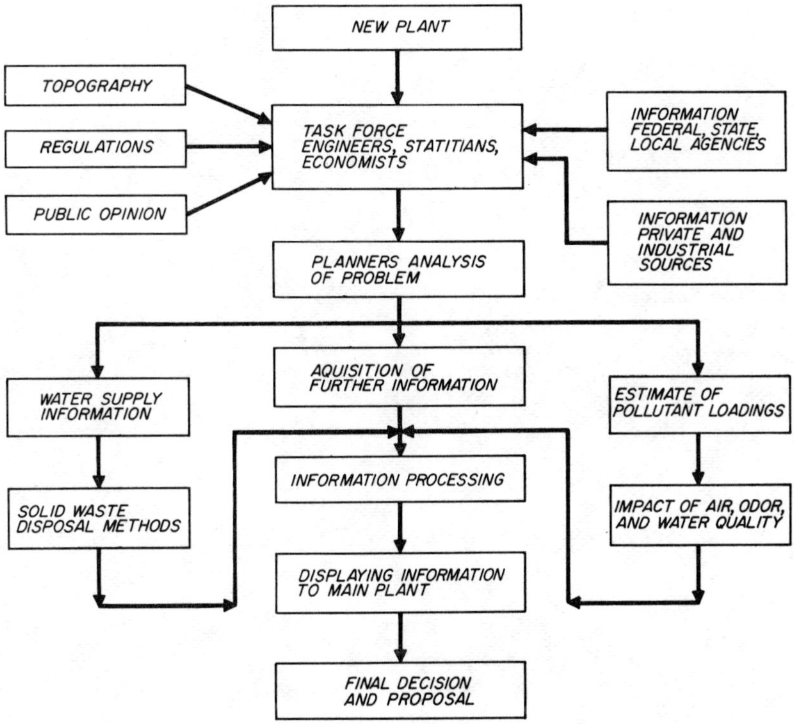

Figure 30-1. Systematic approach to plant site selection.

CHAPTER 31

LEGAL ASPECTS OF ODOR CONTROL

Stanley W. Schroeder, Esq.

Lamb, Eastman & Keats
Washington, D.C.

The treatment of odors under the law has not been uniform and must still be considered in the development stage. Human reactions to marginally disturbing odors tend to be highly subjective, and the law usually hesitates to rush into areas of uncertainty. Judges and lawmakers have also been aware that where odors are not physically harmful, though annoying, they are frequently accepted as the social cost of having industry—and therefore employment—within the area.

The law of odor control has its historic basis in the common law of nuisance, and in many parts of this country that is still the only law available to deal with a problem. Even in many recent statutes, odors are often not treated as a pollutant that is evil per se. Rather, odors are to be controlled only to the extent that they are harmful to the interests of others. As the technology for detection and abatement increases, jurisdictions are able to become specific as to what it is that will be considered a violation and what control measures might possibly be required. Although many of the state and local statutes are seeking objectivity, subjective judgments are still involved where statutes or ordinances require decisions as to what is harmful to health or what should be an actionable annoyance.

THE LAW OF NUISANCE

The statutes in some jurisdictions go no further than to refer to the availability of a nuisance suit as the legal remedy. Others really only incorporate the essential elements of this law in their statutory provisions.

Therefore, it is appropriate to understand the basic aspects of the law of nuisance. Black's *Law Dictionary* defines a nuisance as

> a class of wrongs that arise from the unreasonable, unwarrantable or unlawful use by a person of his own property, either real or personal, or from his own improper, indecent or unlawful personal conduct, working an obstruction or an injury to the right of another or of the public, and producing such material annoyance, inconvenience, discomfort, or hurt that the law will presume resulting damage.

Thus, while a person may use his land in any manner he sees fit, this must be done in a reasonable manner. This reasonableness is called into question if his use causes material injury or annoyance to his neighbor.

In determining whether a given act constitutes a nuisance, the court will consider the act itself, and the place and the circumstances surrounding the act. There are two kinds of nuisances, private and public. A private nuisance may be abated by an injunction brought at the instance of the individual whose property or health is affected. A public nuisance is one which affects an indefinite number of persons in a community. A public official, such as a pollution control officer, normally abates such nuisance. Generally, for the individual to bring a nuisance action he must show that he has sustained damage of a special character, distinct from the injury suffered by the public. Some cases, however, have held that there is a right to recover damages even in a public nuisance case where public rights were affected.[1]

Nuisance actions boil down to balancing the equities between property owners, each asserting his own rights to use his land. Factors that may be relevant in the case, however, include the availability of pollution control devices.[2] Failure to keep pace with the technological advances in pollution controls (as was done by a competitor) resulted in a court ordering adoption of such controls. Of course, in any balancing of the equities the good faith efforts of the polluter, while not absolving him in any way, if they exist, would certainly be a factor if they were absent.

One of the reasons why private nuisance actions have been of limited effectiveness in reducing odors is that the needs of the offending industry to use its property (especially if it is a major local employer) will tend to outweigh the needs of a few individual complainants to use their property.

To overcome this differential, class actions have been attempted, to emphasize to the court the composite social cost to the victims compared with the cost of abating the pollution. *Snyder v. Harris*[3] (a stockholder suit), however, limited the usefulness of bringing class actions by not allowing the separate and distinct claimants to aggregate their respective claims to meet the $10,000 jurisdictional amount necessary to get into federal court. This jurisdictional amount may be more readily met in the future as the range of recognizable pollutant damage is broadened by legislation such as the secondary ambient air quality standards of the Clean Air

Amendments of 1970, which are based on the concept of "welfare" that is expanded to include "effects on economic values and on personal comfort and well-being."[4]

There is an old axiom of law that one who voluntarily places himself in a situation whereby he suffers injury will not prevail. This is referred to as coming to a nuisance, where an individual purchases property with the knowledge of the existence of a nuisance on the adjacent property. There is a possibility in some jurisdictions that the purchaser will be precluded from recovery of damages.

Many other cases, however, hold this not to be applicable, and cases often are involved in factual questions of

1. whether the purchaser was aware of the intensity of the nuisance when he made the purchase
2. whether the intensity has increased since he made the purchase
3. whether the pollution is essential to the industrial activity of the community
4. whether abatement efforts are feasible.

Questions of fact may involve such items as

1. whether a facility is constructed or operated effectively
2. whether noises or smells are extraordinary or unnecessary
3. whether they arise from the ordinary proper course of the business, the character of the neighborhood, the extent and frequency of the injury, the actual effect on the enjoyment of life, health, and property, and the prior use of the property
4. whether the odor is incidental to the use of the land, such as farming or agricultural uses
5. whether the plaintiff is abnormally sensitive to the nuisance
6. whether the enterprise is permitted under a zoning ordinance.

The most widely cited expression in all cases is that a nuisance case cannot be determined by exact rules, but must necessarily depend upon the circumstances of each case. This obviously offers a great deal of flexibility, as well as uncertainty, in the nuisance approach to the abatement of odors.

GENERAL PROBLEMS OF LITIGATION

The question in court cases is not only one of assertions, but of proof. Often it is only in the preparation of a lawsuit that it becomes evident to the complainant that his proof may be lacking or threadbare in areas. Identification of the source of pollution may be easy in a remote area, but can be extremely difficult where there are several possible sources of the pollution. How do you prove that the odors you are complaining of come from factory X, not factory Y or factory Z, or that the odor you complain of is not a combination of odors from all three factories? Perhaps different odors reinforcing each other result in an intensity that is objectionable, whereas individually the smells might not be objectionable.

Where state and local laws provide for odor control, it may be possible to request responsible agencies to verify odor emissions. If no such provision exists, it may be necessary to have analyses made privately.

Another problem is specifying the injury to the complainant. Is it mild discomfort? Severe discomfort? Does it cause illness? Physical or mental aggravation? Is it intermittent or constant, or only when the wind blows in a certain direction? Is the complainant the only one claiming injury? Are others in the community similarly affected? Who are they? Would they be willing to stand up in court and say they want the nuisance abated? Are they or their families employed in the nuisance source? Is the smell limited to a certain time of day or night? Or once a week? Has there been any decrease in land value through lower prices in the neighborhood? Would land values increase anyway because of other factors such as desirable proximity to other commercial areas? Has anybody gone to the doctor because of the odors? Did the complainant know of the nuisance before moving into the area? Is the complainant particularly susceptible to the odor, or would the average person in the community be offended or made sick?

A fundamental question is whether suit is brought in the proper jurisdiction. This may involve questions of whether the plaintiff and the defendant live in the same or in different jurisdictions. Does any statutory law apply in either jurisdiction? Is recourse to state law available? Is interstate commerce involved?[5] If the odor moves across state lines, there may be a possibility of involving the federal government in an enforcement action.

With all these factors involved in the normal nuisance action to abate an odor, it is easy to see why various jurisdictions have enacted statutory and regulatory controls over odors in an attempt to bring some certainty to this area. Some of these efforts will be reviewed in the following section.

STATUTORY LAWS TO CONTROL ODORS

The U.S. Environmental Protection Agency has made an effort to collect and summarize existing and proposed odor control regulations. It hired the Copley International Corporation to survey each air pollution control agency listed in the 1970 *Directory of Governmental Air Pollution Control Agencies*—a total of 256 agencies. The results of this survey, as well as the discussion and recommendations for a model odor control ordinance, are available to the public.[6]

Responses were received by the contractor from 203 agencies and tabulated in this document. With the caveat, however, that the information was collected in the fall of 1971 and cannot be considered current with any certainty. But the information is useful in giving some estimate of the type of laws and provisions within these laws that may be found in dealing with odors. Less than half of the states had laws dealing with odor control. A number of regional and local jurisdictions indicated that where there was a state law they were relying on the state law. In many cases there was not a separate odor law—rather, an inclusion of odors in the air pollution control laws.

Among the conditions necessary for violation of the odor ordinance, reference to the number of complaints turned up only a handful of times. In each case the specification was one complaint, verified by the proper local authority within two hours.

The next condition for violation was whether the odor was deemed offensive or objectionable. A number of regulations stated that the measurement must be made at the property line. A number of the regulations specified that the odor must be deemed offensive or objectionable by 30% of a sample of people exposed to it or living in the affected community, or 75% of the people if the sample is less than six. Also, the sample must be taken on, near, or adjacent to residential, recreational, institutional, retail sales, educational, or hotel premises. In another jurisdiction the determination was to be made by the director of the air pollution agency; in another, by the health officer and 25% of 20 people exposed to it or living in the affected community, or five persons if the sample was of less than 20 people.

The city of Indianapolis, Indiana, for example, required the objectionable or offensive determination to be made by 50% of 20 people exposed to it or living in the affected community (75% of the sample must work or reside in the affected area; 25% of the sample must be from the general public, living or working outside the affected area) where the odor is a residential area problem. The detection must be without any dilution of the air sample. In industrial areas or on or adjacent to industrial premises, where light industry is concerned, there must be detection with a dilution of 5 volumes of odor-free air. Where medium or heavy industry is involved, the dilution must be with 20 volumes of odor-free air.

Some jurisdictions require the determination to be made by people exposed to the odor or living in the affected community but not associated with the problem. In another jurisdiction an independent panel must be used. In some areas the authority to make the decision may be delegated to the representative of the air pollution control officer.

In Vermont the determination is made by persons occupying or frequenting places closest to but beyond the property line of the source of odors. The determination must be made by 15% of 20 people exposed to it or living in the affected community, or 75% of the persons if the sample is less than 20 but more than four people (and provided that these persons are not from the same household). In Virginia the determination is made by the agency, the Air Pollution Control Board, on the basis of a review of data obtained by investigation directed by the Board and by holding a public hearing to hear complaints.

Some jurisdictions specify scentometer readings. In residential and commercial areas, Colorado requires the test to be made with a dilution of odor-free air of 7 or more volumes. In industrial areas the dilution is with 15 or more volumes of odor-free air. Two measurements must be made within one hour, and at least 15 minutes apart, when measurement is made at or beyond the property line of the source.

Dade County, Florida, requires detection by a "considerable number of persons or the public at or beyond the property limits of the premises of the person responsible for the source."

Illinois requires the use of the scentometer with dilutions of odor-free air of 8, 16, and 24 volumes, respectively, for residential, industrial, and other areas, and specifies that two positive concurrent determinations must be made by three trained inspectors in one hour and at least 15 minutes apart, when measured at or beyond the property line of the source. A number of jurisdictions merely specify a dilution device reading of "#2 odor strength."

Just as various jurisdictions use various dilution rates, their directions for locating the detector also vary, from those requiring property line measurement to those prescribing distance from an incinerator stack. Arlington County, Virginia, specifically focuses on the burning of refuse in fuel-burning equipment. A violation exists if objectionable odors are detectable beyond the premises on which the installation is located.

A variety of prohibitions and control requirements are specified in odor ordinances or nuisance laws. Some simply do not permit emission of noxious or objectionable odors which are detectable without instruments at or beyond the property line. They require effective control apparatus to be installed and operated to reduce odor-bearing gases or particulate matter emitted into the atmosphere to a reasonable minimum. Many standards are careful to specify, however, that installation or use of any device or measures to conceal or mask emission of an air contaminant shall be unlawful.

Numerous jurisdictions expect any person treating or processing animal or vegetable matter to require all gases and vapors and gas-entrained effluents from such operations, equipment, or contrivance to be ***incinerated at high temperatures*** or processed in a manner to be determined to be equally or more effective for the control of air pollution. A frequent temperature requirement is incineration at more than 1,200°F for more than 0.3 seconds. Maryland requires first cooling to less than 160°F, and then the noncondensible fraction must be incinerated at more than 1,400°F for 0.4 seconds.

Oregon specifies that any hardboard tempering oven must heat all gases and vapors in a fume incinerator capable of raising temperatures to over 1,500°F for more than 0.3 seconds. Lower operating temperatures may be approved if they are effective in preventing odors from being perceived outside the property line.

Allegheny County, Pennsylvania, requires 1,600°F for more than 0.5 seconds of processing in a manner determined by the bureau chief to be equally or more effective. Knox County, Tennessee, prevents construction of incinerators unless they have secondary burners for use when necessary to destroy odors. Vermont simply requires temperatures of 1,600°F for 0.5 seconds.

In many jurisdictions odors are included in the air pollution laws by virtue of being defined as an air contaminant, and the *inclusion* may be by *definition* or *implication.* A number of counties in California merely say that public nuisance law is applied to control odors or repeat the standard phrase in the California Health and Safety Code:

> No person shall discharge from any non-vehicular source whatsoever such quantities of air contaminants or other material which cause injury, detriment, nuisance or annoyance to any considerable number of persons or the public or which endanger the comfort, repose, health or safety of any persons or the public or which cause or have a natural tendency to cause injury or damage to business or property.

The state of Connecticut specifies that odorous matter may not be discharged beyond the owner's property line which is detrimental to members of the public or interferes with the appetite, sleep, or other environmental health factors of such members. In addition, no person shall construct, install, use, or cause to be used any incinerator which will result in odors that either are annoying or harmful to health in any area of normal human use or occupancy.

The city of Greenwich, Connecticut, apparently has less sweeping constraints. It prevents the creation of objectionable odors on or near residential, recreational, retail sales, hotel, or educational premises. This appears to exclude industrial premises and general commercial, nonretail premises.

Metropolitan Dade County, Florida, requires that solvent vapor or odor emissions from plant sites or other sources as measured at the user's property line shall not exceed the established threshold limit values for solvent vapors or odors. The threshold limit for styrene is 10 ppm.

Illinois requires the abatement of any objectionable odor nuisance. In the event that the rendering processes of one or more companies are contributing to the objectionable odor nuisance, abatement shall be deemed effective when the odor concentration from each process is not more than 120 odor units per cubic foot.

Maryland specifies that incinerators should be maintained in good working order and calibration, and that continuous recording devices be used for indicating temperature, pressure, or other operating conditions. Other reasonable precautions suggested to prevent odors from being discharged include proper storage, covered vehicles or containers, water-tight construction of equipment for handling and transporting offal or vegetable oil, and the use of hoods and fans to enclose and vent storage and handling operations.

The city of Flint, Michigan, prohibits the keeping of any dead carcass or other animal or vegetable matter which may cause any unwholesome, obnoxious, or offensive smell, or the confinement of any horse or animal in pens so as to create unwholesome, unsightly, malodorous, obnoxious, or offensive conditions. Any cart or vehicle used for conveying any swill, garbage, or excrement shall be perfectly tight and covered so as to prevent leaking, spilling, or becoming offensive because of odor. Wayne County, Michigan specifies that when the odor of hydrogen sulfide is found to exist beyond the properly line of the source, measurement shall be taken to see whether the hydrogen sulfide is in excess of 0.005 ppm by volume for a maximum period of two consecutive minutes.

The Minnesota Pollution Control Agency draws distinctions based on the height of stacks. Odor sources emitting from well-defined stacks greater than 50 feet above grade elevation and with adequate dispersion characteristics, as determined by the agency, shall not emit odors greater than 150 odor concentration units. Odor sources of less than 50 feet elevation above grade, or otherwise failing to create dispersion conditions, shall not emit more than 25 odor concentration units. No odor source will have an odor emission rate greater than a million odor concentration units per minute. No odor source shall emit air contaminants which cause odor outside the property line in excess of one odor unit in areas zoned residential, recreational, institutional, retail sales, hotel, or educational, or two odor units in areas zoned light industrial, or four odor units in areas zoned other than these.

The Suburban Air Pollution Commission in West Orange, New Jersey, requires that incinerators not allow any odors to be detected by the sense of smell in any area of human use or occupancy. North Carolina requires any processing of animal, mineral, or vegetable matter to be done with control devices, including wet scrubbers, incinerators, or other such devices as may be approved by the state authorities.

The realm of odors to be considered offensive is given some definition with standards, such as the city of Cincinnati, which says that incinerators may not emit offensive odors, *i.e.*, any odor similar to that of burning garbage, paper, or other cellulosic material. In Oregon, the Columbia-Willamette Air Pollution Authority prohibits outdoor fires which contain garbage, asphalt, waste petroleum products, paint, paint-coated metals or wire, rubber products, plastics, or any substance which normally creates a dense smoke or noxious odors or creates a public nuisance.

In spite of the general prohibition in a number of statutes against the *dilution* of masking of odors, in Tennessee the Chattanooga-Hamilton County Air Pollution Control Bureau specifically allows dilution of odors if such odorous matter is diluted with greater than four volumes of odor-free air. The state of Vermont prohibits the installation or use of any equipment to reduce or conceal an emission which otherwise would constitute a violation of their regulations. In the state of Washington the Northwest Air Pollution Authority specifies that ambient air will not contain odorous substances such as (but not limited to) hydrogen sulfide, mercaptans, organic sulfide, and other aromatic and aliphatic compounds in such concentration or of such duration as will threaten health or safety or prevent the enjoyment and use of property.

The Clean Air Authority of Yakima County, Washington, specifies that all air contaminant sources emitting odor-bearing *gases* and odor-bearing *particulates* shall be operated or controlled in such a manner as to eliminate, as far as practicable, noxious odors. Proven technological improvements, to the extent that the same are economically feasible, shall be incorporated in such air contaminant sources to reduce the emission of such noxious odors in the ambient air.

The Air Pollution Control Authority of Spokane County, Washington, also takes a reasonable approach to what can be done. It requires effective control apparatus and measures to be installed and operated to reduce odor-bearing gases and particulate matter emitted into the atmosphere *to a reasonable minimum.* The Board or its control officer may establish *reasonable requirements* that a building or equipment be closed and ventilated in such a way that all the air, gas, and particulate matter are effectively treated for removal or destruction of odorous matter before emission to the atmosphere.

The West Virginia regulations appear not to put restrictions on odors that affect only the private person. Its restriction is against discharge of air pollutants which cause or contribute to an objectionable odor in any location *occupied by the public.*

The state of Wisconsin requires materials *moved* on public roads, railroads, and navigable waters to be secured and protected in such a manner as to minimize the spreading of odor. The state of Wyoming requires that odor-bearing gases, volumes, fumes, or dusts arising from materials in process shall be confined at the point of origin so as to prevent liberation of odorous matter.

From the foregoing examples we can see that the control of odors is accomplished in many different ways, ranging from outright prohibition to reasonable controls, specific incineration requirements, and—in many cases—a certain amount of flexibility in working out problems which may be more severe in one location than another. Depending on location and local industrial requirements, a wide range of exceptions to the odor laws is also found.

Exceptions

Many statutes provide for upset conditions, breakdown of equipment, or scheduled maintenance of equipment, most of which are contingent upon notification to the local agency. Agricultural operation and associated odors are also widely excepted. This includes the growing of crops and the raising of fowls or animals. A number of exceptions may be found for equipment used exclusively for processing food for human consumption. In Colorado a manufacturing or agricultural source using *best practical* control is exempt, unless odor is detected after dilution with greater than 127 volumes of odor-free air.

Fulton County, Georgia, requires diesel engines to be equipped and operated to minimize objectionable odors. Food service establishments are exempted in other jurisdictions. In Des Moines-Polk County, Iowa, there is an exception for any plant subject to and operating under inspection by the Meat Inspection Division of the U.S. Department of Agriculture. The St. Louis County Health Department in Minnesota makes an exception for temporary operational breakdown and for cleaning air pollution equipment, unless an immediate public health hazard results. Local agencies must be notified.

Missouri excepts incinerators on residential premises that are used exclusively to dispose of refuse originating on the same premises, provided the total number of dwelling units on the premises does not exceed four.

A number of jurisdictions except upset conditions and uncontrollable force, provided local agencies are notified. In Oregon the Columbia-Willamette Air Pollution Authority excepts manufacturing processes if the highest and best practicable treatment and control currently available are used to maintain the lowest possible emission of odorous gases. Outdoor fires are allowed for agricultural burning, for recreational purposes, or cooking food for human consumption where such fires are set or permitted by public officials for fire prevention or fire control.

Virginia excepts incinerators with a rated heat input of more than 40 million Btu's per hour. In Maine, the Southwest Air Pollution Authority excepts space heating equipment other than boilers, internal combustion engines, and vehicles for transport of passengers or freight. Yakima County, in Washington excepts fuel burning equipment for domestic housing.

It should be reemphasized that this information was taken from the EPA publication based on a fall 1971 survey. Because of the interest in protecting the public from air pollution, many state and local agencies are in the process of updating their regulations for odor emissions.

MODEL ODOR ORDINANCE

An attempt by EPA to come up with some uniformity of treatment in the form of a model ordinance was made when EPA hired Copley International Corporation to study the social and economic impact of odors. It was the conclusion of that study that a model ordinance cannot be drafted that is compatible with the legalistic approach to controlling air pollution preferred by every jurisdiction. The contractor, therefore, placed his emphasis in the development of a recommended form of model ordinance on providing a conceptual framework that would promote the most efficient solutions to community odor problems, given the existing state of knowledge. The introduction to this study points out:

> Since the 1950s many advances have been made in the development of air pollution controls with legal controls preceding and often forcing the development of technological controls. In this sequence of events the effectiveness of the legal controls is the critical thing. It rests largely on the availability of equipment to measure concentrations of pollutants and the existence of methodology to quantify the impact of these pollutants on human health and welfare. Legal controls are particularly effective when equipment can measure pollutants to concentrations at which impact is determined to be unlikely.
>
> Progress toward the development of controls over all forms of air pollution has not been consistent. For example, the development of odor controls has been encumbered by two major obstacles arising from the very nature of odor perception:
>
> 1. The lack of equipment capable of *measuring low concentrations* of more than a few odorous substances, [and] coupled with this, the *lack* of *methodology* for *translating* known *concentrations* of substances into odor *intensities*, and
> 2. The lack of methodology to *quantify impact in terms of* odor *intensities.*[7]

This study goes on to say on page 3:

> It was expected that the national survey of the odor problem and the study of the social and economic impact of odors would provide a series of relationships which could be used in the development of legal controls over odors. Yet, as this and other research progressed, it

> became evident that the only measurable impact associated with a vast majority of odor problems was *annoyance* and that a proper means for dealing with such problems must consider measurement of annoyance as central to success. *Public attitude surveys* were recommended as the proper means for doing so.

The approach would be to take samples in two areas–one, the affected community, and the other an odor-free community of similar socioeconomic characteristics. The aim of the survey would be to determine whether a significantly greater proportion of respondents in the affected community state that they were bothered by odors than respondents in the odor-free community. This determination that a "community odor problem" exists would be one of three aspects of establishment of a violation. The other two aspects would be the existence of a chronic situation, defined as odors extending beyond the property limits of the source and occurring more than once or lasting more than one day in any three months. The second item would be the existence of any of the three following conditions:

1. If one or more odor complaints are initiated by residents of a community and verified by local authorities, or
2. If 20 or more odor complaints are initiated without verification, or
3. If odors are detected in the community by local authorities on more than one day in any three-month period. (Such odors shall be of sufficient intensity to be detected using a Barnebey-Cheney Model I-3 Scentometer set at 7 dilutions to threshold.)

This multifaceted approach for establishment of a violation is a result of the assessment of the limitations of the normal procedures in existence for attempting to determine whether an odor problem exists under existing technologies and procedures.

EVALUATION OF PROCEDURES

Use of the Scentometer

The EPA study indicated that the scentometer was useful for measuring odor intensity by diluting the ambient air with odor-free air up to the odor threshold level. It could measure odor intensity levels at a particular location and at a particular time at known or suspected sources. The scentometer has the advantage of requiring only one person for its operation. Different results, of course, can be achieved by different persons using the scentometer. Experience has shown that putting one employee on the scentometer on a regular basis may be necessary to achieve fairly consistent results. If the scentometer is used by a different person every day with significantly different results, there may be some basis for legal challenge.

Use of Panels

In order to compare the intensities of odors in the ambient air with those of reference standards, the EPA felt that judgment panels were useful. If an affected area could be covered by enough panelists, documentation of odor distribution by panels throughout the test area over a period of time would be possible. Also, location of the source of the odor in the test area by reference to odor distribution patterns might be achieved. The aim would be to plot isopleths in order to pinpoint the *locations* of the sources as well as the *extent* of the odors at the time the panels were used.

The difficulty, of course, arises when there is more than one odor in an area. Unless the source being investigated has an odor of such quality that it can be positively identified among other odors, the problem becomes extremely difficult. For practice, panelists have used as a reference standard stable nontoxic odorants producing odors similar in quality to those most frequently encountered in the test areas. This is one way of helping the panelists recognize the odors in the field as being from sources under investigation, thus enabling them to compare the intensities of the odors with the reference standard scales.

Where the odor of a reference standard would be similar to other odors found in the area under investigation, it may be impossible to isolate the source being sought for prosecution. In this situation a reference standard scale can be prepared from an odorant unlike any found. A great deal of training is necessary, however, for the panelists to recognize the odor characteristics and intensities of the various sources. This can be done by trips or laboratory exposures to examples of odorous substances typically emitted. The cost of training and deploying a large number of panelists is prohibitive for the normal surveillance activities. However, they may be justified in support of an action where a source verification is necessary.

The difficulty in using odor panelists, quite naturally, is the degree of subjectivity that can be inherent in the participants. Not only are there physical differences in the olfactory senses, there are also differences of opinion as to whether an odor is good or bad for the community. Where friends or relatives are working in the source being tested, the panels invariably come up with negative findings, which undermine their usefulness for examining local conditions.

Usefulness of Public Attitude Surveys

In sparsely settled areas odors usually do not affect the required number of people for use of public attitude surveys. It is questionable whether the results accurately represent the attitudes of the area surveyed. An attitude survey of a randomly selected sample of residents could satisfy a public nuisance law requirement. But while it is a simple step to claim that attitudes expressing annoyance would represent interference with the community, it is a more difficult step to determine whether interference is substantial and unreasonable. The EPA study states:

> Equity is involved. It is unfair to find substantial and unreasonable interference on the basis of an arbitrary percentage of residents who express annoyance. Recent studies have found that, even in virtually odor-free areas, up to 50% of the residents may express annoyance to odors. This is due to a variety of factors, including, for example, residents with unusual sensitivity to background odors, feelings against nearby industry and neurosis.[8]

The problems of getting a truly representative sampling are compounded by the fact that in many areas there is a difference in the length of time people have resided there. For many oldtimers the neighborhood rendering plant may have been a source of jobs and a boon to local industry for many years. For newcomers without this background, it is only a nuisance, a blight on the investment in their newly purchased homes. In a local operation which may not be a health hazard, such as a well-run rendering plant, there may be little sympathy by local judges, who may ask whether anybody has ever died or gotten sick from the plant. If the answer is no, he may dismiss the case rather than close down a local industry. In such a situation, a local ordinance based on a sampling of annoyance would be a very convenient enforcement tool.

In the absence of any other fair test as to what a community should or shouldn't tolerate, the annoyance factor is as good as any—to the extent that it has built into it the prejudices of the residents in the affected area. This, of course, raises questions for the businessman as to whether or not he has any certainty that a public survey that gave him a vote of confidence one year will necessarily come up with the same results the next year.

PROBLEMS OF LEGAL FORECASTING

For these reasons the businessman planning to locate a facility with potential odor problems might seek a location with specific odor control requirements and make his plans to meet those requirements, or he might try to locate in sparsely settled areas where no community objection would exist. Here too, however, he does not know how long it will be before the population will catch up with him. With the EPA actively looking into the possibilities of odor control model ordinances, there is no certainty that there will not be greater activity in this area in the near future. Only the uncertainties and difficulties of attempting to develop a standard for all situations appear to be a deterrent to rapid action.

When one realizes the great effort being made in industry to control the flavors of beverages and foods, as well as the air man breathes in certain environments, it is obvious that, for odor control, public laws and ordinances and the application of nuisance law really only scratch the surface of available knowledge. In the years to come, as we attempt to improve every aspect of our society, including the odors to which we are subjected, the unanswered question is whether we will become more precise in attempting to identify and quantify discomfort, whether we will attack all sources of odors on the assumption that they are all bad, or whether we will continue to employ the concept of nuisance law which attempts to balance the equities involved between the source of an odor and the recipient of the odor.

If the recent increased activity in the air pollution field by citizens is any indicator, increased controls of odors are not likely to be far behind, inasmuch as the average citizen is more likely to perceive odors in the en-environment than he is to perceive those air pollutants presently controlled by the Clean Air Act. To the extent that odors do not generally constitute a health hazard to the same extent as excessive air pollution, the ability to generate public support for uniform odor control legislation may be less than that for air pollution.

To the extent that our concern for the quality of life continues to increase, however, it is likely that we shall witness an increase of odor controls and statutes and ordinances in those areas where proof of injury might otherwise be difficult. Where injury and damages can be proved, it is logical to resort to the common law of nuisance in the courts because of the possibility of recovering damages where a private nuisance is complained of. A recent setback to class action suits dealt by the Supreme Court in the Eisen case[9] (which required notice to all members of the class before they could be included) makes the public nuisance avenue an unlikely measure for seeking damages for a broad segment of the population.

However, as all pollution control statutes tend to include concepts of welfare (not only concepts of health), and this concept of welfare includes freedom from annoyance, it is likely that we will see more use of the injunctive powers of the courts as the balancing of the equities starts tipping in favor of the community resident rather than the industrial base of our society.

It would be a mistake to assume that the only parties interested in abating odors are residents in the community. On the contrary, other industries in the community may also have a stake in an odor-free environment. Where the gases or particulates causing the odors pervade the atmosphere and are deposited and impregnated in the products of other industries, this may especially jeopardize those businesses where consumer-preference products may be impaired because of latent undesirable odors.

One additional factor may be significant in controlling odors: as air pollution regulations of the basic pollutants covered by the Clean Air Act are enforced through the control of gaseous emissions and particulates, sources of odors may thereby be controlled indirectly. EPA may be wise enough not to attempt to control odors on a uniform nationwide basis, inasmuch as any engineer with the responsibility of looking into the plant's air pollution control problems will probably not overlook possible odor problems in attempting to set up environmentally compatible operations. Furthermore, he will try to avoid leaks of odorous gases or other malfunctions that might trigger inspections for violations of the Occupational Safety and Health Act of 1970.[10]

CHECKLIST FOR LEGAL EVALUATION OF AN ODOR SITUATION

1. What is the nature of the problem?
2. Who has authority in the company to control it?
3. Is the situation covered by any law or ordinance covering odors, air pollution, Occupational Safety and Health Act, general nuisance law? Exactly what do pertinent provisions require?
4. Does the problem involve an accident, periodic occurrence, or constant emission? If a violation has been charged, is it substantiated, or capable of being substantiated? Could there have been mistakes in measurement, procedures, interpretation, or conclusions?
5. Background history:
   a. Have there been previous complaints—by private persons or a public agency, by workers, by the community? (Has the scope of the community changed since then?)
   b. Company responses
   c. Resolution of problem
6. Present company measures or equipment to curb odors, if any. Can this be documented?
7. What new measures or equipment would be needed to prevent such an episode or abate odor?
   a. Cost
   b. Lead time to get equipment
8. If a complaint has been filed, how bad is the effect on complainant or complainants?
   a. Mere annoyance
   b. Health effects

   c. So severe as to support a preliminary injunction to cease operations pending trial, or compliance with statute or ordinance.

9. Consequences of shutdown—cost in
   a. Jobs
   b. Production
   c. Start-up costs
   d. Inventory spoilage
   e. Financial stability of company

10. What court or government agency has enforcement jurisdiction?

11. What is the purpose of the action?
    a. Close down the plant
    b. Have company control the odor
    c. Money damages for actual injuries, punitive damages, exemplary damages?

12. What are the steps of the legal procedure?
    a. Who has the burden of proof?
    b. Avenues of appeal
    c. Possibilities of delaying or expediting procedure

13. Estimate of costs in alternative courses of action: paying damages, settlement, legal fees, control equipment costs, production interruption.

14. Is there any relationship between this action and other court cases or enforcement actions in this industry or political subdivision that would affect the outcome?
    a. Court decisions on what constitutes a nuisance
    b. Decisions on availability of abatement technology
    c. Admissions on economically feasible control measures contained in consent decrees.

15. Is legislation pending that will require compliance anyway? Will defiance only assure passage of the legislation? Or will good faith efforts to control the odor give legislators an example of what is economically and technically feasible and thereby possibly help achieve reasonable legislation?

REFERENCES

1. *Robinson v. Westman*, 224 Minn. 105 (1947); *Clinic and Hospital, Inc. v. McConnell*, 241 Mo. App. 223 (1951).
2. *Rankin v. Harvey Aluminum*, 226 F. Supp. 169 (D. Ore. 1963).
3. *Snyder v. Harris*, 394 U.S. 332 (1969).
4. Secondary ambient air quality standards of the Clean Air Amendments of 1970, Section 109(b) (2) and Section 302h of the Clean Air Act as amended by Public Law No. 91-604, Sections 4 and 5.

5. *U.S. v. Bishop Processing Company* (Court of Appeals, Fourth Circuit), decided March 3, 1970.
6. See *A Study of the Social and Economic Impact of Odors, Phase 3: Development and Evaluation of a Model Odor Control Ordinance,* Copley International Corp. Prepared for the Environmental Protection Agency, dated Feb. 1973 (No. PB-223589), and distributed by the National Technical Information Service in the U.S. Department of Commerce.
7. Copley International Corp. study of the social and economic impact of odors, p. 1.
8. EPA study, p. 153.
9. *Eisen v. Carlisle & Jacquelin*, decided May 28, 1974.
10. Public Law 91-596, 84 Stat. 1590.

CHAPTER 32

ODOR CONTROL CASE LAW

Richard A. Young

Pollution Engineering
Barrington, Ill.

Harry A. Young

Neistein, Richman, Hauslinger & Young, Ltd.
Chicago, Ill.

As previously pointed out, today there is an extensive legislative scheme for regulating odor problems. Most commonly this legislation involves a local municipal ordinance, although there are instances where state statutes are involved as well. Such legislation quite often contains definitions and standards whose violation causes a violation of a particular ordinance or statute. Such definitions and legislative standards ease the job of the courts somewhat since they give them something akin to an objective standard on which to reach a decision as to whether particular odor emissions have violated the law.

Because these odor emission statutes and ordinances are relatively recent, most of the appellate court cases involving them tend to deal with whether or not they are constitutional. To date not many appellate court cases (it is in the appellate court cases where one finds written opinions) have dealt with interpretations of statutes and ordinances dealing with such ordinances. In addition, those cases involving ordinance violations by odor emissions tend to be settled at the trial level, either by compliance, compromise, or trial, due to the substantial expense involved in taking an appeal.

As of the writing of this text, the bulk of the written appellate court decisions deals with odor problems under the old common law of nuisance. However, these common law nuisance cases still are important for several reasons. First, they give the courts background on odor problems. Second, courts doubtless will rely on the precedent created by these older cases in construing and interpreting the newer odor statutes and ordinances. Third, they tend to illustrate the types of problems that one encounters in attempting to deal with and regulate odors. The cases set forth below illustrate some of the principles involved in the present law regulating odors and some of the problems encountered in regulating odors.

BOHAN *v.* PORT JERVIS GAS-LIGHT CO.

This is one of the very early pioneering cases in the law of nuisance and odors. Back in 1880 the Port Jervis Gas-Light Co. (the "Gas Co.") manufactured gas for use in utilities. The original process used coal and caused no substantial complaints from the plant's neighbors. However, in 1880 the Gas Co. introduced naphtha into its manufacturing process, and began to cause tremendous odor problems in the immediate vicinity of the plant. In these early times the law tended to be on the side of the manufacturer in almost all respects. The general legal proposition on this aspect of the law was that if the manufacturer had not done anything negligently in its manufacturing process, and if there was nothing intentional or malicious in its actions, then, if its manufacturing operations caused odors, its neighbors would simply have to endure them.

However, the Court of Appeals for the State of New York (the highest reviewing court in the state) and the trial court in this particular case both recognized that a person who owns property adjacent to an odor-producing manufacturer can sustain a tremendous loss to the resale value of his real estate because of such odor emissions, and in fact may not be able to resell his property once such noxious emissions begin.

The trial court (in a decision which later was affirmed by the Court of Appeals) determined that even if there was neither negligence or malice in the manufacture of such gas, if the odors emitted from the Gas Co.'s plant were so noxious as to destroy neighboring property values, then a jury could determine, after hearing all the facts, that the manufacturing operation constituted a nuisance such that neighboring property owners could be awarded damages.

Since a jury had been demanded and the matter was to be heard as a jury case, the trial judge charged the jury that "to constitute a nuisance, it was essential that the smells and odors from the defendant's works should be sufficient to contaminate and pollute the air, and substantially interfere with the plaintiff's enjoyment of her property." The trial court went on to advise the jury that the question it must ask itself was "Did the odor pollute the air so as to substantially render plaintiff's property unfit for comfortable enjoyment?" If the jury found that the plaintiff's property, because of the odors emitted from the Gas Co.'s plant, had become "unfit for comfortable enjoyment," the jury could find the Gas Co. liable in damages and determine the amount of the damages.

One has to stop for a moment and consider that charge, word by word, and then place oneself in the legal context of 1880 in order to see how revolutionary it was. The court did not say that the Gas Co. had to act either negligently or maliciously. The court did not say that the plaintiff's property had to be destroyed or rendered unfit for human habitation. The court did not say that the fair cash market value of the property had to be destroyed or at least impaired. Rather, the court said that the jury only had to find that the Gas Co.'s odors caused the plaintiff's property to become "unfit for *comfortable* enjoyment" in order for the Gas Co. to be liable in damages. "Comfortable enjoyment" is a very, very flexible standard.

Anxious to avoid establishing such a legal precedent, no matter what the cost, it is no wonder that the Gas Co. fought this case all the way to the highest court in the state of New York. However, the Court of Appeals of the state of New York, which over the years has traditionally been a very forward-looking and progressive court, affirmed the trial court in all respects, particularly its charge that all the noxious odors need do is render the plaintiff's property "unfit for comfortable enjoyment," recognizing that because of the elusive nature of noxious odors it was not possible to frame a more precise standard.

The Court of Appeals went even further. It said that while every person has the exclusive use of his property and, in general, may do with it what he will, nevertheless, he must respect his neighbor's property. In particular, the Court of Appeals said that "he must make a reasonable use of his property, and a reasonable use can never be construed to include those uses which produce destructive vapors and noxious smells, and that result in material injury to the property and to the comfort of the existence of those who dwell in the neighborhood."

The Gas Co., rather than establish the proposition that since it was a quasi-public utility it could do with its property what it willed, established the beginning principles of today's rule of law: that, generally speaking, a neighboring property owner has a natural right to have the air diffused over his property reasonably free from contamination, and that other persons do not have the right to use their property so as to cast impurities into the air over adjoining property in amounts materially interfering with its use and enjoyment, even though the commercial or industrial user may be conducting its operation with all proper care, or at least as well as technology will permit. This principle of law has advanced over the years to the point where the Wisconsin Supreme Court even enjoined a city (cities normally are permitted a wide latitude in the operation of utilities), the city of Fond du Lac, from permitting odorous gases to emanate from its sewage disposal plant.

This, then, is the thrust of the law of nuisance: a man can be held liable for the natural consequences of his actions even though those actions are neither malicious nor negligent and are done for a reasonable business purpose.

For those times, when there were no such things as zoning laws, for example, and when the general proposition was that when a man owned land he owned it from the center of the earth clear up to heaven and could do with it what he wished, it was an important and far-reaching protection against noxious odors, and it is on such general principles of fairness and equity that the constitutionality of today's odor-regulating legislation has been sustained. In fact, many lawyers are of the opinion that eventually one or more reviewing courts will go so far as to find that the right to a completely healthy environment is an inherent right actually protected by the United States Constitution and that pollution, including odor pollution, is a violation of that right giving rise to a cause of action.

CITY OF CHICAGO *v.* FRITZ

In 1960 the defendant owned a garbage dump. In fact, the defendant had owned and operated a dump at that location for over 30 years, apparently without any material change in its operations. The dump was located in an unincorporated and rather sparsely settled area of Cook County, Illinois, well outside the city limits of Chicago. Even though the garbage dump was well outside the city of Chicago, the city of Chicago sued to permanently enjoin the use of the property as a garbage dump, alleging, among other things, that it constituted a common law nuisance. The case was unusual in that, due to a peculiarity of Illinois law, a public nuisance was also a crime, and so the violator faced potential criminal sanctions, such as either jail or a fine, as well as being shut down by injunction, and all this after having operated the dump in the same manner for over 30 years.

Since the suit was for an injunction, there was no jury, and the case was heard before a court-appointed hearing officer, who in those days was called a master in chancery. The master found that the dump was used as a deposit for garbage, refuse, rubbish, chemicals, sludge, and other waste material which the defendants burned, and that as a result nauseous odors and smoke were produced which permeated an area two miles or more from the center of the dump. Apparently the noxious gases and smoke were so bad that the B&O Railroad, whose tracks lay about a mile from the dump, when the wind was right, at times had the safe operations of its trains interfered with.

Building on precedents such as the New York case cited above, the Illinois court had no trouble finding that (1) "a business which is otherwise lawful may become a nuisance either because of the locality in which it is carried on or because it is conducted in an improper manner," and (2) "impregnation of the atmosphere with disagreeable or offensive odors or stenches may become a nuisance where the atmosphere is contaminated to such an extent as to substantially impair the *comfort or enjoyment* of adjacent occupants."

Against all this, the defendants mainly argued that the dump had been in operation for over 30 years. The court held that the lapse of time made no difference, stating: "Defendants' dump is full, reeking, hazardous and a nuisance under the relevant statutes and the common law. It should be abated." Again, we see a court coming back to this very flexible damage standard of impairing the "comfort or enjoyment" of adjacent occupants. However, the relief sought was not just damages but the actual shutting down of the dump by injunction. The court granted the injunction and permanently shut down the dump on a finding that the "comfort" and "enjoyment" of adjacent occupants was being violated by the operation of the dump by its owner.

GREER *v.* CITY OF LENNOX

Another unusual case following the principles laid down in the two preceding cases was decided by the Supreme Court of South Dakota. The city of Lennox, South Dakota, operated a public dump in a careless manner. The city had purchased an old gravel pit in 1953 and converted it into a garbage dump. The city then proceeded to dump all kinds of raw garbage into the dump and burned it every day that weather permitted. No effort was made to cover the garbage with dirt. In fact, the only time that a bulldozer saw its way into the pit was twice a year when the city ran a bulldozer into the pit in an effort to pack the garbage down. Not unnaturally, large numbers of rats appeared and the city began a program of poisoning in an effort to keep the rats down.

A truly innocent and unfortunate gentleman, named John Greer, lived with his wife, baby daughter, and stepson on a farm which adjoined the dump. Greer had owned the farm for years prior to the city's purchasing the old gravel pit and converting it into the city dump. One can well imagine poor Greer's horror as he saw this monster being created at his doorstep, with each passing year bringing stronger and stronger odors, more and bigger rats, and a greater and increasing chance of rat-spread disease to his family, particularly his infant daughter, and his livestock.

Greer initially tried to avoid the courts. He did what any law-abiding, decent citizen would do. He went to the city hall and talked to the proper administrative officials. These officials duly reported the matter to the city council. The city council duly discussed and discussed poor Greer's plight at duly convened meetings of the city council. The local newspaper dutifully covered the story. The mayor and various city officials dutifully inspected the damage, which was considerable, to Greer's property. The culmination of all these earnest efforts by all of these proper and doubtless well-meaning officials was that the city bought and gave Greer a supply of rat poison to let him take care of it himself.

Finally, Greer, who doubtless was not a wealthy man, did what anyone who deals with a municipality almost invariably has to do–he sued. At the trial, the evidence was overwhelmingly in Greer's favor. Greer's evidence showed, without substantial contradiction, that before the dump was established his farm was in good repair and free from rats and offensive odors. After the city opened the dump, Greer's evidence showed that his farm became infested with rats, flies, and noxious odors and that the rats became particularly bothersome and destructive. Apparently the rat problem got to the point where the rats undermined the foundations of buildings, causing them to sag and break! They damaged sills, floors, lower siding boards, and door frames. They caused fear to the point where the Greers were afraid to use their own basement. Greer went on to prove that as a result of all this he incurred expense in fighting the rats, repairing damaged buildings, and fanning grain and that he sustained loss of crops, chickens, feed grain, and loss of value to his buildings. Apparently no effort was made to attach a dollar value to the loss of his comfort and enjoyment caused by the odors.

No mention is made of the city's defense. It possibly was along the traditional municipal defense lines that if Greer won this case, tax monies would be taken from the city, that tax monies are a sacred trust that must be prevented from erosion by spurious damage claims, that the city was low on funds and did not have enough money to pay for employees to administer the dump, that citizenship carries with it certain burdens as well as rights and that Greer's burden was to learn to live with the odors and rats, that Greer wasn't hurt as much as he said he was and the odors, rats, and flies were not really as bad as Greer said they were, and the like. Possibly it was along other lines. Whatever it was, it didn't impress the Supreme Court of South Dakota very much because the court didn't even bother to discuss it.

The case is interesting from at least two aspects. First, one wonders what on earth was happening in that town–whether the city truly believed that Greer wasn't hurt, whether the city truly believed either that Greer had no claim or that the city had valid defenses, whether the city believed that Greer either wouldn't have the nerve or didn't have the money to sue, or whether it just didn't like Greer. In any event, here was Greer running a chicken and livestock farm on the outskirts of town, who obviously was hurt, and the city made him fight his case all the way to the supreme court of the state.

The other interesting aspect of the case is that South Dakota had embodied the concept of common law nuisance as a statute. The statute defined a nuisance as "unlawfully doing an act, or omitting to perform a duty, which act or omission either:

> "(1) Annoys, injures or endangers the comfort, repose, health or safety of others;
>
> "(2) Offends decency;
>
> . . . .
>
> "(4) In any way renders other persons insecure in life, or in the use of property."

Greer was having all of that done to him in stages. He was being "smelled" from his property by noxious odors. His animals and family were being taken over by flies. The rats were taking down his buildings board by board. Not too surprisingly, both the trial court and the supreme court found the city guilty of operating a nuisance and awarded Greer damages.

The statute is interesting because of its breadth. The first case showed us the breakthrough occasioned by the law of nuisance. The courts, faced with the intangible nature of odors, created a new judge-made law, the law of nuisance, which said that a man could be awarded money if his reasonable comfort and enjoyment in the use of his property was disturbed.

The second case took us the next step to the point where the courts said, based on general principles of fairness and equity and without the sanction of a statute or ordinance, that not only could money damages be awarded for damage caused by noxious odors but the extreme relief of a permanent injunction could be used, permanently shutting down the source of such noxious odors if they constitute a common law nuisance by disturbing in an extreme manner the comfortable use and enjoyment of adjacent property.

The third case constitutes the third step, the embodiment of the old common law standards of "annoyance," "comfort," and "quiet enjoyment" in a statute and a court's finding a violation of that statute by, among other things, noxious odors.

INDEX

INDEX